Advances in Intelligent Systems and Computing

Volume 1128

The series "Advances in Intelligent Systems and Computing" contains publications on theory, applications, and design methods of Intelligent Systems and Intelligent Computing. Virtually all disciplines such as engineering, natural sciences, computer and information science, ICT, economics, business, e-commerce, environment, healthcare, life science are covered. The list of topics spans all the areas of modern intelligent systems and computing such as: computational intelligence, soft computing including neural networks, fuzzy systems, evolutionary computing and the fusion of these paradigms, social intelligence, ambient intelligence, computational neuroscience, artificial life, virtual worlds and society, cognitive science and systems, Perception and Vision, DNA and immune based systems, self-organizing and adaptive systems, e-Learning and teaching, human-centered and human-centric computing, recommender systems, intelligent control, robotics and mechatronics including human-machine teaming, knowledge-based paradigms, learning paradigms, machine ethics, intelligent data analysis, knowledge management, intelligent agents, intelligent decision making and support, intelligent network security, trust management, interactive entertainment, Web intelligence and multimedia.

The publications within "Advances in Intelligent Systems and Computing" are primarily proceedings of important conferences, symposia and congresses. They cover significant recent developments in the field, both of a foundational and applicable character. An important characteristic feature of the series is the short publication time and world-wide distribution. This permits a rapid and broad dissemination of research results.

**** Indexing: The books of this series are submitted to ISI Proceedings, EI-Compendex, DBLP, SCOPUS, Google Scholar and Springerlink ****

More information about this series at http://www.springer.com/series/11156

Yukio Ohsawa · Katsutoshi Yada ·
Takayuki Ito · Yasufumi Takama ·
Eri Sato-Shimokawara ·
Akinori Abe · Junichiro Mori ·
Naohiro Matsumura
Editors

Advances in Artificial Intelligence

Selected Papers from the Annual Conference of Japanese Society of Artificial Intelligence (JSAI 2019)

Springer

Editors
Yukio Ohsawa
Department of Systems Innovation
University of Tokyo
Tokyo, Japan

Takayuki Ito
Nagoya Institute of Technology
Nagoya, Japan

Eri Sato-Shimokawara
Department of Information
and Communication
Tokyo Metropolitan University
Tokyo, Japan

Junichiro Mori
School of Engineering
The University of Tokyo
Tokyo, Japan

Katsutoshi Yada
Faculty of Business and Commerce
Kansai University
Osaka, Japan

Yasufumi Takama
Graduate School of System Design
Tokyo Metropolitan University
Tokyo, Japan

Akinori Abe
Faculty of Letters
Chiba University
Chiba, Japan

Naohiro Matsumura
Graduate School of Economics
Osaka University
Toyonaka, Osaka, Japan

ISSN 2194-5357 ISSN 2194-5365 (electronic)
Advances in Intelligent Systems and Computing
ISBN 978-3-030-39877-4 ISBN 978-3-030-39878-1 (eBook)
https://doi.org/10.1007/978-3-030-39878-1

This Springer imprint is published by the registered company Springer Nature Switzerland AG
The registered company address is: Gewerbestrasse 11, 6330 Cham, Switzerland

Preface

The JSAI series of Annual Conference is the central event of the Japanese Society for Artificial Intelligence. Every year we collect papers and presentations and have earnest communications toward the progress of sciences, technologies, and philosophies relevant to AI. Traditionally, the JSAI Program Committee has mainly been responsible for organizing the conference program, selecting the invited speakers, and awarding the work. In JSAI2019, with the challenge of partially internationalizing general sessions, the content was organized on the supports by the newly established International Program Committee.

Summarizing the authors' first desired category of oral presentations in the general sessions, machine learning and AI applications increased further among all the submissions, but others decreased in general. One of the reasons is that many presenters who used machine learning technology as the method of each category wanted the machine learning category. The number of participants has increased more than the number of presentations, and there are 2905 participants (the most in history) for 748 presentations, showing the accelerating interest from the industry—the increase in submissions for AI applications is supporting this inference.

This year, for the first time, an international session was held with the same position as the general session. For the international session, five categories, (1) knowledge engineering, (2) machine learning, (3) agents, (4) robots and real worlds, and (5) human interface and education aid, have been chosen considering this as a challenging start of internationalization. Seven International Program Committee members who can cover these categories worked in the selection of papers, so that each paper had two or more reviewers over a peer review period of two months. In consideration of the evaluation by the evaluation committee and chairperson during the conference, we selected the papers to be published by Springer.

We also aimed to improve the quality of the paper. Both Japanese and English clearly indicated to CFP that the outline should include (1) purpose and (2) summary or conclusion of results. The contributors were carefully noted, and 26 of the 80 talks that were finally given were published as selected papers from Springer following the two stages below.

In the first stage, for all the papers presented in the international session (oral or interactive), at least two reviewers from related fields (from Japan or abroad) for each paper for which the program chairperson confirmed the abstract satisfied the conditions specified in the CFP. The reviewers selected by the International Program Committee members (the editors of this book) reviewed them. This peer review was evaluated according to the following criteria, scoring 0–5 for each criterion.

(a) Originality (novelty of the paper)
(b) Significance (impact on sciences and businesses)
(c) Quality (soundness and clarity of the paper)

This first evaluation period was from mid-March to mid-May 2019. Next, there are 80 cases in total, and the award candidates should be 30–45%. Therefore, in each categorized field, first, the top 33% of candidates in the overall score according to the peer reviewers were selected.

In the second stage, the reviewers consisting of all members of the International Program Committee members reassessed the content and the paper points, including the period of the conference.

(1) Contents: Evaluation of research importance and novelty
(2) Paper points: Evaluation of technical quality and clarity of papers
(3) Presentation points: Evaluation of oral presentations

As a result of consideration by the program chair and the head of the International Program Committee based on the second-stage evaluation, 19 long papers and 7 short papers have been chosen for publication in this book. Then, we finally edited this book including these survived papers. In this final step, we reorganized the selected papers into six parts (I) knowledge engineering, (II) agents, (III) education and culture, (IV) natural language processing, (V) machine learning and data mining, and (VI) cyber-physics, considering the balance of the number of papers. Approximately, (I) corresponds to the previously mentioned (1), (II) to (3), (III) to (5), (IV) to the intersection of all (1) through (5), (V) to (2), and (VI) to (4). This change and rearrangement were originally proposed by Yoshiki Kishi and Danilo E. Miura who assisted the program chair in the editing of this book, attending sessions in JSAI2019 and checking details of all the selected papers. Then we, the editors, found this idea realizes a harmony showing out the feature of JSAI where philosophies, i.e., toward a human-centric society with integrating natural and artificial intelligence.

We sincerely express millions of thanks to President Naohiko Uramoto, Vice-President Shusaku Tsumoto (General Chair of JSAI2019) of JSAI, Ex-president Seiji Yamada, Executive Committee Chair Takafumi Koshinaka, Deputy Program Chair Daisuke Katagami, Assistant Program Chair Teruaki Hayashi, Ex-Program Chair Takashi Onoda, Secretary-General Kazuo Sumida, and

Akiko Yamanobe of May Corporation as the Supporting Company, and all the participants of JSAI2019. We would also like to thank Yoshiki Kishi and Danilo E. Miura for their cooperation in the editing of this post-proceedings, including contacting the author, compiling LaTeX, and checking the contents of the extension.

November 2019

Yukio Ohsawa
The Program Chair of JSAI2019

Katsutoshi Yada
Takayuki Ito
Yasufumi Takama
Eri Sato-Shimokawara
Akinori Abe
Junichiro Mori
Naohiro Matsumura
The International Program Committee of JSAI, as Editor

Contents

Knowledge Engineering

Using Sequence Constraints for Modelling Network Interactions

Johannes De Smedt[1], Junichiro Mori[2(✉)], and Masanao Ochi[2]

[1] Management Science and Business Economics Group, The University of Edinburgh, 29 Buccleuch Place, Edinburgh EH8 9JS, UK
johannes.desmedt@ed.ac.uk

[2] The University of Tokyo, 7-3-1 Bunkyo-ku, Hongo, Tokyo, Japan
mori@mi.u-tokyo.ac.jp, ochi@ipr-ctr.t.u-tokyo.ac.jp

Abstract. This is an extension from a selected paper from JSAI2019. The ubiquitous nature of networks has led to a vast number of works dedicated to the study of capturing their information. Various graph-based techniques exist that report on the characteristics of nodes and edges, e.g., author-citation networks, social interactions, and so on. A significant amount of information can be extracted by summarizing the surrounding network structure of nodes, e.g., by capturing motives, or walk patterns. In this work, we present a new way of capturing the interaction between nodes in a network by making use of the sequence in which they occur. (1) The objective of this paper is to make use of behavioural constraint patterns; a concise but detailed report of node's interactions can be constructed that can be used for various purposes. (2) It is shown how the constraint patterns can be mined form interaction data, and how they can be used for various applications. (3) A case study is presented which uses behavioural constraint patterns to profile user interactions in a communication network.

Keywords: Sequence mining · Network analysis

1 Introduction

Networks are often formed by the interaction of various actors. For example, social networks grow based on friendship or interested-based relations, forum posts and emails link users according to their communication patterns, and citation networks are formed through authors referencing peers in their field. Typically, the construction of these networks is based on either undirected, or directed edges with weights. Furthermore, many network techniques focus on static relationships, I.e., the evolution over time is not investigated. However, a range of new techniques emerged recently that focus on the time-aspect of a network. Most notably, the use of motifs [11], and streams [8] allow to capture the evolution of a network over time. In this paper, we describe a new approach based on behavioral constraints, I.e., constraints based on sequence patterns that allow to describe the order of the interactions of nodes.

Y. Ohsawa et al. (Eds.): JSAI 2019, AISC 1128, pp. 3–13, 2020.
https://doi.org/10.1007/978-3-030-39878-1_1

We investigate how they can be constructed from a network dataset, and use the various patterns to describe the evolution of the network over time. In particular, we apply the sequence mining method to the question-and-answer interaction-based network. Our preliminary results show that profiling network interactions patterns with sequence mining enables track the behaviour of nodes in a transactional network without relying on the typical partial-order based results.

This paper is structured as follows. In Sect. 2, the methodology is presented to mine constraints from network data. Next, Sect. 3 reports on the application on a real-life dataset. Finally, Sect. 4 concludes the paper and reports on the future directions.

2 Behavioural Constraints for Sequence Interactions

In this section, an introduction to behavioural constraints is given, which includes a motivation which ones are suitable for network interactions. Next, the method for mining them is discussed. Finally, a number of potential applications are delineated.

2.1 Constraint Set

Behavioural constraint templates have been long used in various areas of computer science. Most notably, a comprehensive set of Linear Temporal Logic (LTL) templates was proposed for the formal verification of program execution [6]. LTL provides an adequate formalism to search for various temporal properties, such as whether something happens eventually, next, and so on, and can be used in conjunction with typical logical operators to construct expressive relations. The initial set was extended to include various other relations, most notably unary ones. These constraints were later adapted towards the case of process modelling [13]. While initially proposed as LTL formulae which are convertible to Büchi automata, finite trace equivalent regular expressions were introduced in [2] and [15]. Models existing of multiple constraints at the same time can be obtained by conjoining the automata to obtain a global language or automaton, over which all constraints hold.

In Table 1, an overview of the most-commonly used constraints in literature. They are organized according to 7 different categories, including unary and binary constraints. Most notably, the binary constraints follow a hierarchy which is reported in [2] and which covers unordered relations up to chain ordered relations (using the next operator) at the top. Besides, the inclusion of negative constraints is unique, as typically only patterns of positive relations are reported. Negative constraints capture behaviour that has not occurred. Including negative behaviour can be used to find relations that are not apparent at first sight, e.g., in Fig. 1, the fact that nodes A and E are both present in the sequence of C, but do not have interactions themselves, still allows the inference of not succession(A, E). While only unary and binary constraints are included, it is also

Table 1. An overview of Declare constraint templates with their corresponding LTL formula and regular expression.

Type	Template	LTL formula [12]	Regular expression [15]
Unary	Existence(A, n)	$\Diamond(A \wedge \bigcirc(\mathit{existence}(n-1, A)))$	.*(A.*){n}
	Absence(A, n)	$\neg \mathit{existence}(n, A)$	[^A]*(A?[^A]*){n-1}
	Exactly(A, n)	$\mathit{existence}(n, A) \wedge \mathit{absence}(n+1, A)$	[^A]*(A[^A]*){n}
	Init(A)	A	(A.*)?
	Last(A)	$\Box(A \implies \neg X \neg A)$	.*A
Unordered	Responded existence(A, B)	$\Diamond A \implies \Diamond B$	[^A]*((A.*B.*) |(B.*A.*))?
	Co-existence(A, B)	$\Diamond A \iff \Diamond B$	[^AB]*((A.*B.*) |(B.*A.*))?
Simple ordered	Response(A, B)	$\Box(A \implies \Diamond B)$	[^A]*(A.*B)*[^A]*
	Precedence(A, B)	$(\neg B\, U\, A) \vee \Box(\neg B)$	[^B]*(A.*B)*[^B]*
	Succession(A, B)	$\mathit{response}(A, B) \wedge \mathit{precedence}(A, B)$	[^AB]*(A.*B)*[^AB]*
Alternating ordered	Alternate response(A, B)	$\Box(A \implies \bigcirc(\neg A\, U\, B))$	[^A]*(A[^A]*B[^A]*)*
	Alternate precedence(A, B)	$\mathit{precedence}(A, B) \wedge \Box(B \implies \bigcirc(\mathit{precedence}(A, B))$	[^B]*(A[^B]*B[^B]*)*
	Alternate succession(A, B)	$\mathit{altresponse}(A, B) \wedge \mathit{precedence}(A, B)$	[^AB]*(A[^AB]*B[^AB]*)*
Chain ordered	Chain response(A, B)	$\Box(A \implies \bigcirc B)$	[^A]*(AB[^A]*)*
	Chain precedence(A, B)	$\Box(\bigcirc B \implies A)$	[^B]*(AB[^B]*)*
	Chain succession(A, B)	$\Box(A \iff \bigcirc B)$	[^AB]*(AB[^AB]*)*
Negative	Not co-existence(A, B)	$\neg(\Diamond A \wedge \Diamond B)$	[^AB]*((A[^B]*) |(B[^A]*))?
	Not succession(A, B)	$\Box(A \implies \neg(\Diamond B))$	[^A]*(A[^B]*)*
	Not chain succession(A, B)	$\Box(A \implies \neg(\bigcirc B))$	[^A]*(A+[^AB][^A]*)*A*
Choice	Choice(A, B)	$\Diamond A \vee \Diamond B$	.*[AB].*
	Exclusive choice(A, B)	$(\Diamond A \vee \Diamond B) \wedge \neg(\Diamond A \wedge \Diamond B)$	([^B]*A[^B]*) |.*[AB].*([^A]*B[^A]*)

possible to use target-branched constraints to model interactions between more than 2 nodes [4]. Motifs can be handy to capture various profiles of directed arcs between 2 or more nodes [10], but with the constraint sets it is possible to create even more intricate profiles of arc relations between nodes.

Despite not being useful for capturing interaction effects, the unary constraints can be used for adding information to a node's feature vector in case any exist. I.e., if a particular node is always occurring first in a sequence, this might signify a particular pattern, e.g., a person reporting recently-occurred disasters.

Not every constraint is suitable for binary interaction within a network context, i.e., not chain succession is, in general, not suitable for profiling behavior, as it holds in many situations. Besides, absence is hard to identify unless a particular

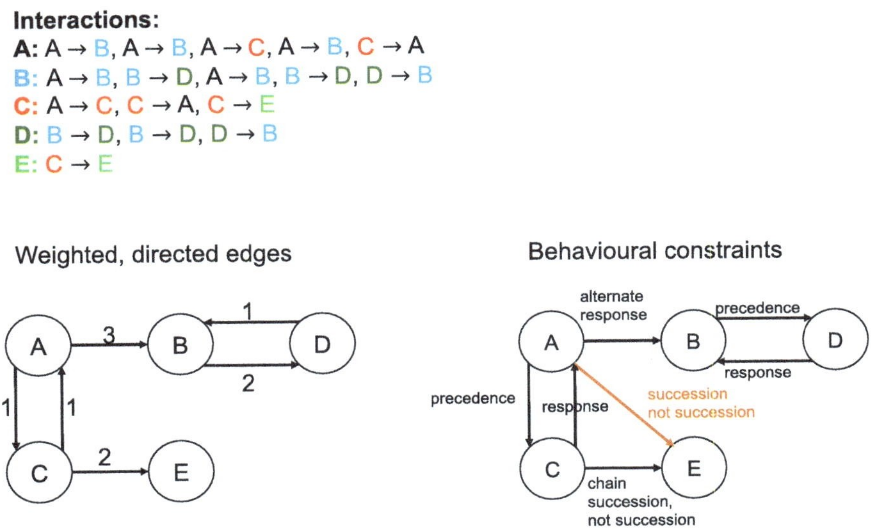

Fig. 1. Running example

node is scrutinized for this behaviour in the sequence of another node. Exclusive choice and not co-existence are similar in this respect, where the latter does not require the presence of either. Similar to not chain succession, this might lead to the discovery of many frequently non-occurring pairs.

2.2 Mining the Patterns

We define transactional network data as an ordered set of interactions T between nodes from the set N, where each transaction is a tuple $(n_1, n_2, ts) \in T$ with $n_1 \in N$ the initiating node, $n_2 \in N$ the receiving node, and $ts \in \mathbb{N}^+$ a timestamp. T can be read sequentially, where each node $n \in N$ has a sequence $s_n \subset 2^{|N|}$ that is extended whenever a transaction $t \in T$ is for that node is witnessed. I.e., s_n gets extended with $\langle n, n_o \rangle$ whenever n is the initiating node, and with $\langle n_o, n \rangle$ when n is on the receiving end given another node $n_o \in N$.

By using the interesting Behavioural Constraint Miner [3], we can mine all patterns in a sequence s_n to obtain a set of constraints C_{s_n}. Note, however, that if a given binary constraint $c(n, n_2) \in C_{s_n}$ holds for n in its own sequence, this still has to be verified with the sequence of the other node. If $c(n, n_2)$ is not present in that sequence, the constraints do not hold. Consider for example the interaction in Fig. 1. Despite the evidence in the sequence of A that there exists an alternate succession relationship between A and B due to the alternating ABABAAB pattern, the sequence of B rather indicates that other occurrences of B happen in between (e.g. B→D), breaking the pattern. Hence, a final step is required to recursively ensure that $C_n = \{c \mid c \in C_n \wedge c \in C_{n_i} \forall n_i \in \mathcal{N}(n) \vee c \notin C_n \wedge c \notin C_{n_i} \forall n_i \in \mathcal{N}(n)\}$ where $\mathcal{N}(n) \subseteq N$ denotes the neighbourhood of node

n to check that all constraint pertaining to n are either both in its constraint set and the constraint set of its neighbours to avoid conflict, or that it is present in an unrelated node (e.g. the connection succession(A, E) in Fig. 1). To conclude the discovery of sequence templates from the network interactions, the sets C_n are pruned according to the constraint hierarchy.

2.3 Applications

The mining of interactions in a network as sequences has several applications. Most notably, the sequence information can be used for analysing the patterns that exist between nodes, and their evolution over time. By tracking what patterns exist, and whether they return over time gives an overview of how certain relations change and what the underlying sequential behaviour is. In this case, unary constraints might not be useful, but especially the hierarchy of binary constraints can pin down how strong the relationship between two nodes is.

Next, the sequence patterns can be used as features of a node to obtain vector representations of nodes or relations between nodes. The presence of relations can be stored in a binary vector for relations, or the number of relations of each type can be stored for nodes. In this case, also unary constraints help define the node in terms of where in a sequence (init/last), how often (existence/exactly), besides with what other nodes the node is interacting (binary constraints). The features can be used towards node classification [1] or node relation prediction [9].

Given that all constraints have formal semantics in LTL or regular expressions, it is possible to exploit these in order to simulate the sequence behaviour later. These languages support state machine models which can generate strings of nodes by creating a global state machine that models all the constraints' behaviour. This can potentially feed into techniques such as node2vec [7], which typically employ random walks [14] for getting node representations.

Finally, by using the transitivity properties of the constraints, link inference/prediction [9] can also be made.

3 Application

In this section, the approach is applied to a real-life dataset in order to retrieve the evolution of constraints over time to profile the network interactions from a behavioural perspective.

3.1 Data

We apply the sequence method to the Math Overflow dataset, as used in [11]. On the Overflow web sites, users post questions and receive answers from other users, and users may comment on both questions and answers. We derive a transactional network by creating an edge (u, v, t) if, at time t, user u: (1) posts an answer to user v's question, (2) comments on user v's question, or

(3) comments on user v's answer. The data contains 24,818 nodes with 506,550 interactions over 2,350 days and deals with question-and-answer data from users regarding mathematical problems.

We retrieve the constraints over the dataset by splitting the interactions into contingent blocks of a varying time length. In this case, we used blocks of 4 h (14,102 blocks), 2 days (1,175 blocks), 100 days (23 blocks), and 1,000 days (2 blocks) in order to track the evolution of the constraints. The evolution is captured by tallying the shift in constraint type between nodes present in subsequent blocks. For this analysis, we limit the constraint set to the 7 most common sequence patterns. The following coding was used:

- 1: not succession
- 2: precedence
- 3: alternate precedence
- 4: chain precedence
- 5: response
- 6: alternate response
- 7: chain response

In the columns, 0 stands for the absence of a constraint in subsequent time blocks between nodes that both reoccur.

In order to understand the difference in behaviour of various user types, we analyse the evolution of various node type and look into nodes that have a degree less than or equal to 3 (low involvement), a degree of 4–10 (medium involvement), and a degree of more than 10 (high involvement - 5,821 nodes). The results can be found in Tables 2 and 3 for incoming and outgoing relations respectively. Besides, the node with the highest authority score [5] is included as well, in order to illustrate how the most important node in terms of the dispersion of information acts within the web site. The results can be found in Table 4. The colours denote the place in the distribution, where red is higher and green is lower. Scores with different colours and equal scores indicate a difference in value behind the significant digits.

3.2 Interpretation

Influence of Degree. From Tables 2 and 3, we can learn that most relationships between nodes in subsequent blocks vanish, however, this seems to be more of an issue for blocks that are longer (>4 h) especially for nodes with a lower degree ($\leq$10). It seems a majority of the activity is relatively one-off. In general, all alternate relations happen sparsely given their very low share in the overall constraint evolution tally, meaning most nodes only interact once during a particular time block.

More interesting patterns can be observed when looking at how different constraint types evolve over time. A vast majority of chain precedence relations, a result of uninterrupted interaction between nodes, are replaced by not succession (4→1) and vice versa (1→4), and a shift from not succession to chain response

Table 2. An overview of the proportion of incoming constraints that shift from one sequence pattern into another for nodes with different degrees between different time blocks of varying lengths.

		4 hours								2 days							
IN		0	1	2	3	4	5	6	7	0	1	2	3	4	5	6	7
≤3	1	0.072	0.025	0.025	0.000	0.100	0.007	0.000	0.086	0.214	0.022	0.028	0.000	0.040	0.013	0.000	0.028
	2	0.063	0.013	0.011	0.000	0.011	0.001	0.000	0.012	0.144	0.018	0.022	0.000	0.004	0.008	0.000	0.004
	3	0.000	0.001	0.000	0.000	0.001	0.000	0.000	0.001	0.000	0.001	0.001	0.000	0.005	0.000	0.000	0.001
	4	0.076	0.064	0.032	0.001	0.038	0.008	0.001	0.034	0.079	0.038	0.030	0.001	0.010	0.014	0.001	0.015
	5	0.058	0.023	0.004	0.001	0.008	0.007	0.000	0.011	0.116	0.018	0.007	0.000	0.006	0.021	0.000	0.003
	6	0.000	0.001	0.000	0.000	0.001	0.000	0.000	0.001	0.000	0.000	0.000	0.000	0.000	0.000	0.000	0.000
	7	0.050	0.042	0.011	0.000	0.012	0.015	0.000	0.075	0.027	0.021	0.006	0.000	0.004	0.008	0.000	0.019
4–10	1	0.086	0.018	0.037	0.000	0.047	0.011	0.000	0.059	0.229	0.014	0.038	0.001	0.011	0.021	0.000	0.016
	2	0.106	0.013	0.028	0.000	0.011	0.007	0.000	0.021	0.201	0.018	0.027	0.000	0.005	0.011	0.000	0.008
	3	0.001	0.000	0.000	0.000	0.000	0.000	0.000	0.000	0.003	0.000	0.003	0.000	0.001	0.000	0.000	0.001
	4	0.062	0.023	0.031	0.000	0.018	0.010	0.000	0.035	0.042	0.010	0.014	0.000	0.003	0.011	0.000	0.008
	5	0.091	0.010	0.010	0.000	0.007	0.014	0.000	0.013	0.164	0.014	0.007	0.000	0.003	0.019	0.000	0.006
	6	0.001	0.000	0.000	0.000	0.000	0.000	0.000	0.001	0.001	0.000	0.001	0.000	0.000	0.000	0.000	0.000
	7	0.062	0.019	0.012	0.000	0.008	0.021	0.000	0.103	0.038	0.007	0.006	0.000	0.001	0.011	0.000	0.019
>10	1	0.143	0.028	0.028	0.001	0.024	0.017	0.001	0.027	0.218	0.014	0.018	0.001	0.002	0.017	0.000	0.002
	2	0.146	0.024	0.024	0.001	0.008	0.011	0.000	0.013	0.274	0.017	0.023	0.001	0.001	0.007	0.001	0.002
	3	0.003	0.001	0.001	0.000	0.000	0.000	0.000	0.000	0.007	0.000	0.001	0.000	0.000	0.001	0.000	0.000
	4	0.046	0.028	0.015	0.000	0.010	0.010	0.000	0.015	0.016	0.002	0.002	0.000	0.001	0.002	0.000	0.001
	5	0.142	0.022	0.008	0.000	0.006	0.021	0.001	0.009	0.278	0.017	0.006	0.001	0.001	0.024	0.001	0.002
	6	0.003	0.001	0.001	0.000	0.000	0.000	0.000	0.000	0.006	0.000	0.001	0.000	0.000	0.001	0.000	0.000
	7	0.050	0.026	0.009	0.000	0.005	0.019	0.000	0.050	0.021	0.002	0.002	0.000	0.000	0.004	0.000	0.003

		100 days								1000 days							
		0	1	2	3	4	5	6	7	0	1	2	3	4	5	6	7
≤3	1	0.330	0.022	0.024	0.001	0.001	0.017	0.000	0.000	0.277	0.022	0.027	0.000	0.000	0.018	0.000	0.000
	2	0.283	0.020	0.028	0.001	0.001	0.002	0.000	0.000	0.312	0.031	0.025	0.001	0.000	0.003	0.000	0.000
	3	0.002	0.000	0.000	0.000	0.000	0.000	0.000	0.000	0.001	0.000	0.001	0.000	0.000	0.001	0.000	0.000
	4	0.011	0.001	0.001	0.000	0.000	0.001	0.000	0.000	0.005	0.001	0.001	0.000	0.000	0.001	0.000	0.000
	5	0.212	0.013	0.008	0.001	0.000	0.013	0.000	0.000	0.230	0.022	0.008	0.001	0.000	0.011	0.000	0.000
	6	0.001	0.000	0.000	0.000	0.000	0.000	0.000	0.000	0.000	0.000	0.000	0.000	0.000	0.000	0.000	0.000
	7	0.001	0.000	0.000	0.000	0.000	0.000	0.000	0.000	0.001	0.000	0.000	0.000	0.000	0.000	0.000	0.000
4–10	1	0.309	0.023	0.032	0.001	0.000	0.029	0.000	0.000	0.256	0.024	0.040	0.002	0.000	0.032	0.001	0.000
	2	0.264	0.021	0.027	0.001	0.000	0.005	0.000	0.000	0.265	0.027	0.039	0.002	0.000	0.006	0.001	0.000
	3	0.005	0.000	0.000	0.000	0.000	0.000	0.000	0.000	0.008	0.001	0.001	0.000	0.000	0.001	0.000	0.000
	4	0.003	0.000	0.001	0.000	0.000	0.000	0.000	0.000	0.001	0.000	0.000	0.000	0.000	0.000	0.000	0.000
	5	0.231	0.017	0.007	0.000	0.000	0.015	0.000	0.000	0.231	0.022	0.011	0.001	0.000	0.023	0.001	0.000
	6	0.003	0.000	0.000	0.000	0.000	0.000	0.000	0.000	0.004	0.000	0.000	0.000	0.000	0.000	0.000	0.000
	7	0.001	0.000	0.000	0.000	0.000	0.000	0.000	0.000	0.000	0.000	0.000	0.000	0.000	0.000	0.000	0.000
>10	1	0.194	0.024	0.033	0.001	0.000	0.034	0.001	0.000	0.160	0.032	0.046	0.001	0.000	0.047	0.001	0.000
	2	0.252	0.032	0.047	0.001	0.000	0.012	0.001	0.000	0.216	0.043	0.062	0.001	0.000	0.019	0.001	0.000
	3	0.005	0.001	0.001	0.000	0.000	0.001	0.000	0.000	0.004	0.001	0.001	0.000	0.000	0.001	0.000	0.000
	4	0.001	0.000	0.000	0.000	0.000	0.000	0.000	0.000	0.000	0.000	0.000	0.000	0.000	0.000	0.000	0.000
	5	0.254	0.033	0.012	0.001	0.000	0.048	0.001	0.000	0.222	0.043	0.017	0.001	0.000	0.068	0.002	0.000
	6	0.005	0.001	0.001	0.000	0.000	0.001	0.000	0.000	0.005	0.001	0.001	0.000	0.000	0.001	0.000	0.000
	7	0.001	0.000	0.000	0.000	0.000	0.000	0.000	0.000	0.000	0.000	0.000	0.000	0.000	0.000	0.000	0.000

Table 3. An overview of the proportion of outgoing constraints that shift from one sequence pattern into another for nodes with different degrees between different time blocks of varying lengths.

		100 days								1000 days							
OUT		0	1	2	3	4	5	6	7	0	1	2	3	4	5	6	7
	1	0.136	0.058	0.017	0.000	0.059	0.014	0.002	0.068	0.133	0.021	0.030	0.000	0.017	0.016	0.001	0.030
	2	0.080	0.041	0.013	0.000	0.005	0.015	0.001	0.006	0.216	0.025	0.020	0.000	0.004	0.009	0.000	0.007
	3	0.000	0.000	0.000	0.000	0.000	0.000	0.000	0.000	0.001	0.000	0.000	0.000	0.000	0.000	0.000	0.000
	4	0.013	0.086	0.009	0.000	0.013	0.008	0.000	0.010	0.018	0.019	0.004	0.000	0.004	0.004	0.000	0.006
	5	0.082	0.021	0.004	0.000	0.005	0.011	0.001	0.008	0.236	0.012	0.004	0.000	0.003	0.025	0.001	0.005
≤3	6	0.002	0.002	0.000	0.000	0.000	0.000	0.000	0.000	0.002	0.001	0.000	0.000	0.000	0.000	0.000	0.000
	7	0.022	0.114	0.001	0.000	0.002	0.029	0.000	0.048	0.028	0.040	0.003	0.000	0.002	0.027	0.001	0.023
	1	0.167	0.060	0.018	0.001	0.028	0.013	0.001	0.026	0.152	0.013	0.021	0.000	0.007	0.017	0.001	0.008
	2	0.116	0.038	0.017	0.000	0.007	0.007	0.001	0.007	0.253	0.022	0.018	0.001	0.003	0.009	0.002	0.004
	3	0.001	0.001	0.000	0.000	0.000	0.000	0.000	0.000	0.004	0.000	0.000	0.000	0.000	0.000	0.000	0.000
	4	0.032	0.063	0.010	0.000	0.008	0.011	0.000	0.010	0.022	0.014	0.003	0.000	0.002	0.007	0.001	0.002
	5	0.119	0.036	0.003	0.000	0.007	0.017	0.001	0.007	0.268	0.022	0.004	0.001	0.003	0.023	0.002	0.005
4–10	6	0.003	0.001	0.001	0.000	0.001	0.001	0.000	0.001	0.006	0.001	0.000	0.000	0.000	0.000	0.000	0.000
	7	0.033	0.061	0.005	0.000	0.004	0.023	0.001	0.034	0.033	0.015	0.004	0.000	0.001	0.018	0.002	0.009
	1	0.138	0.026	0.029	0.001	0.026	0.017	0.000	0.030	0.220	0.014	0.018	0.001	0.002	0.017	0.000	0.002
	2	0.144	0.023	0.025	0.001	0.008	0.010	0.000	0.014	0.272	0.017	0.023	0.001	0.001	0.008	0.000	0.002
	3	0.003	0.001	0.001	0.000	0.000	0.000	0.000	0.000	0.007	0.000	0.001	0.000	0.000	0.001	0.000	0.000
	4	0.048	0.026	0.017	0.000	0.011	0.010	0.000	0.017	0.017	0.002	0.003	0.000	0.001	0.002	0.000	0.001
	5	0.139	0.021	0.008	0.000	0.006	0.020	0.001	0.010	0.275	0.016	0.006	0.001	0.001	0.024	0.001	0.002
>10	6	0.002	0.000	0.001	0.000	0.000	0.000	0.000	0.000	0.006	0.000	0.001	0.000	0.000	0.001	0.000	0.000
	7	0.052	0.023	0.010	0.000	0.005	0.019	0.000	0.054	0.021	0.002	0.002	0.000	0.000	0.004	0.000	0.003

		100 days								1000 days							
		0	1	2	3	4	5	6	7	0	1	2	3	4	5	6	7
	1	0.180	0.010	0.014	0.000	0.000	0.017	0.002	0.001	0.211	0.007	0.023	0.000	0.000	0.028	0.002	0.000
	2	0.297	0.012	0.023	0.000	0.000	0.008	0.000	0.000	0.265	0.016	0.022	0.000	0.000	0.010	0.000	0.000
	3	0.000	0.000	0.000	0.000	0.000	0.000	0.000	0.000	0.001	0.000	0.000	0.000	0.000	0.000	0.000	0.000
	4	0.002	0.000	0.000	0.000	0.000	0.000	0.000	0.000	0.001	0.000	0.000	0.000	0.000	0.000	0.000	0.000
	5	0.359	0.016	0.003	0.000	0.000	0.037	0.001	0.001	0.338	0.020	0.002	0.000	0.000	0.043	0.001	0.000
≤3	6	0.002	0.000	0.001	0.000	0.000	0.001	0.000	0.000	0.004	0.001	0.000	0.000	0.000	0.000	0.000	0.000
	7	0.007	0.001	0.000	0.000	0.000	0.001	0.000	0.001	0.002	0.000	0.000	0.000	0.000	0.000	0.000	0.000
	1	0.167	0.009	0.015	0.000	0.000	0.018	0.001	0.000	0.179	0.014	0.022	0.000	0.000	0.027	0.001	0.000
	2	0.305	0.021	0.029	0.000	0.000	0.009	0.000	0.000	0.271	0.024	0.032	0.000	0.000	0.015	0.000	0.000
	3	0.004	0.000	0.000	0.000	0.000	0.000	0.000	0.000	0.005	0.000	0.000	0.000	0.000	0.001	0.000	0.000
	4	0.002	0.000	0.000	0.000	0.000	0.000	0.000	0.000	0.000	0.000	0.000	0.000	0.000	0.000	0.000	0.000
	5	0.332	0.024	0.007	0.000	0.000	0.039	0.001	0.000	0.302	0.029	0.008	0.000	0.000	0.050	0.002	0.000
4–10	6	0.007	0.000	0.001	0.000	0.000	0.001	0.000	0.000	0.011	0.001	0.001	0.000	0.000	0.002	0.000	0.000
	7	0.004	0.000	0.000	0.000	0.000	0.001	0.000	0.000	0.001	0.000	0.000	0.000	0.000	0.000	0.000	0.000
	1	0.198	0.025	0.034	0.001	0.000	0.034	0.001	0.000	0.168	0.033	0.047	0.001	0.000	0.048	0.001	0.000
	2	0.251	0.032	0.047	0.001	0.000	0.012	0.001	0.000	0.215	0.044	0.063	0.001	0.000	0.018	0.001	0.000
	3	0.005	0.001	0.001	0.000	0.000	0.001	0.000	0.000	0.005	0.001	0.001	0.000	0.000	0.001	0.000	0.000
	4	0.001	0.000	0.000	0.000	0.000	0.000	0.000	0.000	0.000	0.000	0.000	0.000	0.000	0.000	0.000	0.000
	5	0.251	0.032	0.012	0.001	0.000	0.047	0.001	0.000	0.214	0.043	0.017	0.001	0.000	0.066	0.001	0.000
>10	6	0.005	0.001	0.001	0.000	0.000	0.001	0.000	0.000	0.005	0.001	0.001	0.000	0.000	0.001	0.000	0.000
	7	0.001	0.000	0.000	0.000	0.000	0.000	0.000	0.000	0.000	0.000	0.000	0.000	0.000	0.000	0.000	0.000

Table 4. An overview of the proportion of constraints that shift from one sequence pattern into another, both for incoming and outgoing constraints of the node with the highest authority score in the network.

	4 hours								2 days							
IN	0	1	2	3	4	5	6	7	0	1	2	3	4	5	6	7
1	0.172	0.020	0.019	0.000	0.007	0.019	0.000	0.009	0.195	0.020	0.026	0.001	0.001	0.025	0.001	0.001
2	0.219	0.021	0.026	0.001	0.003	0.008	0.001	0.005	0.267	0.028	0.035	0.001	0.001	0.011	0.001	0.002
3	0.006	0.000	0.000	0.000	0.000	0.000	0.000	0.000	0.006	0.000	0.000	0.000	0.000	0.000	0.000	0.000
4	0.040	0.007	0.006	0.000	0.003	0.005	0.000	0.005	0.010	0.001	0.001	0.000	0.000	0.001	0.000	0.000
5	0.224	0.020	0.007	0.000	0.003	0.026	0.001	0.004	0.264	0.028	0.009	0.001	0.001	0.036	0.001	0.002
6	0.002	0.001	0.000	0.000	0.000	0.000	0.000	0.000	0.005	0.000	0.000	0.000	0.000	0.000	0.000	0.000
7	0.062	0.010	0.008	0.000	0.002	0.012	0.001	0.015	0.014	0.001	0.001	0.000	0.000	0.002	0.000	0.001

OUT	0	1	2	3	4	5	6	7	0	1	2	3	4	5	6	7
1	0.195	0.018	0.026	0.001	0.009	0.025	0.000	0.010	0.215	0.020	0.029	0.001	0.001	0.030	0.001	0.001
2	0.201	0.016	0.021	0.001	0.005	0.010	0.001	0.008	0.249	0.025	0.033	0.001	0.001	0.011	0.001	0.001
3	0.006	0.001	0.001	0.000	0.000	0.001	0.000	0.000	0.007	0.001	0.001	0.000	0.000	0.001	0.000	0.000
4	0.046	0.006	0.011	0.000	0.002	0.011	0.000	0.004	0.009	0.001	0.002	0.000	0.000	0.001	0.000	0.000
5	0.203	0.017	0.007	0.001	0.005	0.027	0.002	0.006	0.263	0.025	0.007	0.001	0.001	0.038	0.001	0.001
6	0.005	0.001	0.001	0.000	0.000	0.001	0.000	0.000	0.005	0.001	0.001	0.000	0.000	0.001	0.000	0.000
7	0.051	0.005	0.008	0.000	0.001	0.011	0.000	0.014	0.008	0.001	0.001	0.000	0.000	0.001	0.000	0.000

	100 days								1000 days							
IN	0	1	2	3	4	5	6	7	0	1	2	3	4	5	6	7
1	0.099	0.023	0.040	0.001	0.000	0.051	0.002	0.001	0.011	0.027	0.013	0.000	0.000	0.047	0.001	0.001
2	0.185	0.043	0.087	0.001	0.000	0.032	0.001	0.001	0.077	0.059	0.049	0.000	0.003	0.121	0.003	0.001
3	0.001	0.000	0.001	0.000	0.000	0.002	0.000	0.000	0.000	0.000	0.000	0.000	0.000	0.000	0.000	0.000
4	0.001	0.000	0.001	0.000	0.000	0.000	0.000	0.000	0.000	0.001	0.001	0.000	0.000	0.000	0.000	0.000
5	0.209	0.053	0.023	0.001	0.000	0.121	0.003	0.001	0.138	0.101	0.039	0.000	0.000	0.285	0.005	0.000
6	0.005	0.001	0.003	0.000	0.000	0.003	0.000	0.000	0.001	0.003	0.003	0.000	0.000	0.008	0.000	0.000
7	0.002	0.000	0.001	0.000	0.000	0.001	0.000	0.000	0.001	0.000	0.000	0.000	0.000	0.001	0.000	0.000

OUT	0	1	2	3	4	5	6	7	0	1	2	3	4	5	6	7
1	0.175	0.059	0.069	0.001	0.000	0.060	0.000	0.000	0.070	0.026	0.219	0.006	0.000	0.100	0.000	0.001
2	0.182	0.056	0.083	0.002	0.000	0.017	0.001	0.000	0.061	0.025	0.269	0.015	0.000	0.035	0.001	0.000
3	0.003	0.001	0.001	0.000	0.000	0.001	0.000	0.000	0.003	0.000	0.001	0.000	0.000	0.000	0.000	0.000
4	0.001	0.001	0.000	0.000	0.000	0.000	0.000	0.000	0.000	0.001	0.001	0.000	0.000	0.000	0.000	0.000
5	0.152	0.049	0.026	0.001	0.000	0.053	0.000	0.000	0.019	0.015	0.060	0.001	0.000	0.061	0.000	0.000
6	0.001	0.000	0.000	0.000	0.000	0.000	0.000	0.000	0.000	0.000	0.001	0.000	0.000	0.000	0.000	0.000
7	0.000	0.000	0.000	0.000	0.000	0.000	0.000	0.000	0.000	0.000	0.001	0.003	0.000	0.003	0.000	0.000

(1→7) mainly for a degree lower or equal to 3 and 4 h blocks and incoming relationships. The same trend holds for 2–1000 days, but here there is a higher rate of precedence to precedence (2→2). This indicates that in the short term, mostly the absence of response of the node (1) is eventually replaced by a very close interaction (4) in the short run, and that for the longer run (100–1000 days), the 2→2 indicates that the other party initiates a message often without there being a strict conversion towards other constraints such as response (5).

For nodes with a degree between 4 and 10, the evolutions look different with a strong persistence of chain response (7→7) in the short run (4 h), and a similar

(1/2→1/2) in the longer run. The former indicates that nodes tend to maintain a close interaction over short time spans, which is substituted by more sporadic interchange for longer time spans.

For nodes with a high degree, 7→7 for 4 h, and 5→5 relations are present for a large share of the constraint evolutions. Hence, these nodes tend to maintain close contact at first, and afterwards remain the consequent in a response relationship indicate being especially forthcoming in terms of responding to other users.

Overall, users with a high degree tend to be more involved in mutually positive relationships (not 1), and reciprocate by being mostly involved in response relationships (5–7).

For outgoing constraints, we see a different picture, where incoming chain responses (7) are often converted into not succession (1) by the receiving party, which persists for 4 h and 2 days at a degree of 3. For longer periods, 5→5 or, response to response is prevalent.

For nodes with a degree between 4 and 10, most initial relations end in not succession, later converging to a similar profile as nodes with a lower degree, and a high proportion of 5→5. The same holds for nodes with a high degree, where initially 7→7 is prevalent.

These observations clarify which nodes tend to be the source of cutting of contact in short time blocks (mostly with degree ≤10), and explain the interpretation of the incoming nodes earlier. In this respect, the evolution of constraints can be tracked between different types of nodes in terms of degree

Authority. The node with the highest authority score which has a degree of 2,680 does not follow the trends of the other nodes discussed above. Again, the high proportion of 5→5 evolution is present for incoming relationships, but instead of materialising later on, this happens even for shorter time blocks (4 h). This indicates that the authority will respond regardless of any established relationship.

4 Conclusion and Future Work

In this paper, we have shown how mining network interaction patterns can be profiled using sequence mining techniques. We apply the sequence mining method to the question-and-answer interaction-based network. Our preliminary results show that employing sequence patterns enables us track the behaviour of nodes in a transactional network and summarize their interactions without relying on the typical partial-order based results that are offered in sequence mining, while still going beyond the typical general nature of motifs that focus on directed arcs between 2 or 3 actors [11] . In a small experimental evaluation, we demonstrate the usefulness of the approach in the context of message board analysis.

For future work, we envision to focus on testing the patterns in the context of feature engineering, and link inference.

References

1. Bhagat, S., Cormode, G., Muthukrishnan, S.: Node classification in social networks. In: Social Network Data Analytics, pp. 115–148. Springer (2011)
2. Ciccio, C.D., Mecella, M.: A two-step fast algorithm for the automated discovery of declarative workflows. In: IEEE Symposium on Computational Intelligence and Data Mining, CIDM 2013, Singapore, 16–19 April, 2013, pp. 135–142 (2013)
3. De Smedt, J., Deeva, G., De Weerdt, J.: Behavioral constraint template-based sequence classification. In: ECML/PKDD (2). LNCS, vol. 10535, pp. 20–36. Springer (2017)
4. Di Ciccio, C.D., Maggi, F.M., Mendling, J.: Efficient discovery of target-branched declare constraints. Inf. Syst. **56**, 258–283 (2016)
5. Ding, C.H.Q., Zha, H., He, X., Husbands, P., Simon, H.D.: Link analysis: hubs and authorities on the world wide web. SIAM Rev. **46**(2), 256–268 (2004)
6. Dwyer, M.B., Avrunin, G.S., Corbett, J.C.: Patterns in property specifications for finite-state verification. In: ICSE, pp. 411–420. ACM (1999)
7. Grover, A., Leskovec, J.: node2vec: scalable feature learning for networks. In: Proceedings of the 22nd ACM SIGKDD International Conference on Knowledge Discovery and Data Mining, pp. 855–864. ACM (2016)
8. Latapy, M., Viard, T., Magnien, C.: Stream graphs and link streams for the modeling of interactions over time. Social Netw. Analys. Min. **8**(1), 61:1–61:29 (2018)
9. Liben-Nowell, D., Kleinberg, J.: The link-prediction problem for social networks. J. Am. Soc. Inform. Sci. Technol. **58**(7), 1019–1031 (2007)
10. Milo, R., Shen-Orr, S., Itzkovitz, S., Kashtan, N., Chklovskii, D., Alon, U.: Network motifs: simple building blocks of complex networks. Science **298**(5594), 824–827 (2002)
11. Paranjape, A., Benson, A.R., Leskovec, J.: Motifs in temporal networks. In: WSDM, pp. 601–610. ACM (2017)
12. Pesić, M.: Constraint-based work on management systems: shifting control to users. Ph.D. thesis, Eindhoven University of Technology (2008). 26
13. Pesic, M., van der Aalst, W.M.P.: A declarative approach for flexible business processes management. In: Business Process Management Workshops. LNCS, vol. 4103, pp. 169–180. Springer (2006)
14. Pons, P., Latapy, M.: Computing communities in large networks using random walks. In: International Symposium on Computer and Information Sciences, pp. 284–293. Springer (2005)
15. Westergaard, M., Stahl, C., Reijers, H.A.: Unconstrainedminer: efficient discovery of generalized declarative process models. BPM Center Report BPM-13-28, BPMcenter.org, p. 28 (2013)

Prediction of Onset of Lifestyle-Related Diseases Using Regular Health Checkup Data

Mitsuru Tsunekawa[1](✉), Natsuki Oka[1], Masahiro Araki[1], Motoshi Shintani[2], Masataka Yoshikawa[3], and Takeshi Tanigawa[4]

[1] Kyoto Institute of Technology, Matsugasaki, Sakyo-ku, Kyoto, Japan
m-tsune@ii.is.kit.ac.jp
[2] SG Holdings Group Health Insurance Association, 637, Suiginnya-cho, Shimogyo-ku, Kyoto, Japan
[3] Japan System Techniques Co., Ltd., 2-3-18, Nakanoshima, Kita-ku, Osaka, Japan
[4] Juntendo University, 2-1-1, Hongo, Bunkyo-ku, Tokyo, Japan

Abstract. This is an extension from a selected paper from JSAI2019. This study proposes a method to predict the onset of lifestyle-related diseases using periodical health checkup data. In this study, we carefully examine insurance claims data to identify onset of diseases and use the data for supervised learning. We aim to predict whether lifestyle-related diseases, except cancer, will develop within a year. We adopt the undersampling and bagging approach to address the class imbalance problem. The precision and recall of the proposed method are found to be 0.32 and 0.89, respectively. Compared with a baseline that sets thresholds for each examination item and considers their logical sum, our method achieves much higher precision while maintaining the recall; this allows suppression of the number of targets for health guidance, without increasingly neglecting those who are likely to become severely ill.

Keywords: Machine learning · Class imbalance · Medical informatics · Undersampling · Bagging

1 Introduction

Many people have recently begun using Internet mail orders, which has greatly increased the number of deliveries. Consequently, social interest in the work environment and health management of courier drivers is increasing. If appropriate health guidance can reduce the occurrence of lifestyle-related diseases (except cancer) among drivers and prevent severe idiopathic illnesses during driving, medical expenses and traffic accidents can be decreased. We, therefore, aim to use regular health checkup data of employees from a home delivery company to predict the onset of lifestyle-related diseases within a year as accurately as possible so that appropriate health guidance can be provided.

Y. Ohsawa et al. (Eds.): JSAI 2019, AISC 1128, pp. 14–26, 2020.
https://doi.org/10.1007/978-3-030-39878-1_2

There have been many studies on the prediction of diseases using medical data. For example, [1] created a risk equation using the Cox proportional hazards model to estimate colorectal cancer risk using the results of electronic medical records of a unified database in a specific country or region. Statistical analysis techniques have been used for cohort studies that predict the occurrence probability of myocardial infarction and cerebral infarction from health examination results [2]. The prediction target of [2] overlaps with this study, but the data and prediction methods are different. [3] proposed a model that predicts pneumonia hospitalization using the Lasso logistic regression of regular health checkup data; similar to our study, periodical health checkup data of healthy people are used for prediction here. However, our study differs in that it uses machine learning techniques according to data characteristics. Recently, many studies have used machine learning and data mining techniques to predict disease onset from medical data. [4] highlighted the superiority of machine learning techniques to predict cardiovascular risk from routine clinical data. [5] developed a model that uses a neural network to predict ventricular tachycardia one hour before occurrence. Because ventricular tachycardia can lead to sudden cardiac death, predicting the occurrence is useful to treat it immediately. There was also a study that developed a probability-based artificial neural network model, called NORF, for predicting risks of oyster norovirus outbreaks [6]. Similarly, prediction research using deep learning has been actively conducted [7,8]. In this study, to emphasize the explainability of prediction results, we did not use deep learning because its grounds for judgment tend to be black boxes. The data used in this study includes data on occupations other than drivers.

2 Target Data

2.1 Data Overview

We used insurance claim and regular health checkup data of employees from the SG Holdings Group. Health insurance claim data is created when a person is injured or ill and visits a medical institution, whereas health checkup data is created regularly (typically once a year). These two datasets were anonymized and linked with a hash code for unique identification of patients. In this study, disease onsets extracted from insurance claim data were used as correct answers for onset prediction; health checkup data was used as input for onset prediction. Health insurance claim data includes basic patient information such as sex, age, medical treatment date, diagnosed disease, medical procedure, and prescribed medicine. On the other hand, health checkup data includes information such as patient height, weight, blood pressure, and red blood cell count.

In this study, we analyzed insurance claim data from 2006 to 2018 and health checkup data from 1996 to 2017 for individuals aged 15–74 years. There were 961,906 sheets of health checkup data for 156,145 people, and 1,617,078 sheets of insurance claim data for 108,581 people.

2.2 Disease Names as Prediction Targets

An individual's diagnosed disease names were obtained by searching through the disease name codes in the health insurance claims data and cross-referencing with those in the ICD-10, which is the international statistical classification code of diseases and related health problems created by the World Health Organization. Table 1 presents the ICD-10 codes and the corresponding disease names to be predicted (hereinafter, referred to as "severe disease names").

Table 1. ICD-10 codes of severe disease names.

ICD-10	Disease name
E10	Insulin dependent diabetes mellitus
E11	Non-insulin dependent diabetes mellitus
E14	Diabetes mellitus other than the above
I20	Angina pectoris
I21, I22	Acute myocardial infarction
I42	Cardiomyopathy
I44 I49	Arrhythmia, Conduction defects
I60, I690	Subarachnoid hemorrhage
I61, I691	Intracerebral hemorrhage
I63, I693	Cerebral infarction

2.3 Feature Values for Prediction

The following examination items from health checkup data were used as feature values for prediction. Health examination data included not only the numerical data of inspection results, but also the results of a questionnaire on lifestyle habits and the judgment results of six levels derived from the examination data by medical institutions. The six levels were normal, almost normal, guidance required, reexamination required, close examination required, and treatment required. Factors of abdominal girth and visual acuity judgment, heart rate, visual acuity judgment, fundus judgment, and metabolic judgment were removed because 50% or more values were missing. Health examination data also included findings freely described by doctors; however, we excluded them because natural language understanding is required for using free description.

Health examination data items used as input features are as follows:

Sex; Age; Height; Weight; Body fat percentage; Systolic blood pressure; Diastolic blood pressure; Number of red blood cells; Hemoglobin; Hematocrit; Platelet count; GOT; GPT; γ-GTP; Total cholesterol; HDL cholesterol; LDL cholesterol; Neutral fat; Uric acid; Creatinine; eGFR; HbA1c; Questions regarding insulin injections or medicines to lower blood sugar, medicines to lower blood pressure, ameliorate dyslipidemia, stroke, chronic renal failure, and anemia; Lifestyle questions related to smoking habits, weight change from the age of 20 years, exercise habits, walking habits, walking speed, weight change over the past year, eating speed, meals just before going to bed, after dinner snacks, skipping breakfast, drinking habits, amount of alcohol consumed, sleeping time, willingness to improve lifestyle habits, and willingness to receive health guidance; Judgments on urinary protein and urine sugar; Representative judgment; Judgments on physical measurements, hearing ability, blood pressures, anemia, liver function, renal function, uric acid and gout, blood sugar, sugar metabolism, and urinalysis; Examination judgment.

2.4 Characteristics of Data

The data typically had two characteristics. First is considerable imbalance; for example, in 2017, the proportion of people diagnosed with severe diseases was only 4.5%. Learning of such unbalanced data may be greatly affected by the properties of many negative examples, i.e., persons not diagnosed with severe diseases. Therefore, a method that can successfully learn this imbalanced data must be adopted.

Second, classifying data as positive or negative is not easy. In this study, we aimed to predict whether a person who is healthy at the time of a regular health checkup will be diagnosed with a severe diseases within a year of the checkup. Consequently, data was classified as positive if the person would fall sick within a year and as negative if not.

The point at which the target disease name first appeared in an employee's insurance claim data is not necessarily the point when he/she first developed the disease. It is not uncommon for individuals with previously diagnosed diseases to join a health insurance association, especially, in an industry with large personnel flow. However, because the data used in this study was from a health insurance association, it was only for the period of joining the association; the insurance claim data before entering the association could not be confirmed. The next section describes a way to address this problem.

3 Data Selection and Machine Learning

3.1 Data Selection

We address the classification problem of predicting whether individuals will suffer severe illness within one year by using medical examination data. The selection method for positive and negative data was as follows.

First, to address the previously mentioned data availability problem, we determined whether an individual's first-time diagnosis of a severe disease as per the insurance claim data was actually the first. We first calculated the hospital visit interval for the disease after the diagnosis. If the hospital visit interval was shorter than the interval between the day of joining the health insurance association and the day of first-time diagnosis of a severe disease, the diagnosis was considered to be the first; alternatively, if the visit interval was greater, it was considered not to be first. The visit interval was calculated using three interval data, which was considered as sufficient. The specific procedure was as follows (see also Fig. 1):

1. Extract the oldest data item (*) containing a severe disease name from the individual's insurance claim data.
2. Select three consecutive data items with the same disease name that are newer than the extracted data and calculate the hospital visit intervals.
3. Retrieve the individual's oldest insurance claim data (**) and calculate the interval between data (**) and data (*).
4. If the maximum of the three values calculated in "2." is smaller than that calculated in "3.," regard data (*) as the first-diagnosis data of the disease.

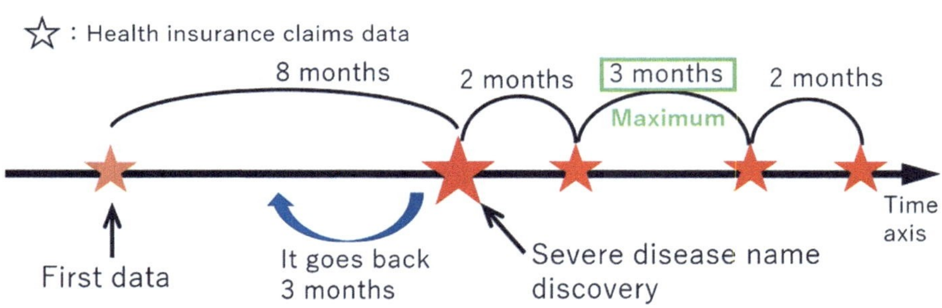

Fig. 1. Judging the genuineness of the first diagnosis.

Next, positive health checkup data was chosen from the data included within one year or less before the first appearance of the severe disease. When there were multiple data items in the range, we considered the oldest one. We also considered the amount of changes in the health checkup data, calculated the differences between the chosen data and the previous data and between the chosen data and the previous two data, and added them to the feature set (Fig. 2). Because some items in the health checkup data would have changed with the development of the disease, we considered that the discrimination accuracy would improve by explicitly adding the change to the feature set.

For negative data, we excluded data of individuals who had been diagnosed with a severe disease even once; only the remaining data was used. Furthermore, if there is no insurance claim data more than one year after a health checkup of an individual, we excluded the health heckup data from negative data because

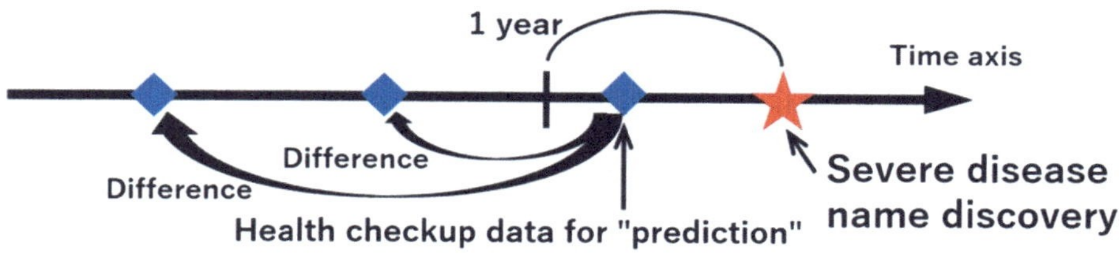

Fig. 2. Selection of positive data.

the person may have left his job and may have been diagnosed with a severe disease within a year. As with positive data, we calculated the differences using three consecutive health examination data and added them to the feature set (Fig. 3).

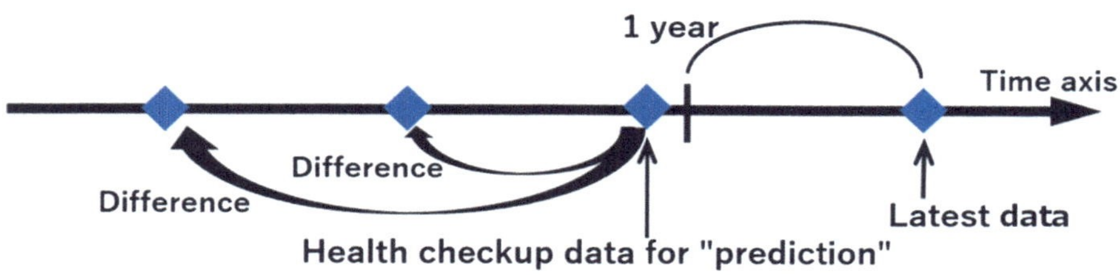

Fig. 3. Selection of negative data.

There were cases in which an individual had multiple medical examination data that fit the selection criteria. This was true for both positive and negative data. However, we used data for only one medical examination per individual to prevent data imbalance. If we did not select the oldest data for positive data, we used data after diagnoses of one of the severe diseases; however, negative data could be selected for any point in time.

Thus, we obtained 1255 positive data and 37664 negative data, with 132 features. The missing values were filled with median values.

3.2 Machine Learning for Imbalanced Data

This study deals with imbalanced data with only approximately 3% positive cases. Many methods for learning imbalanced data have been proposed such as cost-based strategies, undersampling, and oversampling. Cost-based strategies penalize misclassifications of the minority class more heavily than those of the majority class. Undersampling considers equal number of examples from both positive and negative classes by discarding training data from the majority class; whereas, oversampling achieves this by introducing duplicate instances from the minority class. Because duplication tends to cause overfitting, Synthetic Minority

Over-sampling Technique (SMOTE) [9] was proposed, which creates synthetic examples of the minority class by combining neighboring examples.

Wallace, Small, Brodley, and Trikalinos [10] argued that "undersampling + bagging" is the best strategy for handling imbalance. From the probabilistic interpretation of class imbalance, undersampling should be used to handle an imbalance in most scenarios; further, bagging should be used to reduce the variance of the approach. They also showed that "undersampling + bagging" should outperform cost-based strategies and SMOTE on datasets with high dimensionality, low prevalence, or small training set size. Based on their claim, we adopted "undersampling + bagging", and used decision trees, without pruning, as weak learners because they were unstable; combining them led to better performance.

4 Results and Discussion

4.1 Comparison with a Baseline Method

Five hundred weak learners were used in the proposed method. If the number of weak learners was changed to anywhere between 100 and 500, there was only slight change in recall and precision; however, if the number was less than 100, the precision reduced. Recall and precision did not change even when the number of classifiers was increased beyond 500; hence, the number of weak learners was set to 500. Because the decision tree used for the weak learners was an algorithm that is not affected by scale, data scaling was not performed. We adopted stratified 10-fold cross-validation as an evaluation method. Table 2 presents the confusion matrix of the proposed method. The positive precision and recall were 0.32 and 0.89, respectively.

Table 2. Confusion matrix of the proposed method.

		Predicted class	
		Positive	Negative
Actual class	Positive	1120	135
	Negative	2366	35298

We used a judgment category table[1] officially released by the Japan Society of Ningen Dock as the baseline method. This is because notifications to the examinees of regular health checkup are usually made based on similar standards in Japan. A threshold value for each item was set and discrimination was carried out by the logical OR operation on each item. We only used the items that were common to the input features of this study. The total number of items used was

[1] https://www.ningen-dock.jp/wp/wp-content/uploads/2013/09/Dock-Hantei2018-20181214.pdf.

13. The table classified health checkup data into four categories: No abnormality, Mild abnormality, Follow-up required, and Medical treatment required.

The precision-recall curves of the proposed method and the baseline method are shown in Fig. 4. The thresholds of three categories, excluding "No abnormality", were used to plot the precision-recall curve of the baseline method, plotted from left to right, in the order of Medical treatment required, Follow-up required, and Mild abnormality. For the proposed method, the ratio of the positive and negative examples in undersampling was changed from left to right as follows: 1:16, 1:8, 1:4, 1:2, 1:1, 1:0.5, and 1:0.25.

Two precision-recall curves are plotted for the proposed method: one with all 132 available features, and the other with the same 13 features as the baseline method.

The difference between the baseline method (13 features) and the proposed method (13 features) shows improvement when using the machine learning method, and the difference between the proposed method (13 features) and the proposed method (132 features) shows the improvement due to the increase of the features.

Because the graph of the proposed method lies clearly above that of the baseline method, the proposed method can be considered superior to the baseline method. The results demonstrate that a much higher precision can be obtained by the proposed method when the recall is approximately the same degree as the baseline method; increasing precision while maintaining recall is important as it allows suppression of the number of targets for health guidance without further neglecting those who are likely to be severely ill. In other words, the proposed method will provide more effective health guidance within the limited number of public health nurses. Moreover, recall or precision can be set to each customer's desired value by changing the ratio of positive and negative cases in undersampling.

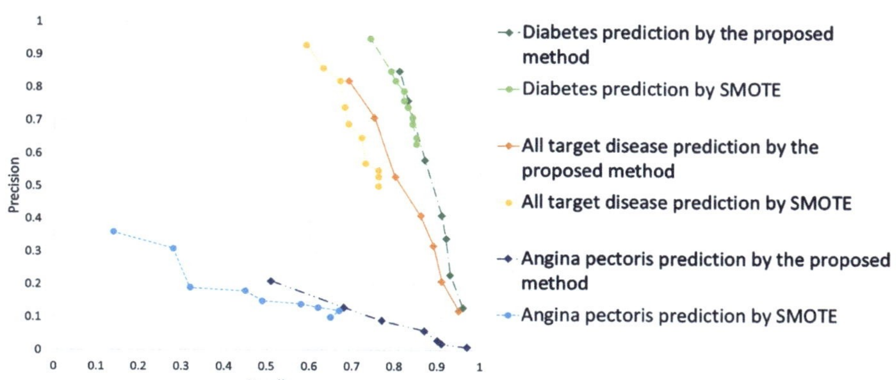

Fig. 4. Precision-recall curves of proposed and baseline methods.

4.2 Important Features and Influence of Disease Frequency

We identified the features that were important for classification. The top three were HbA1c, metabolism judgment, and the question regarding insulin injections or a drug that lowers blood glucose. Metabolic judgment refers to the judgment of the accumulation of risk factors leading to lifestyle-related diseases in six levels from the health checkup data. HbA1c is one of the indicators of diabetes, and prescribing insulin injection and medication to lower blood glucose is a treatment for diabetes. The fact that there were two diabetes-related features among the top three seems realistic when considering the number of diabetes mellitus, which was as high as 74% among the positive data. Besides, diabetes seems to be easy to predict using the two features; therefore, we hypothesized that because the onset of diabetes could be accurately predicted, the overall prediction of lifestyle-related diseases was successful.

To confirm this hypothesis, we performed several analyses. First, the results in Table 2 were reanalyzed for each disease name. There were 1255 positive cases; Table 3 shows a breakdown by disease name. The recall rate gradually decreases as the number of positive examples decrease. Acute myocardial infarction and intracerebral hemorrhage had very few positive cases (three); hence, the recall is not reliable.

Table 3. Breakdown by disease name for prediction by the proposed method.

Disease names	Positive example	True positive	False negative	Recall
Diabetes mellitus	857	797	60	0.930
Angina pectoris	208	175	33	0.841
Arrhythmia, Conduction defects	104	83	21	0.798
Cerebral infarction	66	49	17	0.742
Cardiomyopathy	14	10	4	0.714
Acute myocardial infarction	3	3	0	1.000
Intracerebral hemorrhage	3	3	0	1.000
Total	1255	1120	135	0.89

In the above analysis, we could discuss the relationship between recall and disease frequency, but not with precision because we did not make predictions for each disease. Therefore, next, by making predictions for each disease, we will examine the changes in recall and precision depending on the frequency of the diseases. We will first focus solely on diabetes and try to predict it. Positive cases were defined as persons diagnosed with diabetes mellitus; negative cases comprised patients with severe diseases other than diabetes and healthy people. The created dataset comprised of 921 positive and 37998 negative cases. As a result of prediction with 1:1 undersampling, the positive precision and recall were 0.34 and 0.92, respectively. The confusion matrix is shown in Table 4.

Table 4. Confusion matrix for diabetes prediction using the proposed method.

		Predicted class	
		Positive	Negative
Actual class	Positive	836	85
	Negative	1812	36186

Next, for comparison, we tried to predict the second most frequent disease, angina, using the proposed method. Again, positive cases were defined as persons diagnosed with angina pectoris, and negative cases consisted of patients with severe diseases other than angina pectoris and healthy people. The created dataset comprised of 229 positive and 38690 negative cases. Prediction with 1:1 undersampling showed positive precision and recall of 0.03 and 0.90, respectively. Table 5 shows the confusion matrix. It shows that the recall did not change much, but the precision dropped significantly. Figure 5 in the next section shows the change in recall and precision when the ratio of positive and negative cases in undersampling is varied. It can be confirmed that the precision of predicting angina was low from the figure.

Table 5. Confusion matrix for angina pectoris prediction by the proposed method.

		Predicted class	
		Positive	Negative
Actual class	Positive	204	25
	Negative	5133	33557

4.3 Comparison with an Oversampling Method

In this section, we compare the proposed method with an oversampling method. SMOTE [11] was used to augment positive data and random forests were used for learning. When this method was used to identify all target disease names as positive examples, positive precision and recall were 0.52 and 0.76, respectively The confusion matrix is shown in Table 6.

Table 6. Confusion matrix using SMOTE and random forests.

		Predicted class	
		Positive	Negative
Actual class	Positive	952	303
	Negative	866	36798

The precision-recall curves for this method when all target disease names are identified as positive examples, when only diabetes is identified as positive examples, and when only angina pectoris is identified as positive examples, are shown in Fig. 5. For SMOTE, we changed the ratio of the positive and negative examples in oversampling from left to right as follows: 0.1:1, 0.2:1, 0.3:1, 0.4:1, 0.5:1, 0.6:1, 0.7:1, 0.8:1, 0.9:1, and 1:1. For comparison, the precision-recall curves of the proposed method for the same three targets are also plotted in the figure. The ratio of the positive and negative examples in undersampling was changed in the same manner as in Sect. 4.1.

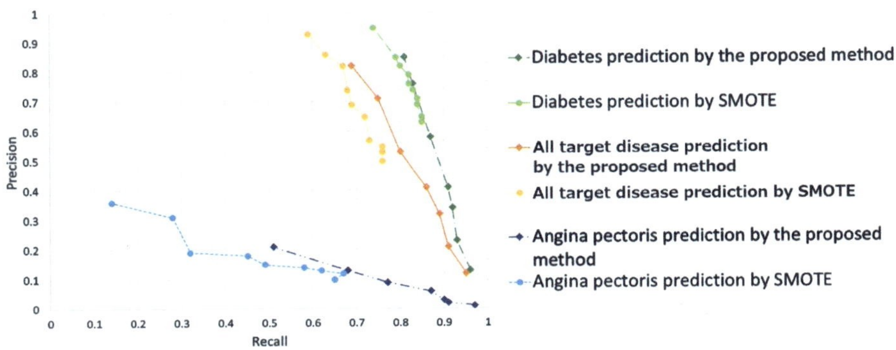

Fig. 5. Precision-recall curves for all lifestyle-related diseases and for each of the top two diseases with a large number of patients: Comparison between the proposed method and SMOTE.

The figure shows that there is not much difference between the proposed method and SMOTE when diabetes is the target for prediction. However, SMOTE has a slightly lower performance than the proposed method when all lifestyle-related diseases or angina are considered. This is because the prediction of all lifestyle-related diseases and angina are more difficult when using SMOTE than diabetes considering the proposed method as a standard.

5 Conclusion and Future Work

In this study, we proposed a method to predict the onset of lifestyle-related diseases (except cancer) using periodical health checkup data; learning data was selected based on insurance claim data. When all target disease names were identified as positive cases, positive precision and recall values of 0.32 and 0.89, respectively, were obtained. Compared to the judgment category table (used by the Japan Society of Ningen Dock) as a baseline, much higher precision could be obtained when the recall was about the same degree.

We excluded doctor findings from the health examination data; however, applying natural language processing to these findings will allow the data to be used. As another method of coping with imbalanced data, we plan to use an anomaly detection method that constructs a model using data of healthy people as normal data and then detecting data that does not fit the model. We also plan to predict the onset of lifestyle-related diseases after one year of a regular medical examination.

A major limitation of this study was that although insulin injection and drugs that lower blood glucose should not be prescribed before receiving a diagnosis of diabetes, the item "Do you take insulin injection or a drug that lowers blood glucose" was used as one of the main items to predict the onset of diabetes. This means that the selection process of positive and negative data needs to be reconsidered. There may be cases in which over-the-counter medicines were purchased without a visit to the doctor.

References

1. Hippisley-Cox, J., Coupland, C.: Identifying patients with suspected colorectal cancer in primary care: derivation and validation of an algorithm. Br. J. Gen. Pract. **62**(594), e29–e37 (2012). https://doi.org/10.3399/bjgp12X616346
2. Yatsuya, H., Iso, H., Li, Y., Yamagishi, K., Kokubo, Y., Saito, I., Sawada, N., Inoue, M., Tsugane, S.: Development of a risk equation for the incidence of coronary artery disease and ischemic stroke for middle-aged Japanese, Japan public health center-based prospective study. Circ. J. **80**(60), 1386–1395 (2016). https://doi.org/10.1253/circj.CJ-16-0081
3. Uematsu, H., Yamashita, K., Kunisawa, S., Otsubo, T., Imanaka, Y.: Prediction of pneumonia hospitalization in adults using health checkup data. PLoS One **12**(6) (2017). https://doi.org/10.1371/journal.pone.0180159
4. Weng, F. S., Reps, J., Kai, J., Garibaldi, M. J., Qureshi, N.: Can machine-learning improve cardiovascular risk prediction using routine clinical data? PLoS One **12**(4), (2017). https://doi.org/10.1371/journal.pone.0174944
5. Lee, H., Shin, S.Y., Seo, M., Nam, G.B., Joo, S.: Prediction of ventricular tachycardia one hour before occurrence using artificial neural networks. Sci. Rep. **6**, 32390 (2016). https://doi.org/10.1038/srep32390
6. Wang, J., Deng, Z.Q.: Modeling and prediction of oyster norovirus outbreaks along gulf of Mexico coast. Environ. Health Perspect. **124**(5), 627–633 (2016). https://doi.org/10.1289/ehp.1509764
7. Celik, G., Baykan, O.K., Kara, Y., Tireli, H.: Predicting 10-day mortality in patients with strokes using neural networks and multivariate statistical methods. J. Stroke Cerebrovasc. Dis. **23**(6), 1506–1512 (2014). https://doi.org/10.1016/j.jstrokecerebrovasdis.2013.12.018
8. Peng, J.C., Ran, Z.H., Shen, J.: Seasonal variation in onset and relapse of IBD and a model to predict the frequency of onset, relapse, and severity of IBD based on artificial neural network. Int. J. Colorectal Dis. **30**(9), 1267–1273 (2015). https://doi.org/10.1007/s00384-015-2250-6
9. Buda, M., Maki, A., Mazurowski, M.A.: A systematic study of the class imbalance problem in convolutional neural networks. Neural Netw. **106**, 249–259 (2018). https://doi.org/10.1016/j.neunet.2018.07.011

10. Wallace, C.B., Small, K., Brodley, E.C., Trikalinos, A.T.: Class imbalance, Redux. In: IEEE 11th International Conference on Data Mining. IEEE Xplore (2011). https://doi.org/10.1109/ICDM.2011.33
11. Nitesh, V.C., Aleksandar, L., Lawrence, O.H., Kevin, W.B.: SMOTEBoost: improving prediction of the minority class in boosting. In: 7th European Conference on Principles and Practice of Knowledge Discovery in Databases, pp. 107–119. Springer, (2003). https://doi.org/10.1007/978-3-540-39804-2_12

Feasible Affect Recognition in Advertising Based on Physiological Responses from Wearable Sensors

Taweesak Emsawas[1(✉)], Ken-ichi Fukui[2], and Masayuki Numao[2]

[1] Graduate School of Information Science and Technology, Osaka University, Osaka, Japan
taweesak@ai.sanken.osaka-u.ac.jp
[2] The Institute of Scientific and Industrial Research (ISIR), Osaka University, Osaka, Japan

Abstract. This is an extension from a selected paper from JSAI2019. Recent studies in affective computing have facilitated and stimulated the development of systems and sensors that can recognize and interpret human affects. Affective computing has been applied in various domains, and one of the applied domains is in the marketing area to increase the consumers' appeal and attraction. In particular, advertisements (ads) can convey amounts of information in a short time. Therefore, using physiological responses can help to acquire a user's feedback and obtain an advantage. This study proposes non-invasive affect recognition in each scene of an advertising video using electroencephalogram (EEG), electrocardiogram (ECG) and eye-tracking. The preliminary analysis of EEG shows the relationship between scene feeling score and emotional affects regarding physiological responses. Hence, we also trained two types of recognition models: window recognition and sequence learning. The models learned from the physiological responses and questionnaires on a user's preference in each ad scene.

Keywords: Affect recognition · Physiological response · Wearable sensor

1 Introduction

In the marketing area, Marketer always tries to reach maximum customer for getting their attraction and sales. Consequently, the ads have been created in various ways, such as articles, billboards, videos. Nowadays, TV commercial plays a role in human life, and consumer commonly watches many ad videos through various channels. The goal of creating ads video is to convey amounts of information in a short time. For this reason, all scene in the ads is essential information and should make an impression on consumers. To comprehend feedback and obtain an advantage, consumers' response is needed, and there can gather in many ways such as questionnaires, external appearances, and internal appearances. However, the questionnaires can not be guaranteed reliability.

Y. Ohsawa et al. (Eds.): JSAI 2019, AISC 1128, pp. 27–36, 2020.
https://doi.org/10.1007/978-3-030-39878-1_3

Also, in external appearances, the subject might pretend to conceal their real emotions like a facial expression and speech response. The acquired responses can be biased for any reason in these circumstances. In contrast, the internal responses—the physiological signals are directly in response to the nervous system of humans. Cannon's theory [1] mentioned that emotional state changes according to the nervous system of human body. Thus far, many efforts have been taken into gathering the internal responses for analyzing genuine emotions and acquiring informative knowledge.

Recently, the studies are interested in understanding the user's preferences regarding human affect or emotion. Hence, emotion recognition based on physiological signals has been a hot topic and applied in various domains such as health care, game and commercial [2,3]. The emotions can be represented in various ways, which are principally discrete categories and multiple dimensions. In psychology, human affect is a concept to describe the experience of feeling or emotion, and affect transmits physiological responses with stimuli (e.g., pictures, audio and videos). To acquire the data from a subject, sensors have been developed for different purposes such as for laboratory or wearable depend on the feasibility and accurate circumstances [4].

In recognition method based on physiological data [5], after extracting and selecting the features, various researches utilize these features to train a model and classify different emotional states. However, each model applies a different method to represent input data or physiological signals. General techniques train with the whole signal data and classify labels such as multilayer perceptron (MLP) [6], support vector machine (SVM) [7] and so on. While another technique, which studies continuous time interval of sequential behavior, or sequence learning [8] in affective computing recognizes the continuous human affects appropriately. Long-short term memory (LSTM) networks [9] are efficient sequence learner that suits to recognition tasks. Moreover, the learning alternatively considers the classification tasks, which are subject-independent to generalize a model spaciously, or subject-dependent to specific a model individually [10].

This study focuses on continuous ads-affect recognition that considers how the strength of changing states while watching ads. We propose subject-independent recognition using non-invasive wearable sensors, and a 15-s ad video was selected as stimuli. The data acquisition and preprocessing are described in Sect. 2, including electroencephalogram (EEG), electrocardiogram (ECG), and eye-tracking. In this study, we study the subject affects as negative or positive (low or high) emotional affects each scene interval. In Sect. 3, we present our research methodology and experiment setting. The data trend is presented, which represents an observation of the relationship between physiological data and feeling score. Furthermore, The learning techniques, including MLP, SVM, and LSTM networks, were performed to recognize these emotional human affects. In Sect. 4, the result of the preliminary experiment shows a relation between the scene-feeling score of questionnaires and the physiological responses. Moreover, recognition and comparison results are shown in this section. The results of continuous affect recognition can support the interpretation of the scene interesting and how subjects interact with ad scenes. Finally, in Sect. 5. We present the study's discussion and conclusion.

2 Data Acquisition and Preprocessing

In this study, 130 subjects between 20 and 50 years of age including half of males and females participated in the experiments, and all subjects had never watched the ad video before, to assure a genuine reactions. The 15-s ad video of toothpaste was chosen as stimuli to elicit emotional affect from the subjects. This ad video contains 11 scenes which respectively present tooth problems, bacteria illustration, brushing teeth, cleaning up and toothpaste products. The subjects were asked to watch the video while wearing Neurosky's sensors. Then, after watching the video, all subjects were asked to fill out the questionnaires concerning scene feeling based on 7 scales of preference (-4 to 4). We represent the score of 0 to 4 and -1 to -4 as positive and negative labels respectively.

2.1 Electroencephalogram (EEG)

The frontal-brain activity plays a crucial role in emotion recognition [11] and distinguishes emotional affects, which were known to vary in affective valence such as positive and negative. In this research, EEG data were acquired from one electrode in the frontal-brain area. Various techniques have been proposed in the feature extraction step to estimate the emotional state from the acquisition signals [12,13]. Frequency domain features as power spectral density (PSD) seem to be the most commonly used method [5]. Consequently, PSD feature extraction was employed to extract features from EEG signal into five frequency bands: delta (0–4 Hz), theta (4–8 Hz), alpha (8–13 Hz), beta (13–30 Hz), and gamma (>30 Hz) from low to high frequencies. Neurosky's sensor calculated and recorded five frequency bands in each second.

2.2 Electrocardiogram (ECG)

CardioChip/BMD10X-based devices of Neurosky's sensors recorded and converted to ECG signals. ECG signal consists of P-QRS-T waves in each one cardiac cycle. The features were extracted throughout the whole P-QRS-T segment as the standard deviation of the RR intervals (SNDD), the number of pairs of successive RR intervals that differ by more than 50 ms (NN50) and the proportion of NN50 that divides by the total number of NN intervals (PNN50) [14,15].

2.3 Eye-Tracking

Video-based method of eye tracking is widely used in commercial eye trackers which measures horizontal and vertical components of the movements of both eyes [16]. The distance of each X and Y pairs was calculated and used as an eye-tracking feature.

To synchronized all physiological features, the lower-frequency feature was duplicated multiple times unit it matches the timestamp with the higher-frequency feature. For example, in the case of all features, EEG and ECG features were duplicate until they match with eye-tracking features.

For the sake of simplicity, the targeted labels of affect recognition were extracted from questionnaires using mapping and rearrangement depend on the scene period. Thus, each second which involves more than 2 scenes would calculate the average of time-weighting by

$$f_t = \sum_{i}^{N_t}(sf_i * t_i) \tag{1}$$

where f_t is feeling score at time t, sf_i is feeling score at scene i, t_i is time weight at scene i, N_t is the number of scene which relates to time t and t is observed timestamp $(1, 2, 3, ..., 15)$.

3 Research Methodology

In this study, we focused on the sequence-to-sequence learning between physiological signals and emotional affect. Preliminarily, we presented the trend value to comprehend the relation of physiological data over time. For affect recognition, the emotional affect was classified as positive and negative. The techniques can be divided into two types: window recognition and sequence learning. The input and output data were constructed from extracted features and questionnaires, which are shown in Fig. 1. This research also experiments to compare the features used as only EEG features and all features because some physiological responses, in 15-s ads-video watching tasks, can not gain explicit responses.

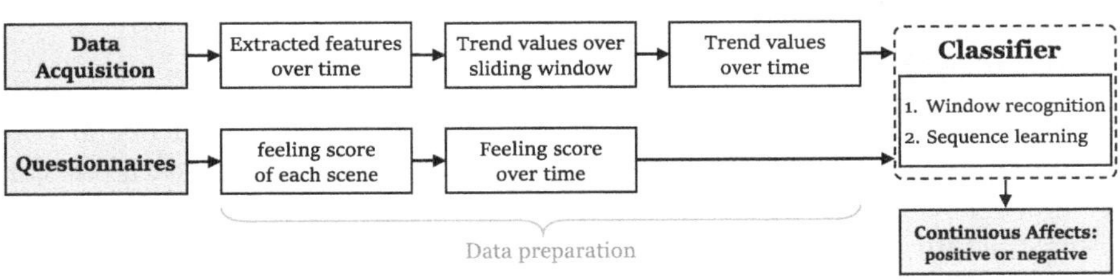

Fig. 1. Affect recognition process using physiological responses

3.1 Data Trends

To observe the relationship of physiological data and feeling score, representation of the same axis is required to indicates whether each particular band is increasing or decreasing over time, and how strong each data values. We present the data trend by using trend value or r which represents the trend of data over time. Each trend value calculated by using the 4-second sliding window technique. The positive value is a direct variation of data values over time, and the negative value is an inverse variation. The trend value can be calculated by

$$r = \frac{\sum_i^n (x_i - \bar{x})(t_i - \bar{t})}{\sqrt{\sum_i^n (x_i - \bar{x})^2} \sqrt{\sum_i^n (t_i - \bar{t})^2}} \tag{2}$$

where n is sliding window size, x is data value, t is time in second and i is running number of sliding window.

These trend values were only applied for PSD features of EEG data. While ECG and Eye-tracking, their features were directly used as the input of classifier.

3.2 Window Recognition

Window recognition recognizes emotional affects each second in which this study applied the 4-second sliding window for segmentation technique to analyze temporal data and recognize emotional affects over time. The model input was constructed from the trend values in Sect. 3.1, and the shifting size is 1 s. Then, the values were mapped to the time axis for data and label consistency. Meanwhile, the label or output was the feeling score value at each second. In our attempt to compare with the standard models, MLP and SVM are ubiquitous techniques were chosen. Two layers of feed-forward neural networks were trained with 100 hidden units and softmax activation in each layer. Besides, we trained SVM models with a linear kernel. For illustrating the data representation, Fig. 2 (left) shows the window as red-rectangle and shading, which is the information level of the sliding window at the final time step. The classifiers merely learn the information from the individual window. However, these techniques are not able to transmit useful information between each window entirely but can learn along with the values according to the sliding window size.

3.3 Sequence Learning

We also applied sequence learning [8] as LSTM networks [9] which improve from a recurrent neural network (RNN). The main idea of the RNN is to memorize previous information in the network's internal states by using self-connections. The benefit of an RNN is the ability to use contextual information from the whole data sequence, and the dependencies of inputs can remain in the network. Figure 2 (right) shows the information level of learning techniques. The LSTM architecture preserves information by using memory cells and gate units which allows the exhibiting of temporal dynamic behavior for a time series data. In

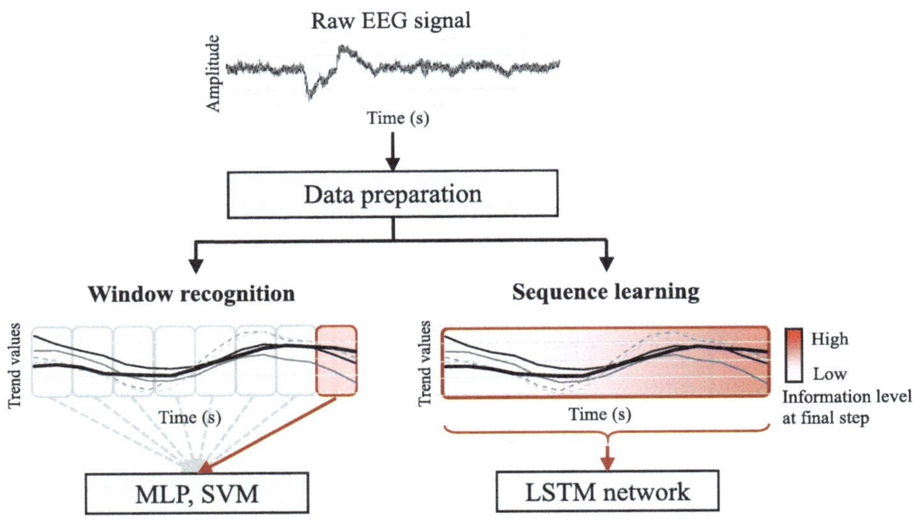

Fig. 2. Information level of window recognition and sequence learning techniques

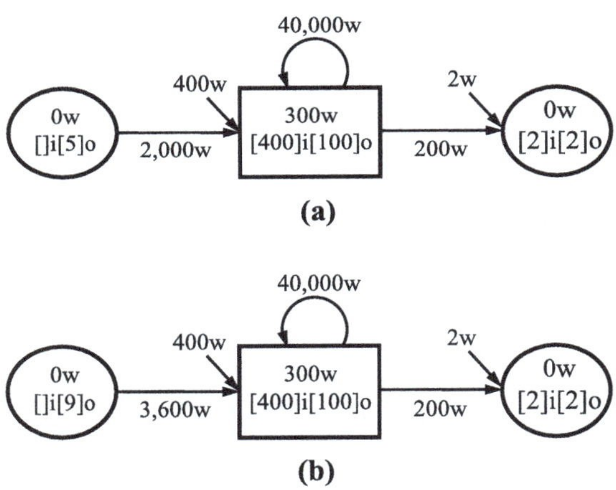

Fig. 3. The structure of the LSTM networks in our experiments. (a) For only EEG features experiment, The network has one node in the input layer (left circle) for the input vector of size 5 and the hidden layer (center rectangles), which represents 100 memory blocks. One node in the output layer (right circle) represents the output vector of size 2 (positive and negative). The number of weights between the input layer to memory block is 2,000. The self-loop connections in memory block have 40,000 weights and 400 bias weights. The number of weights between memory block to the output layer is 200 and there are 2 bias weights at the output layer. (b) the input layer has the input vector of size of 9 and the number of weights between the input layer to memory block is 3,600. The others are the same as (a)

the experiment of sequence learning, the LSTM networks were used to learn the dependencies of physiological responses. After constructing from the trend values in Sect. 3.1, we perform a 1-second sliding window that contains the data values, and each feeling score in a second was used as the label. The sliding window is shifted along the data waves to construct an input vector for each window. Then, for only EEG features experiment, we constructed the fully connected LSTM networks with 5 input nodes according to frequency domains, 100 memory blocks, and 2 output nodes according to the target output. Besides, we adjusted the input layer to 9 input nodes for all features experiment. The network setup for both only EEG features and all features used experiments are shown in Fig. 3. After having been trained, the network can be used to predict the label sequence of feeling score.

4 Experiments and Results

4.1 Preliminary Experiment

In our preliminary experiment, we calculated and investigated the trends considering the five frequency bands of only EEG data over time. The subject's averages between EEG data over time are shown in Fig. 4, and the average time-weighting of the feeling score is shown in Fig. 5, respectively.

Considering these illustrations, we found evidence that shows the relation between EEG responses and scene feeling scores. Fortunately, we can indicate the unusual point around 3 to 7 s, which is the disgusting scene of bacteria. Hence, we continued to examine the recognition experiment based on physiological responses using machine learning techniques.

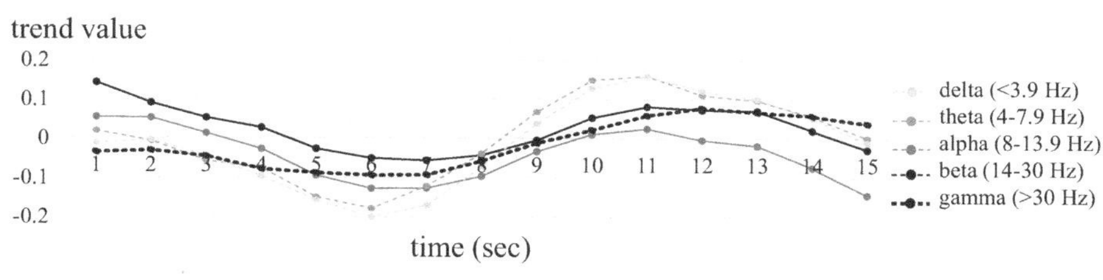

Fig. 4. Average data trends of EEG data over time for all subjects

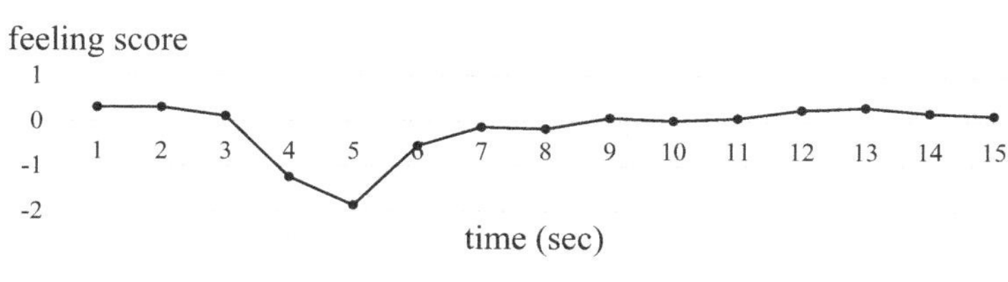

Fig. 5. Average feeling score overtime for all subjects

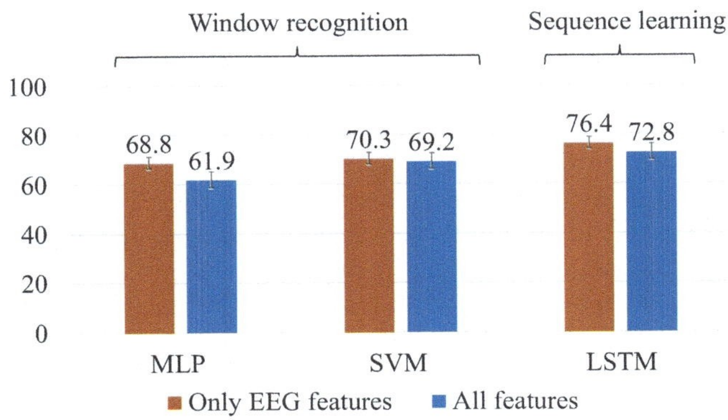

Fig. 6. Accuracy of affect recognition

4.2 Recognition Experiment

In this section, we divide the experiments, according to the features used, into two experiments: only EEG features and all features. These two experiments applied MLP, SVM and LSTM network to recognize the output label or feeling score each second. 5-fold cross-validation was performed for all subjects. The trend values and feeling score over time were used as the dataset and labels. This dataset was proportionally divided into 5 fold and used for train and test the models. In addition, MLP and LSTM networks need the validation set in the training process, so the training sets of each fold were divided into the training set and validation set with 80% and 20% respectively. The results are shown in Fig. 6. In the experiment of only EEG features, LSTM network achieved the highest accuracy at 76.4%, and SVM and MLP achieved 70.3% and 68.8% respectively. Unfortunately, all features experiment achieved an accuracy of 72.8%, 69.2%, 61.9% respectively.

To illustrate the comprehensive of continuous recognition, we plot two user results of LSTM network prediction, which are shown in Fig. 7. The blue bold-line shows the target feeling score, and the orange dot-line shows the predicted feeling score. The x-axis represents the time domain, and the y-axis represents the binary emotional affects: positive and negative. The results can appropriately capture the emotional affect over time.

5 Discussion and Conclusion

In this study, we present a study of feasible emotional affect recognition based on physiological responses from wearable sensors. The models constructed through this study approach can be used for detecting the emotional affects change over time. For using physiological signals, based on the recognition results, the best results are achieved in recognition when only EEG features are used. Considering ECG features and according to previous researches, we found that ECG

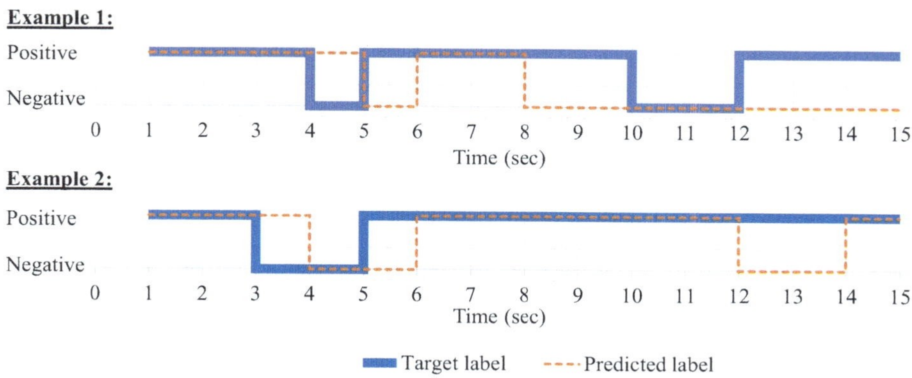

Fig. 7. The examples of target and predicted results using LSTM network along the duration

features could not perform good results in a short duration. While eye-tracking features in this experiment could not denote the emotional affects, the subjects just watched the interesting points in each scene. In recognition techniques, we present general recognition and sequence learning techniques. The sequence learning outperformed accuracy through the learning of data sequences. The results could gain a continuous affect along with the duration and apply to comprehend the feedback and obtain an advantage from the interesting points in ads. Machine learning techniques can detect the user's response to ads watching by using wearable sensors. The acquired responses can be utilized as informative knowledge without concealing.

References

1. Cannon, W.: The James-Lange theory of emotions: a critical examination and an alternative theory. Am. J. Psychol. **39**, 106–124 (1927)
2. Lee, N., Broderick, A.J., Chamberlain, L.: What is 'neuromarketing'? A discussion and agenda for future research. Int. J. Psychophysiol. **63**(2), 199–204 (2007)
3. Shu, L., Xie, J., Yang, M., Li, Z., Li, Z., Liao, D., Xu, X., Yang, X.: A review of emotion recognition using physiological signals. Sensors **18**(7), 2074 (2018)
4. Ragot, M., Martin, N., Em, S., Pallamin, N., Diverrez, J.M.: Emotion recognition using physiological signals: laboratory vs. wearable sensors. In: Ahram, T., Falcão, C. (eds.) Advances in Human Factors in Wearable Technologies and Game Design, vol. 608, pp. 15–22. Springer, Cham (2017)
5. Alnafjan, A., Hosny, M., Al-Ohali, Y., Al-Wabil, A.: Review and classification of emotion recognition based on EEG brain-computer interface system research: a systematic review. Appl. Sci. **7**, 1239 (2017)
6. Popescu, M.C., Balas, V.E., Perescu-Popescu, L., Mastorakis, N.: Multilayer perceptron and neural networks. WSEAS Trans. Circuits. Syst. **8**(7), 579–588 (2009). http://dl.acm.org/citation.cfm?id=1639537.1639542
7. Hearst, M.A.: Support vector machines. IEEE Intell. Syst. **13**(4), 18–28 (1998). https://doi.org/10.1109/5254.708428

8. Graves, A.: Supervised sequence labelling with recurrent neural networks, vol. 385 (2012)
9. Hochreiter, S., Schmidhuber, J.: Long short-term memory. Neural Comput. **9**, 1735–1780 (1997)
10. Chen, J., Hu, B., Wang, Y., Moore, P., Dai, Y., Feng, L., Ding, Z.: Subject-independent emotion recognition based on physiological signals: a three-stage decision method. BMC Med. Inform. Decis. Mak. **17**, 167 (2017)
11. Schmidt, L.A., Trainor, L.J.: Frontal brain electrical activity (EEG) distinguishes valence and intensity of musical emotions. Cogn. Emot. **15**(4), 487–500 (2001). https://doi.org/10.1080/02699930126048
12. Kim, M., Kim, M., Oh, E., Kim, S.P.: A review on the computational methods for emotional state estimation from the human EEG. Comput. Math. Methods Med. **2013**, 573734 (2013)
13. Shukla, J., Barreda-Ángeles, M., Oliver, J., Nandi, G., Puig, D.: Feature extraction and selection for emotion recognition from electrodermal activity. IEEE Trans. Affect. Comput. **PP**, 1 (2019)
14. Karpagachelvi, S., Arthanari, M., Sivakumar, M.: ECG feature extraction techniques - a survey approach. Int. J. Comput. Sci. Inf. Secur. 8 (2010)
15. Dewangan, N.: A survey on ECG signal feature extraction and analysis techniques. Int. J. Innov. Res. Electr. Electron. Instrum. Control Eng. **3**, 12–19 (2017)
16. Chennamma, H., Yuan, X.: A survey on eye-gaze tracking techniques. Indian J. Comput. Sci. Eng. 4 (2013)

Deep Markov Models for Data Assimilation in Chaotic Dynamical Systems

Calvin Janitra Halim(✉) and Kazuhiko Kawamoto

Chiba University, 1-33, Yayoicho, Inage Ward, Chiba-shi, Chiba 263-8522, Japan
calvinjh@chiba-u.jp, kawa@faculty.chiba-u.jp

Abstract. This is an extension from a selected paper from JSAI2019. Recently, the use of deep learning in data assimilation has been gaining research attention. For instance, the time-series deep Markov model has been proposed along with an inference network trained using variational inference. However, the original proposal did not fully leverage the model ability for data assimilation. Therefore, we aim to evaluate the suitability of a deep Markov model and its inference network for a chaotic dynamical system, which is a common problem in data assimilation. We evaluate the model under various generative conditions. The results show that when information about part of the target model is known, the model has comparable performance to a smoothed unscented Kalman filter, even under the presence of process and observation noise.

Keywords: Data assimilation · Chaotic system · Deep Markov model · Variational inference

1 Introduction

The advancement of computer processors has allowed the simulation of complex real-world processes that are modelled by dynamical systems and numerically. However, hurdles such as chaos and numerical errors prevent the use of these models for long-term forecasting. To overcome this problem, data assimilation uses both observations and numerical models as inputs to statistical methods to reduce estimation error.

Data assimilation was motivated by the need to accurately predict/forecast natural events. For example, chaos in the theoretical models to forecast weather causes high susceptibility even to small changes or errors in the variables. This phenomenon is obvious when we use observations from sensors as initial values. Such observations can be affected by electromagnetic noise, inaccurate positioning, and other sources of statistical noise. When observations are used to forecast the system evolution, as measurements are affected by sensing errors, the consequent chaos in the system may lead to increasing divergence between the forecasted and real values with longer prediction horizons. The effectiveness

Y. Ohsawa et al. (Eds.): JSAI 2019, AISC 1128, pp. 37–44, 2020.
https://doi.org/10.1007/978-3-030-39878-1_4

of data assimilation depends on several factors such as the underlying models, minimization methods of estimation errors, and methods to handle unobservable variables. The mainstream method to perform data assimilation is to consider the problem as a Bayesian inference problem [1]. In fact, most methods for data assimilation are based on this type of inference, including the Kalman filter [2] and its nonlinear approximations (ensemble Kalman filter [3], extended Kalman filter [4]), and particle filters, also known as sequential Monte Carlo methods. Other popular methods include variational methods (3D-Var and 4D-Var), which aim at minimizing a cost function over the available data.

On the other hand, the success of deep learning has led researchers to focus on its combination with conventional data assimilation. Gu et al. [5] and Kutschireiter et al. [6] proposed particle filters augmented using principles of neural networks and with their unique characteristics. Additionally, Cintra et al. [7] used a multilayer perceptron for data assimilation. Loh et al. [8] proposed a modified long short-term memory model to predict the gas flow rates in mature gas wells, and then used an ensemble Kalman filter to update the prediction according to observations.

Krishnan et al. [9] derived the *deep Markov model* (DMM) by using neural networks to extend the Gaussian state-space model, which is essentially a hidden Markov model with continuous latent variables. They also proposed a method based on variational inference [10] to determine a state-space model by using a structured inference network to approximate the posterior latent distribution. As data assimilation problems are often represented with a state-space model and solved using Bayesian inference, we believe that the DMM approach is suitable to solve them.

Although Krishnan et al. [9] showed the potential of the model over simple synthetic data, polyphonic music generation and clinical health records, we think that more comprehensive experiments are needed to evaluate the model over chaotic systems. Furthermore, Krishnan et al. did not fully test the model ability when only part of the generative model (i.e., transition and emission functions) is available. Thus, we aim to assess the method on chaotic systems under varying availability of the transition or emission function during training.

2 Deep Markov Model

Krishnan et al. [9] proposed the DMM that consists of a generative model and an inference network trained with variational inference. The DMM uses the evidence lower bound as objective function by comparing the posterior distributions of latents generated by the inference network with their prior distributions and maximizes the expected log-likelihood of observations retrieved from the generative network.

2.1 Generative Model

The generative model takes a sequence of latents and produces the corresponding observations using a Gaussian state-space model, whose transition and emission

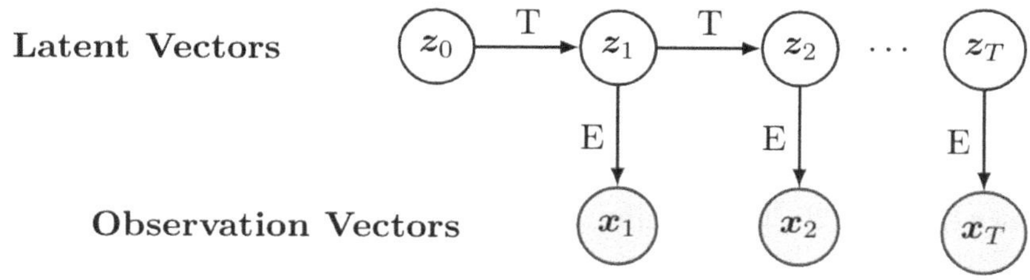

Fig. 1. Generative model (T, transition function; E, emission function)

functions can be substituted by neural networks. The transition function uses a gated transition function, which is essentially a gated recurrent unit [11] not conditioned by observations, similar to the Markovian properties of the latents.

2.2 Inference Network

The inference network takes an observation sequence and infers its corresponding latents. This network is structured upon the factorization of the posterior latent distribution [9]. Krishnan et al. [9] followed the factorization using a backward recurrent neural network (RNN), which outputs a hidden unit per timestep and then uses a *combiner* function, with the hidden unit and previous latent as input, to retrieve (sample) the approximate latent for the current timestep. As the RNN is propagated to a previous instant, we adopt the notation ST-R, *STructured-Right*, like in [9]. The structures for both the generative model and inference network are respectively shown in Figs. 1 and 2.

3 Experiments and Results

3.1 Dataset and Evaluation

We trained and evaluated the proposed approach on the Lorenz 96 model [12], which exhibits chaos and is often used to evaluate data assimilation methods. This model provides atmospheric-like dynamics, resembling atmospheric weather models, and is defined by the following differential equation:

$$\frac{dy^{(i)}}{dt} = (y^{(i+1)} - y^{(i-2)})y^{(i-1)} - y^{(i)} + F, \tag{1}$$

where $y^{(i)}$ is the i-th state of the model, $y^{(-1)} = y^{(N-1)}, y^{(0)} = y^{(N)}, y^{(N+1)} = y^{(1)}$, $\frac{dy^{(i)}}{dt}$ is the derivative of $y^{(i)}$ w.r.t. time t, and F is the forcing constant, set to 8.0 to elicit chaotic movement [12]. We then set the transition and emission functions of the generative model according to the following stochastic processes:

$$\textbf{Transtion}: \boldsymbol{z}_t \sim \mathcal{N}(\text{Lorenz96RK4}(\boldsymbol{z}_{t-1}), 5I), \tag{2}$$

$$\textbf{Emission}: \boldsymbol{x}_t \sim \mathcal{N}(\boldsymbol{z}_t, 5I), \tag{3}$$

where I is the identity matrix, function Lorenz96RK4(.) represents the 4th order Runge–Kutta integration of the Lorenz 96 model with timestep of 0.01, and $\mathcal{N}(\boldsymbol{\mu}, \Sigma)$ denotes a multivariate Gaussian distribution with mean vector $\boldsymbol{\mu}$ and covariance matrix Σ. Both the latent and observation sizes were set to 20. The training and validation data have 5000 and 500 datapoints, respectively, with sequence length of 25. The training and evaluation parameters used in the experiments are listed in Table 1.

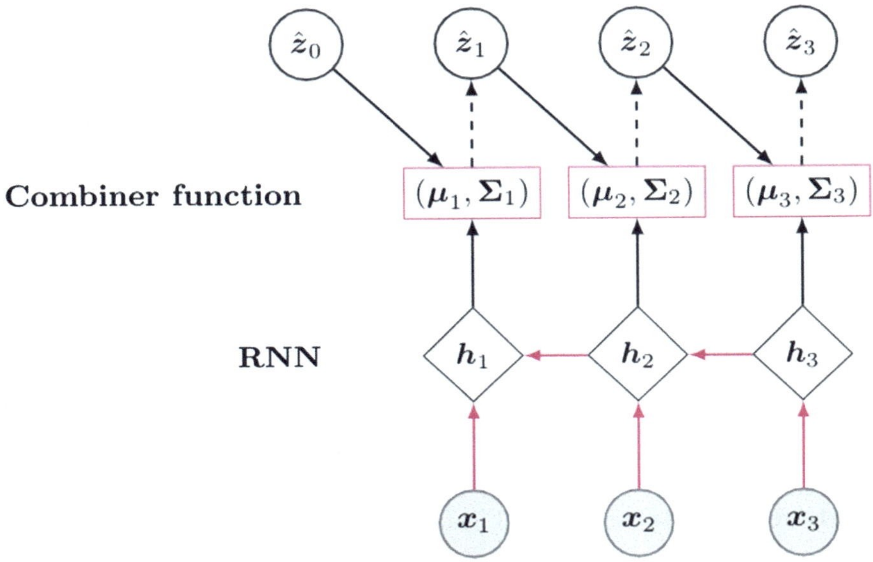

Fig. 2. Inference network for sequence length of 3

We used a smoothed unscented Kalman filter (UKF) as baseline and obtained the root-mean-square error (RMSE) between the true latents, observations, and their reconstructed counterparts to measure the model performance. We also evaluated the model on four conditions imposed on the generative model:

1. Condition 1: *fixed transition and emission*
2. Condition 2: *unknown transition and fixed emission*
3. Condition 3: *fixed transition and unknown emission*
4. Condition 4: *unknown transition and emission*

In the conditions, *fixed* indicates that the corresponding function of the generative model is set as the corresponding Eq. 2 or 3, and *unknown* indicates that the function is substituted by a neural network that is inferred during training. In addition, note that the smoothed UKF is only evaluated on condition 1, as it is not an adaptive filter.

Table 1. Training and evaluation parameters

Parameter	Value	Parameter	Value
RNN layers	1	Emission layers	2
RNN size	40	Emission hidden units	40
Transition type	Gated transition	Regularizer	l_2 Norm
Transition layers	2	Optimizer	Adam
Transition hidden units	80	Learning rate	0.0008
Emission type	MLP	Batch size	250

3.2 Experiment Modifications

Here, we describe the training variants we implemented for evaluating conditions 3 and 4.

Condition 3. During the experiment for evaluating condition 3 with the parameters shown in Table 1, the estimation of the Kullback–Leibler (KL) divergence on the training algorithm used a prior estimated from the transition function, whose inputs were retrieved from the posterior calculated by inference network, instead of directly using the latent vectors from the dataset. This is useful when backpropagating on the gated transition function.

However, during training, the inference network did not learn the proper posterior distribution and increasingly diverged from the true value over time. We believe that chaos from the Lorenz 96 model combined with added transition noise causes this divergence. As chaotic models are very sensitive to small perturbations on the state variables, even slight variations in the posterior can cause the calculated priors to diverge largely from true priors, resulting in inaccurate KL divergence.

Therefore, to prevent this problem, as we considered the transition function as fixed in this condition, we used the latent priors directly from the dataset. This decision was successful for the inference network to approximate the posterior latent distribution, as reported in Sect. 3.3.

Condition 4. As inferring both transition and emission is very difficult and considering the transition function is a chaotic system with added noise, we increased the annealing rate of the KL divergence to reach 1 from 0 after 50,000 parameter updates, aiming to increase the accuracy of the inferred posterior latents. The results after increasing the annealing rate are reported in Sect. 3.3.

3.3 Results

The evaluation results are shown in Fig. 3. *ST-R*, *UKF*, and *Observations* denote the RMSE (vertical axis) of the DMM (with ST-R), UKF, and observations, respectively, and the horizontal axis shows the training epochs. The RMSE of the DMM converges to that of the UKF as training proceeds on conditions 1 to 3. Hence, the DMM achieves state estimation comparable to the UKF and can infer part of the generative model when unavailable, showing the ability of the DMM as adaptive filter. However, the model cannot estimate the system state when only observations are given, even after increasing the annealing rate of the KL divergence (condition 4). We believe this occurs by the insufficient information of the sampled observations to infer the unknown generative network.

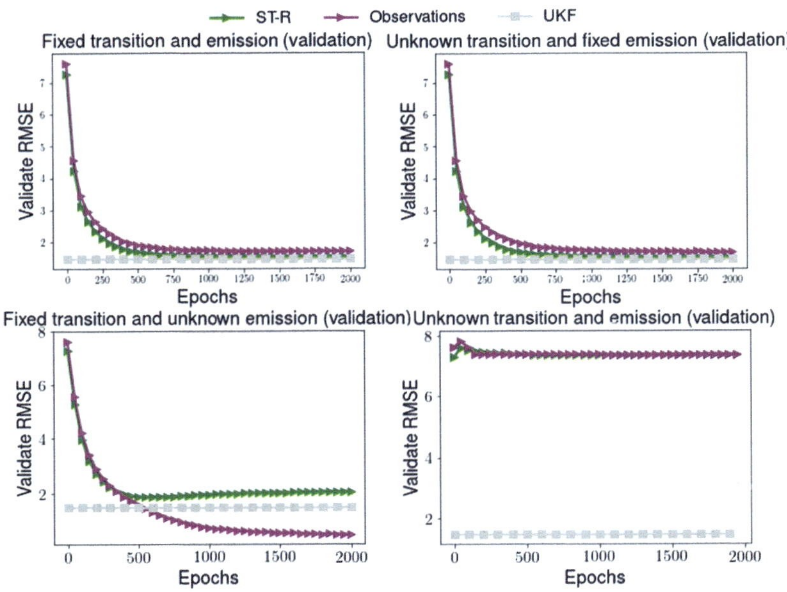

Fig. 3. RMSE from four evaluated conditions

Consider the results for condition 3. Although training converges to the RMSE of the UKF, the convergence performance is poorer than for conditions 1 and 2, which use the unmodified training algorithm. The modification described in Sect. 3.2 causes overfitting to the training data, failing to capture the inherent process noise of the transition function, as shown by the slightly larger RMSE obtained from condition 3 compared to conditions 1 and 2. In addition, the observation RMSE becomes very small as training proceeds. This might also be the result of the model overfitting the observations, similar to the results from condition 3. Therefore, we can conclude that the original training algorithm performs better when both the transition and emission functions are known, because it can better infer the process error encoded into the transition function. However, when only the transition function is known, the proposed modification can be used for training, given that the unmodified algorithm fails to learn this function. In addition, the model may be fine-tuned to suitably capture process noise after training. Still, more research is required on how to do this model fine-tuning.

4 Conclusion

We evaluated the structured inference network and DMM proposed by Krishnan et al. on a chaotic dynamical system over four conditions imposed to the generative network. The evaluation results show that the inference network can suitably approximate the true latent variables, converging close to the results from the smoothed UKF and suggesting the model suitability for state estimation of real-world models. The model performance was confirmed when the transition and emission functions are given, or when only either of them is available, as the network has the ability to infer the transition network when trained with the latent variables. Nevertheless, when both the emission and transition functions are unknown (i.e., learning an unknown system), the structured inference network and DMM fail to capture the dynamics of the system. This might be caused by the difficulty in learning a variational inference network.

We also verified that modifying the calculation of the KL divergence term by using the latent variables directly from the dataset can reduce the complexity of training when the emission function is unknown, resulting in performance comparable to that from the smoothed UKF. However, when the emission function is known, it is better to train using the original method proposed by Krishnan et al., possibly by the tendency of overfitting to the dataset retrieved by our approach, as shown in the results from condition 1. A more robust approach might be devised for this specific condition, such as taking the weighted average of the proposed and conventional KL divergences to capture the stochasticity of the transition function. This development is part of our future work. Still, the proposed approach can be used for learning a chaotic dynamical system with unknown emission function.

Overall, the DMM is suitable for data assimilation even over a chaotic dynamical system (with the addition of stochasticity). We plan to further evaluate harder grid problems and improve the model inference capability to outperform existing data assimilation methods.

Acknowledgements. This work was supported by JSPS KAKENHI under Grant Number JP16K00231.

References

1. Wikle, C.K., Berliner, L.M.: A Bayesian tutorial for data assimilation. Phys. D: Nonlinear Phenom. **230**, 1–16 (2007). https://doi.org/10.1016/j.physd.2006.09.017
2. Kalman, R.E.: A new approach to linear filtering and prediction problems. J. Basic Eng. **82**(1), 35–45 (1960)
3. Evensen, G.: Sequential data assimilation with a nonlinear quasi-geostrophic model using Monte Carlo methods to forecast error statistics. J. Geophys. Res. Oceans **99**(C5), 10143–10162 (1994)
4. Groves, P.D.: Principles of GNSS, inertial, and multisensor integrated navigation systems (2nd edn.). IEEE Aerosp. Electron. Syst. Mag. **30**(2), 26–27 (2015)

5. Gu, S.S., Ghahramani, Z., Turner, R.E.: Neural adaptive sequential Monte Carlo. In: Advances in Neural Information Processing Systems, vol. 28, pp. 2629–2637. Curran Associates, Inc. (2015)
6. Kutschireiter, A., Surace, S.C., Sprekeler, H., Pfister, J.-P.: Nonlinear Bayesian filtering and learning: a neuronal dynamics for perception. Sci. Rep. **7**(1), 8722-1–8722-13 (2017)
7. Cintra, R.S., Campos Velho, H.F.: Data assimilation by artificial neural networks for an atmospheric general circulation model. In: Advanced Applications for Artificial Neural Networks, chap. 14 (2018)
8. Loh, K., Omrani, P.S., Linden, R.V.: Deep learning and data assimilation for real-time production prediction in natural gas wells. arXiv:1802.05141 [cs.LG] (2018)
9. Krishnan, R.G., Shalit, U., Sontag, D.: Structured inference networks for nonlinear state space model. In: AAAI (2017)
10. Jordan, M.I., Ghahramani, Z., Jaakkola, T.S., Saul, L.K.: An introduction to variational methods for graphical models. Mach. Learn. **37**(2), 183–233 (1999)
11. Cho, K., van Merriënboer, B., Gülçehre, Ç., Bahdanau, D., Bougares, F., Schwenk, H., Bengio, Y.: Learning phrase representations using RNN encoder-decoder for statistical machine translation. In: Proceedings of the 2014 Conference on EMNLP, pp. 1724–1734 (2014)
12. Lorenz, E.N.: Predictability: a problem partly solved. Semin. Predict. **1**(1), 1–18 (1995)

Reducing the Number of Multiplications in Convolutional Recurrent Neural Networks (ConvRNNs)

Daria Vazhenina(✉) and Atsunori Kanemura

LeapMind Inc., Tokyo 150-0044, Japan
{daria,atsu-kan}@leapmind.io

Abstract. This is an extension from a selected paper from JSAI2019. Convolutional variants of recurrent neural networks, ConvRNNs, are widely used for spatio-temporal modeling, since they are well suited to model sequences of two-dimensional inputs. Similar to conventional RNNs, the introduction of gating architecture, such as ConvLSTM, brings additional parameters and increases the computational complexity. The computation load can be an obstacle in training efficient models and putting ConvRNNs in operation in real-world applications. However, the correspondence between ConvRNN unit complexity and its performance is not well investigated. We propose to reduce the number of parameters and multiplications by substituting some convolutional operations with the Hadamard product. We evaluate our proposal using the task of next video frame prediction and the Moving MNIST dataset. The proposed method requires 38% less multiplications and 21% less parameters compared to the fully convolutional counterpart. In price of the reduced computational complexity, the performance measured by structural similarity index measure (SSIM) decreased about 1.5%. ConvRNNs with reduced computations can be used in a wider range of situations like in web applications or embedded systems.

This paper is an extension of a selected paper from JSAI2019 [10].

Keywords: Neural networks · Convolution · Hadamard product · Recurrent neural networks · Deep learning

1 Introduction

Convolutional recurrent neural networks (ConvRNNs) are widely used because of their ability to model temporal information using two-dimensional input. ConvRNNs were design by stacking a convolutional neural network (CNN) layer and a long-short term memory (LSTM) layer [2] and successfully applied for sequential video processing and other tasks involving two-dimensional input that has temporal dependency [2,6]. It allowed to obtain informative representation that improve overall model performance. Representations (or features) of a video are

Y. Ohsawa et al. (Eds.): JSAI 2019, AISC 1128, pp. 45–52, 2020.
https://doi.org/10.1007/978-3-030-39878-1_5

useful in video contents description, activity recognition, and other tasks. ConvRNNs parse video frames sequentially and encode frame-level information.

While lots of work were done on speeding up and improving performance of conventional (i.e. non-convolutional) RNNs, less attention was paid to ConvRNNs. Similar to conventional RNNs, gated variants were proposed for their convolutional counterparts, such as ConvLSTM [6] with three gates, ConvGRU [1] with two gates, and reduced-gate ConvLSTM [3] with one gate, which is a variant of non-convolutional JaNet [9]. Gates are sensitive to short-term and long-term patterns in the input and help to overcome vanishing gradient problem. Other improvements for ConvRNNs focused on fitting better to their target task and increased overall model performance, but they resulted in increasing memory footprint, the number of parameters, and the number of floating point multiplications [7,11].

Recently, for conventional RNNs, Li et al. [4] proposed to reduce the numbers of multiplications and parameters by substituting matrix multiplication between a weight matrix and an input vector to the Hadamard product. It allowed to build a deeper network of up to 21 layers and slightly outperformed state-of-the-art models for three different tasks.

We investigate the influence of substituting convolution with the Hadamard product in ConvLSTM, ConvGRU, and reduced-gate LSTM. It is expected that such substitution would result in a large performance drop. Then, instead of replacing all the convolution operations in a network, we used the Hadamard product only in some parts of a network, in order to keep a good balance between reducing the numbers of parameters and multiplications and performance drops.

As an example of using ConvRNN, we selected the task of next video frame prediction, which learns video sequence representations of individual video frames in an unsupervised manner [8]. The next frame prediction problem is useful because (1) we don't need to obtain lots of labeled data, which is often a difficult task before adopting deep learning, and (2) whereas a system trained for one specific task will learn representations for that specific task, internal representations learned by the model for predicting next frames will be re-usable for other tasks [8]. Also, targets in supervised learning contain much less information than input data, especially in terms of video action recognition, where there is only one label per many frames. That is why learning how to forecast the future of an image sequence requires the prediction model to understand and efficiently encode the content and dynamics for a certain period of time.

2 ConvRNNs

2.1 Basic ConvRNN Architecture

The most widely used variant of ConvRNN is ConvLSTM, proposed in [6], where conventional LSTM is modified by replacing matrix multiplication with convolution and changing the shape of input X_t to 2D from 1D. Gating architecture

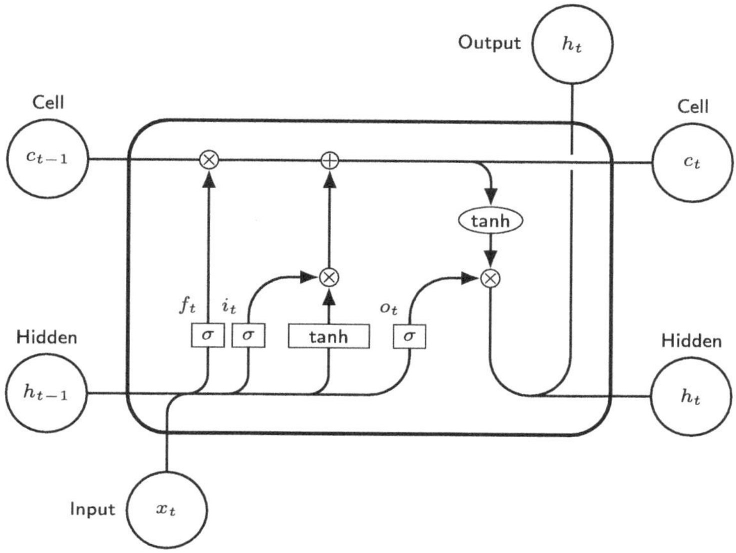

Fig. 1. ConvLSTM unit architecture.

allows to control information flow within ConvLSTM unit, as shown in Fig. 1. A ConvLSTM unit is described by the following equations:

$$i_t = \sigma(X_t * W_{xi} + h_{t-1} * W_{hi} + b_i), \tag{1}$$
$$f_t = \sigma(X_t * W_{xf} + h_{t-1} * W_{hf} + b_f), \tag{2}$$
$$o_t = \sigma(X_t * W_{xo} + h_{t-1} * W_{h0} + b_o), \tag{3}$$
$$c_t = f_t \odot c_{t-1} + i_t \odot \tanh(X_t * W_{xc} + h_{t-1} * W_{hc} + b_c), \tag{4}$$
$$h_t = \tanh(c_t \odot o_t). \tag{5}$$

Here $*$ means convolutional operation and $\odot$ is the Hadamard (element-wise) product. $W_{..}$ is a set of convolutional kernels. Subscript is used to distinguish sets of kernels, it doesn't indicate any dependency between them. h_t and h_{t-1} are current and previous hidden states, respectively and c_t is cell output. i_t, f_t, and o_t are input, forget, and output gates, respectively, all of which has the same size as hidden state. As gates values lie in the $(0, 1)$ interval, they allow the memory cells to store and access information over many training examples. For example, as long as the input gate i_t has an activation value near zero, the activation of the cell c_t will not be overwritten by the new inputs arriving in the network, and can therefore be made available to the network much later in the sequence, by opening the output gate o_t.

2.2 Next Video Frame Prediction Model Using ConvRNN

In unsupervised video representation model, we use an encoder-decoder pipeline with two different decoders termed a predictor and a reconstructor. This model was proposed in [8] as a composite model, which allowed to train the most reliable representations. In the composite model, several video frames are fed into the recurrent encoder and then its final hidden state, so called a learned representation, is used as the initial hidden state in the predictor decoder and the reconstructor decoder, while the two decoders do not share all other parameters (e.g. weight matrices). Learning the two tasks of prediction and reconstruction using the same encoder allows us to improve overall model performance, rather than using different encoders. The loss function is defined to be the sum of one for the reconstructor and the other for the predictor, and we can do backpropagation on it.

Our experiments in this paper are based on the network architecture proposed in [5], where the next frame prediction pipeline has been improved by stacking recurrent layers on the top of the CNN layers in the encoder part and reversing those recurrent and CNN layers in the decoder part, changing the CNN layers into so-called deconvolution layers. This stacking technique reduces the resolution of input image frames with convolutional layers before feeding it to the recurrent ones, thus reducing the number of parameters in the recurrent layers. This significantly reduce mean square error (MSE) of the model for both prediction and reconstruction tasks.

3 Proposed Improvements

We investigated and analyzed replacing convolutional operation with Hadamard product in ConvRNN units. This idea was inspired by IndRNN [4], which reduces the number of multiplications in the vanilla recurrent cell. It was proposed to use Hadamard product instead of matrix multiplication for the connection to the previous hidden state h_{t-1}. Since IndRNN was proposed for RNNs without gates, we cannot adopt it for our purpose of improving ConvLSTM, which includes many gates. Reduction of the number of multiplications by substituting convolution with Hadamard product has not been investigated so far for gated convolutional recurrent units.

We expect that changing all connections with the previous hidden state h_{t-1} will reduce the performance of the overall model significantly. Therefore, we retain one convolutional operation when connecting with h_{t-1}. We design the following three types of recurrent units with different degrees of replacement with Hadamard product:

- Conv*: The baseline model with standard ConvRNN structure,
- ConvIndConv*: The proposed combination of convolution and Hadamard product,
- ConvInd*: The model where convolution is replaced with Hadamard product.

Here the asterisk sign "*" is a placeholder of LSTM, GRU, or Janet. That is, "ConvInd*" can be any of ConvIndLSTM, ConvIndGRU, or ConvIndJanet. Figure 2 shows our expectations on the computational complexity (lower is better) and the performance measured by MSE (lower is better).

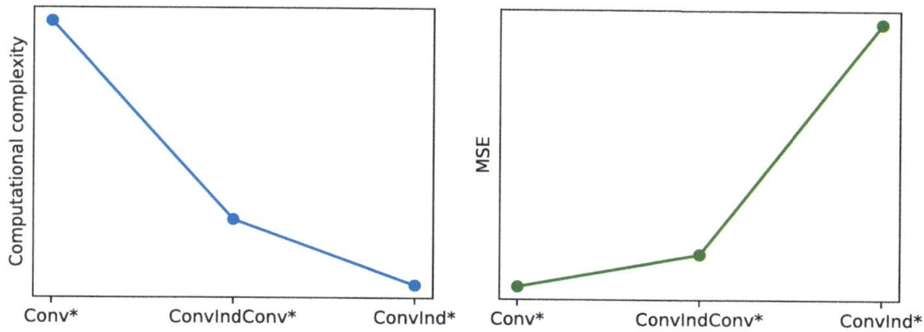

Fig. 2. Expected computational reduction and MSE increase due to substitution of convolution operation with Hadamard multiplication.

The ConvInd* models replace the convolution operations at the gates and cell units of the Conv* models with the Hadamard product. Similar to the original paper we apply it only to connect to the previous hidden state h_{t-1}. The equations behind the ConvIndLSTM unit are as follows:

$$i_t = \sigma(X_t * W_{xi} + h_{t-1} \odot W_{hi} + b_i), \tag{6}$$
$$f_t = \sigma(X_t * W_{xf} + h_{t-1} \odot W_{hf} + b_f), \tag{7}$$
$$o_t = \sigma(X_t * W_{xo} + h_{t-1} \odot W_{h0} + b_o), \tag{8}$$
$$c_t = f_t \odot c_{t-1} + i_t \odot \tanh(X_t * W_{xc} + h_{t-1} \odot W_{hc} + b_c), \tag{9}$$
$$h_t = \tanh(c_t \odot o_t). \tag{10}$$

ConvIndGRU and ConvIndJanet were changed in the similar manner.

In the ConvIndConv* models, the Hadamard product is used only for the gates units, and the cell unit, c_t, is calculated using the convolution, as follows:

$$c_t = f_t \odot c_{t-1} + i_t \odot \tanh(X_t * W_{xc} + h_{t-1} * W_{hc} + b_c). \tag{11}$$

In case of GRU and Janet architecture we use convolution for the final hidden state computation.

The changes we have introduced are highlighted in Fig. 3. The term, that is used in the cell output computation, is calculated using convolution in the ConvIndConv* unit.

4 Experimental Results

4.1 Database Description

We use the moving MNIST dataset, which has been widely used to evaluate video frame prediction models. We used the same train and test settings as in [8]. Each sequence in this dataset consists of 20 image frames with two random digits from

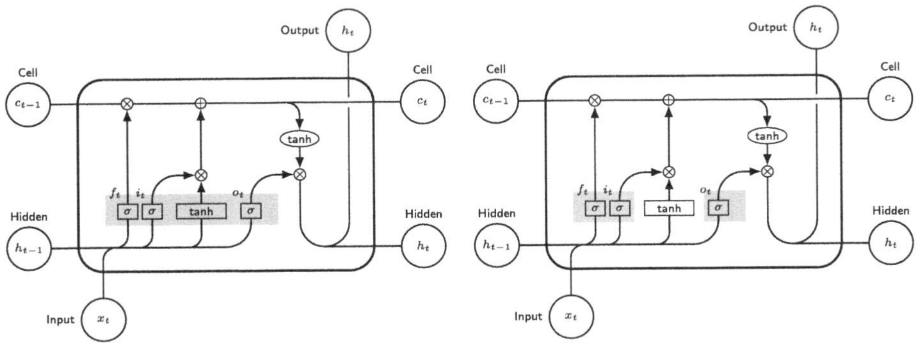

Fig. 3. ConvLSTM unit architecture with highlighted operations, where Hadamard multiplication is used.

the MNIST dataset moving in the patches of size 64×64 bouncing at the edges. The performance in next frame prediction was measured by MSE (the lower this value the better) and SSIM (the higher this value the better) between prediction and the ground truth [12].

4.2 Performance Evaluation

We evaluated three types of gated ConvRNNs and their ConvInd* and ConvInd-Conv* variants described in Sect. 3. We considered the number of multiplications and the number of parameters for complexity evaluation.

Figure 4 compared the prediction performance of the three types of ConvRNNs, presented in Sects. 2.1 and 3, in terms of MSE and SSIM. We can observe that the model with original ConvLSTM cell without the Hadamard product showed the best performance in terms of SSIM, while the model with ConvGRU cell was slightly better in terms of MSE. All the models with ConvInd* cell showed significant drop in performance of about 30% relative to their fully convolutional counterparts. Keeping the convolution operation in (11), not changing to the Hadamard product, allowed to reduce performance drop to about 5% relative to their fully convolutional counterparts.

Figure 5 shows that the best performing model with ConvLSTM cell requires the largest number of multiplications and parameters. The model with ConvInd-ConvLSTM cell showed the second best SSIM value, requiring about 37.5% less multiplications and about 21% less parameters. Its relative drop of SSIM value is 1.54% and MSE is 4.2%. This shows that it is possible to use less parameters and multiplications for gates calculations without big loss in performance.

Units ConvGRU and ConvIndConvLSTM have almost equal number of parameters, while the number of multiplications is fewer for the later one. This advantage is also reflected in better SSIM measure of ConvIndConvLSTM unit over ConvGRU one. These results show that reducing computational complexity of gated ConvRNNs by reducing number of gates in the recurrent unit, i.e. use

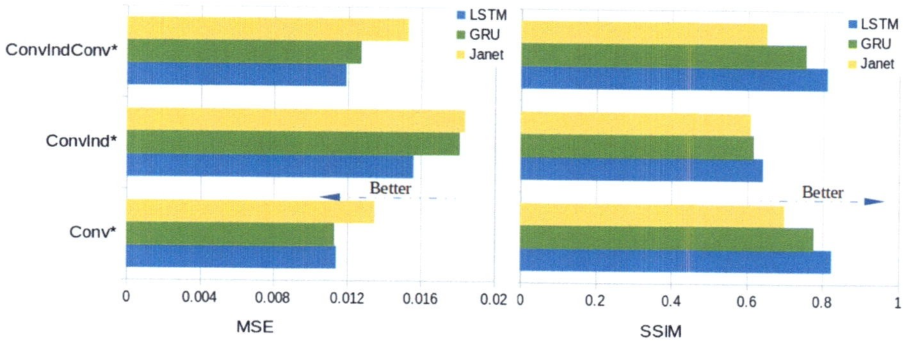

Fig. 4. ConvInd* and ConvIndConv* models performance evaluation in terms of MSE and SSIM.

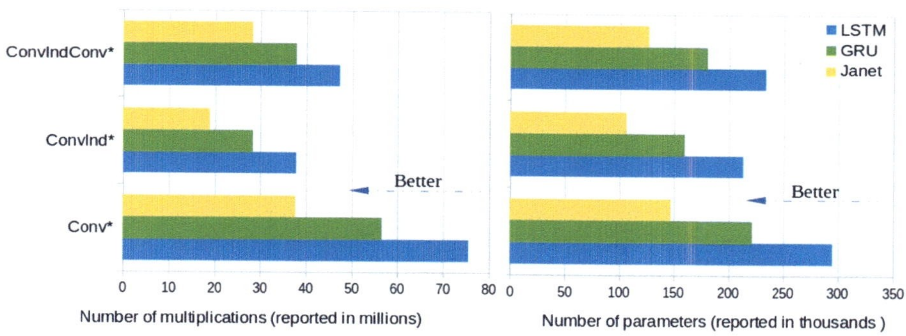

Fig. 5. ConvInd* and ConvIndConv* models performance evaluation in terms of number of parameters and multiplications.

GRU with two gates instead of LSTM architecture with three gates, does not reflect in improved performance and requires more sophisticated way of complexity reduction.

5 Conclusions

In this work, we presented computational and memory load reduction method and compare three gated ConvRNN variants using next video frame prediction task. The usage of the ConvLSTM unit with three gates constantly allowed to achieve higher results, than fewer gated variants of ConvRNNs. We propose to reduce computational and memory load by introducing Hadamard product, when computing connection to the previous hidden state. We showed that it is possible to reduce the number of parameters and multiplications in the ConvRNN architecture with a small drop in the performance of the overall model. ConvInd* models provide significant reduction in the number of parameters and multiplications, but drop in performance of those models is pretty large. Proposed ConvIndConv* models, where convolution operation is kept in the cell computation, allowed to achieve minor drop in performance compared to ConvInd* models, and also kept number of parameters and multiplications smaller compared to Conv* models. We demonstrated, that proposed way of reducing memory and computational load is more efficient, than reducing number of gates, in terms of SSIM and number of multiplications.

References

1. Ballas, N., Yao, L., Pal, C., Courville, A.: Delving deeper into convolutional networks for learning video representations. In: International Conference Learning Representations (ICLR) (2016)
2. Donahue, J., Hendricks, L.A., Guadarrama, S., Rohrbach, M., Venugopalan, S., Saenko, K., Darrell, T.: Long-term recurrent convolutional networks for visual recognition and description. In: IEEE Conference Computer Vision and Pattern Recognition (CVPR), pp. 2625–2634 (2015)
3. Elsayed, N., Maida, A.S., Bayoumi, M.: Reduced-gate convolutional LSTM using predictive coding for spatiotemporal prediction. arXiv:1810.07251 (2018)
4. Li, S., Li, W., Cook, C., Zhu, C., Gao, Y.: Independently recurrent neural network (IndRNN): building a longer and deeper RNN. In: IEEE Conference Computer Vision and Pattern Recognition (CVPR), pp. 5457–5466 (2018)
5. Sautermeister, B.: Deep learning approaches to predict future frames in videos. Master's thesis, Technishe Universität München (2016)
6. Shi, X., Chen, Z., Wang, H., Yeung, D.Y., Wong, W., Woo, W.: Convolutional LSTM network: a machine learning approach for precipitation nowcasting. In: Advances in Neural Information Processing Systems (NIPS), pp. 802–810 (2015)
7. Shi, X., Gao, Z., Lausen, L., Wang, H., Yeung, D.Y., Wong, W., Woo, W.: Deep learning for precipitation nowcasting: a benchmark and a new model. In: Advances in Neural Information Processing Systems (NIPS), pp. 5617–5627 (2017)
8. Srivastava, N., Mansimov, E., Salakhudinov, R.: Unsupervised learning of video representations using LSTMs. In: International Conference Machine Learning (ICML), pp. 843–852 (2015)
9. van der Westhuizen, J., Lasenby, J.: The unreasonable effectiveness of the forget gate. arXiv:1804.04849 (2018)
10. Vazhenina, D., Kanemura, A.: Reducing the number of multiplications in convolutional recurrent neural networks (ConvRNNs). In: Annual Conference of the Japanese Society for Artificial Intelligence (JSAI) (2019)
11. Wang, Y., Gao, Z., Long, M., Wang, J., Yu, P.S.: PredRNN++: towards a resolution of the deep-in-time dilemma in spatiotemporal predictive learning. In: International Conference Machine Learning (ICML) (2018)
12. Wang, Z., Bovik, A.C., Sheikh, H.R., Simoncelli, E.P.: Image quality assessment: from error visibility to structural similarity. IEEE Trans. Image Process. **13**(4), 600–612 (2004)

Agents

Rewards Visualization System Promotes Information Provision

Fujio Toriumi[1(✉)], Hitoshi Yamamoto[2], and Isamu Okada[3]

[1] The University of Tokyo, 7-3-1 Hongo, Bunkyo-ku, Tokyo, Japan
tori@sys.t.u-tokyo.ac.jp
[2] Rissho University, 4-2-16 Osaki, Shinagawa-ku, Tokyo, Japan
[3] Soka University, 1-236, Tangimachi, Hachioji-shi, Tokyo, Japan

Abstract. This is an extension from a selected paper from JSAI2019. If users participated in a Consumer Generated Media (CGM) sites on the web do not provide sufficient information to the site, the site would be insignificant. Despite this point, many users have lower incentives to provide something than those to receive something. This is why such the information-sharing sites face the so-called free-riding issues. Therefore, the information behaviors of CGM are regarded as a kind of a public good game. In this paper, we model such behaviors to a restricted meta-reward game and analyze some effects of their information behaviors on CGM to be activated using agent-based simulations. Our simulation shows that CGM can be activated if at least half agents observe the other agents' meta-rewards information. Moreover, our results show that agents evolutionarily refer individual meta-reward rates to keep cooperation when they can learn their reference strategies.

Keywords: Public goods game · Consumer Generated Media · Agent-based simulation

1 Introduction

We can see many varieties of Consumer Generated Media (CGM) including Wikipedia, TripAdviser, and questions/answer forums such as Yahoo Answers, on the information-sharing platforms operated on the web nowadays. In those platforms, users' voluntary participation bears values, and thus they have network externality in which the more active users are, the more the values of the CGM are. Although users' motivations to generate contents are not only rational thinking but also intrinsic psychological minds including self-disclosure [1], brand image [2] and communication [3], institutional designs from the perspective of the rational and logical incentives should be considered because inactivate and unsuccessful CGM are not so rare.

CGM rely on user-provided information and thus fail if information is not provided. Getting users to provide information generally requires effort costs including time costs and click costs [4]. Therefore, CGM users are given incentives

Y. Ohsawa et al. (Eds.): JSAI 2019, AISC 1128, pp. 55–65, 2020.
https://doi.org/10.1007/978-3-030-39878-1_6

to discourage free riding, a situation in which users receive information, but do not provide it. While huge CGM never worry about freeriding, many managers of small-sized CGM pay attention to it. CGM can be regarded as a kind of public goods game–a social dilemma game in which users may refrain from paying costs (that is, free riding), although they could benefit substantially if they contributed.

To avoid the free-rider problem, many CGM adopt incentive systems for users to receive comments as appreciation for posting articles. These comments are considered rewards for contributing to the public goods game. Moreover, many real CGM systems provide Like buttons to react to comments, which can be regarded as meta-rewards. This is because comments also give psychological benefits to original article providers as well as Like buttons give psychological benefits to their receivers.

The public goods game framework is a strong analytical tool for understanding the contents generate behaviors of users. Some studies insist that information sharing services in online networks have public goods game features [5–7]. Empirical studies have analyzed influences on the cooperative behaviors of reciprocity and network structures in social media [8–10] and some data analyses consider the effects of reputation systems on online markets [11,12].

In public goods game research, kinship and reciprocity promote cooperation [13]. Sanctions, including punishments for free riding and rewards for cooperative behavior, also encourage cooperation [14–18]. Theoretical analyses have pointed out that cooperation through sanctions cannot be maintained due to second-order free riders, who cooperate but shirk sanction behaviors to non-cooperative others [19]. To avoid second-order free riders, Axelrod [20] introduced a meta punishment to punish those who did not participate in punishing the second-order free riders. Despite this approach, some studies have pointed out that meta punishments alone cannot maintain a stable cooperative regime [21–23]. Okada et al. [19] extended the meta punishment concept and exhaustively explored all combinations of meta incentive systems, including meta rewards as well as meta punishments.

Toriumi et al. [24] used a public goods game model to show that meta rewards are required to maintain cooperation. A meta reward is a reward for those who gave a reward to cooperative users. Many CGMs implement a function that allows other users to express their gratitude to those who provided information, and the users who expressed their gratitude can also be given something as a reward.

Those models are based on a linkage hypothesis, there is a positive correlation between the probability of imposing the first-order sanctions and that of imposing the second-order sanctions. However, experimental studies have no consensus on this linkage hypothesis. Some experiments support the linkage between the first-order sanctions and cooperative behaviors [25,26] while others deny it [27,28]. The linkage hypothesis between the first-order and second-order sanctions is partially supported by an experiment of a one-shot public goods game [29].

Toriumi et al. [30] have discovered conditions to promote cooperation even if there is no linkage between the first-order sanctions and the second-order ones. In thier study, it is effective to visualize players' second-order sanctions. Their study, however, assumes that all players can observe all of others' second-order sanctions. Here we extend their study by relaxing that assumption and we explore some features. Our main purposes of this study are to clarify

1. how much users observing the other users' second-order sanctions are needed for promoting cooperation, and
2. whether users observe the other users' second-order sanctions or not if they are available.

2 Models and Methods

To analyze the effects of article-providing behaviors on players' reference behaviors in real CGM, we developed a restricted meta reward game model in Toriumi et al. [24]. Here, we introduce a summary of the model.

2.1 A Restricted Meta Reward Game Model

We consider N agents playing a restricted meta reward game. The game is run for a discrete time and each period is referred to as a round. In each round, all agents play three sequential steps in serial order. Using the case of Agent i as an example, Agent i has its own strategy denoted by (b_i, r_i, rr_i), which we will explain later.

In the first step, the agent provides its own token into a public pool with probability b_i and otherwise does not. In CGM, a contribution and a non-contribution are, respectively, regarded as an information-providing behavior and a non-providing behavior. If a token is provided by Agent i, i must pay a cost κ_0, also the other $N-1$ players receive a benefit, ρ_0.

In the second step, rewards for providing a public good may occur. In CGM, posting a comment to an information provider is regarded as a reward. If and only if Agent i provides a token, the other $N-1$ agents consider whether or not they will give a reward to Agent i. Agent $j(\neq i)$ gives a reward to Agent i with probability $p_{r_{i \to j}}$ and otherwise does not. This probability is calculated as $p_{r_{i \to j}} = \delta \cdot r_j$, where r_j is j's own reward parameter and δ is an expected rate of meta rewards newly introduced in this model to consider the third challenge of the above-mentioned prior studies. If a reward is given, Agent i gains a constant benefit, ρ_1, while Agent j must pay a constant cost, κ_1.

In the third step, meta rewards for giving rewards may occur. In our model, meta rewards from contributors are possible in the first step only to consider the second challenge of the previous studies, thus making this model a restricted game. In CGM, a reply to comments is regarded as a meta reward. If and only if Agent i received a reward from Agent j, Agent i can decide whether to give a meta reward to Agent j with probability rr_i, and otherwise not. While Toriumi et al. [31] assumes that $r_i = rr_i$, our model assumes that these are independent of each other to consider the linkage hypothesis. If a meta reward is given, Agent j gains a constant benefit, ρ_2, while Agent i must pay a constant cost, κ_2.

Each agent plays the above three steps four times in each round. When all agents complete these steps, each agent's final payoff at each round is regarded as its fitness value.

Figure 1 illustrates the conceptual diagram of the model.

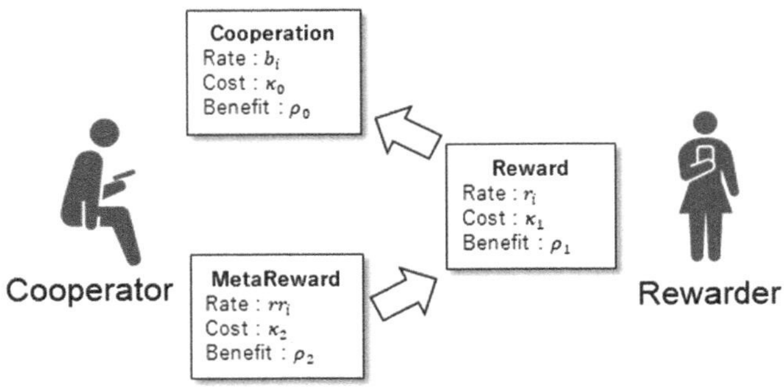

Fig. 1. Outline of extended CGM model [30]

2.2 Adaptive Process of Strategy

At the end of each round, each agent evolves their own strategy. Although many evolution algorithms have been tested [21] we employ roulette selection as a selection mechanism because it is one of the most basic selection methods adopted by Axelrod [32].

To do so, a strategy, (b_i, r_i, rr_i), is coded as a binary code and is regarded as a locus. Agent i randomly selected as a parent agent by roulette selection. The probability that Agent i is randomly chosen is defined as

$$\Pi_i = \frac{exp(\frac{v_i - \bar{v}}{\sigma}) + \varepsilon}{\sum exp(\frac{v_j - \bar{v}}{\sigma}) + \varepsilon}, \tag{1}$$

where $v_i, \bar{v}$, and σ are, respectively, Agent i's fitness value, the average values of fitness of all agents, and the standard deviation of the fitness values of all agents. Value ε is set to 0.0001 to avoid division by zero. This probability function shows that a strategy with a higher payoff tends to spread in the next generation. Next, the strategy parameters are converted to binary code as in Axelrod's procedure [32].

After adopting this method, each binary code in the new locus may reverse one's value (either 0 to 1 or 1 to 0) with a constant probability, 1%. The focal agent has this new locus as their own strategy in the next round.

2.3 Simulation Scenarios

In the restricted meta reward game, there is no incentive to give meta rewards, and thus players never provide meta rewards. To consider this point, we introduce players' expectations of meta rewards. We then explore how these expectations are reflected in the probability of providing rewards using the following three scenarios that are different values of expected rates of meta rewards, λ.

1. No reference ($\lambda = 1.0$): players do not use any reference
2. Social reference ($\lambda = \frac{1}{N}\sum_k rr_k$): players use the average rate of meta rewards in the group
3. Individual reference ($\lambda = rr_i$): players use cooperator i's probability of meta rewards

Scenario 1 is a baseline. Scenario 2 describes a situation that players can get information on a providing rate of meta rewards in CGM. For instance, we suppose a system in which seeing all meta rewards for rewards by others is possible. Scenario 3 describes a situation that visualizes a providing rate of meta rewards for information provided in CGM. In this scenario, we assume that players can decide whether or not to provide meta rewards to a cooperator after they check the providing rate of meta rewards of the focal cooperator.

2.4 Parameter Setting

For simplicity, we set the values of the parameters above by installing two new intervening parameters: δ and μ. Table 1 shows the values of the other parameters in the simulation.

$$\kappa_0 = 1.0 \tag{2}$$
$$\rho_n = \mu \cdot \kappa_n \tag{3}$$
$$\kappa_n = \delta \cdot \kappa_{n-1}, \tag{4}$$

where $n = 1, 2$.

3 Simulation Results

3.1 Summary of Comparing Three Scenarios

As the previous study [30] demonstrated, three scenarios described in the previous section have different performances. Figure 2 shows the simulation results of the cooperation rates in the three cases where we simulated 100 runs with different random seeds in each scenario and the vertical axis represents the step numbers. This figure supports that the case of Scenario 3 can promote and maintain a cooperative regime while the others (Scenarios 1 and 2) cannot. This is because the unsucceeded cases cannot keep the reward rates and they decrease immediately after the beginning. No reward never bears cooperation, and thus

Table 1. Simulation Parameters

Param	Value
N	100
Simulation steps	1000
μ (benefit-cost ratio)	2.0
δ (discount ratio)	0.8
ρ_0 (benefit of cooperation)	2.0
κ_0 (cost of cooperation)	1.0
ρ_1 (benefit of reward)	1.6
κ_1 (cost of reward)	0.8
ρ_2 (benefit of meta-reward)	1.28
κ_2 (cost of meta-reward)	0.64

they do not reach a cooperative regime. In Scenario 3, on the other hand, the cooperation rate increases from the beginning, then the meta reward rate also increases and, finally, the reward rate increases. Therefore, the case maintains a stable cooperative regime despite the parameter value δ is 1 in Scenario 3 while it is less than 1 in Scenario 1.

The essential point of the difference is that players with high meta reward rates tend to receive more rewards than those with low meta reward rates. In both Scenarios 1 and 2, a cooperator's meta reward rate is independent of the rate of receiving rewards, and thus players with high meta reward rates do not always have higher probabilities of receiving rewards than those with low meta reward rates. Despite this, the former players must give meta rewards by paying costs. Therefore, neither scenario has an incentive for raising meta rewards. In Scenario 3, on the other hand, rewards are given by referring to the individual rate of meta rewards and, indirectly, players with low meta reward rates receive a kind of punishment; i.e., they are not given rewards. Players with low meta reward rates gain relatively small amounts of payoffs, and thus they are distinct sooner or later under the selection pressure. At last, players with high meta reward rates become the majority. In order to receive meta rewards, players tend to post rewarding actions, and in order to receive rewards, players tend to cooperate. Therefore, these traits indicate that a cooperative regime emerges if players can observe the other individual's meta reward rate.

3.2 Effects on Cooperation of the Rate of Browseable Users

As mentioned before, it is quite important for promoting cooperation to depend cooperative behaviors on the meta-reward rates of article-providers. However, players must cost for capturing the mera-reward rates generally. Here, we test effects of the rates of browseable players on the rates of posting articles. The browseable players can gain the meta-reward rates of their focal article-providers.

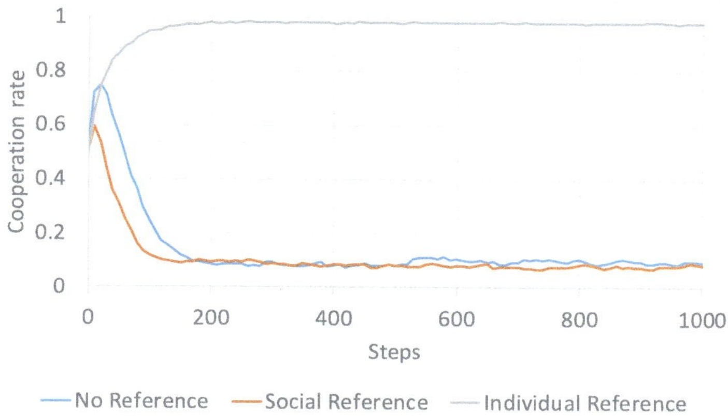

Fig. 2. Cooperation rates in three scenarios

To do so, we introduce a new parameter, $r (0 \leq r \leq 1)$, which means the rate of the browseable agents in the population and we set Scenarios 4 and 5. In a simulation with $r = r_0$, nature chooses the fraction r_0 of the agents at the beginning and the chosen players can gain the meta-reward rate of all players a la Scenario 3. In Scenario 4, the rest of the players have no reference a la Scenario 1. In Scenario 5, the rest gain the average rate of the meta-reward rates a la Scenario 2. The following list is the summary.

4. No reference ($\lambda = 1.0$) + Individual Reference ($\lambda = rr_i$)
5. Social reference ($\lambda = \frac{1}{N} \sum_k rr_k$) + Individual Reference ($\lambda = rr_i$)

Figures 3 and 4 show the results of cooperation rates, reward rates and meta-reward rates with r in Scenario 4 and Scenario 5, respectively. Figure 3 shows that a cooperative regime can be feasible when more than 60% of agent can refer the other agents' meta-reward rates. On the other hand, Fig. 4 shows that it can even if the rate of the browseable agents is 40%.

These results support that cooperative regimes emerge easier when CGM users consider a society-wide meta-reward rate than when they consider nothing about the meta-rewards even if the browseable users are not so many. However, at least 40% of users must refer the other agents' meta-reward rates in promoting cooperation. CGM may provide the individual meta-reward rates using their previous behaviors, and thus CGM can be expected to activate when the system provides an information-provider's estimated meta-reward rate to information-receivers.

3.3 Adaptation of Reference Strategy

We then test an adaptive process of agent's reference strategy. Whether an agent refers information on the other agents or not depends, and thus its reference strategy should intrinsically be a parameter an agent has individually.

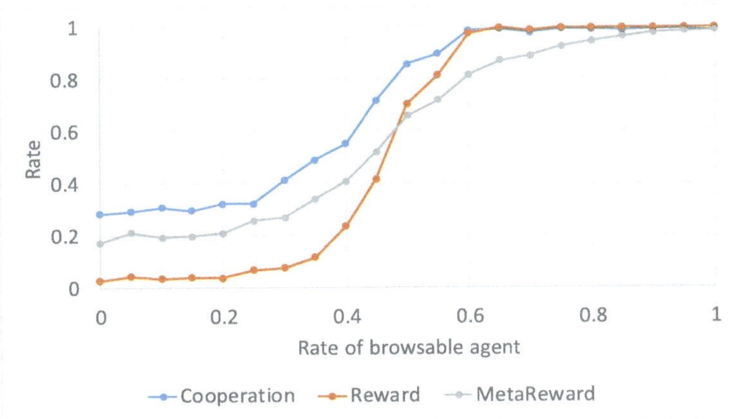

Fig. 3. Results of scenario 4

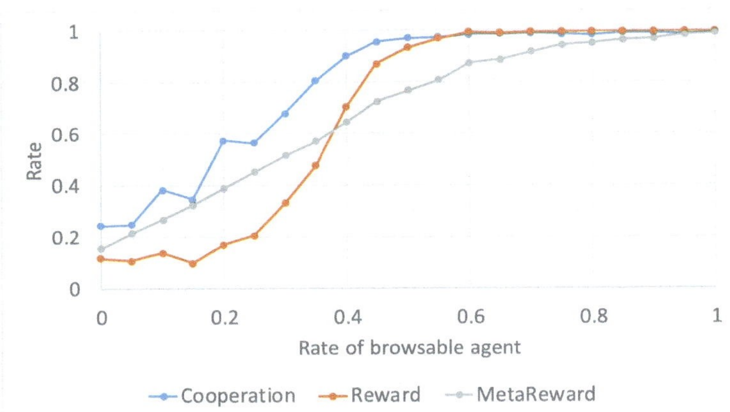

Fig. 4. Results of scenario 5

Here, we added a new parameter on reference rate into an agent's strategy which is changeable in its adaptive process. An agent refer one's focal player's meta-reward rate with a probability of the value of the parameter when one considers one's reward action. If one does not refer it, one sets $\lambda = 1.0$ as well as Scenario 1 to determine one's reward action.

Figure 5 shows the cooperation rates, the reward rates, the meta-reward rates, and the reference rates in the situation. The reference rates keep about 0.6 while the cooperate rates goes to 1. As shown in Fig. 3 of the case of Scenario 4, the cooperative regime can be feasible if the reference rate exceeds about 0.6. This result supports that cooperation can emerge when agents decide their reference behaviors through the adaptive process on their strategies. It also implies that agents' autonomous individual reference behaviors are likely to be cooperation without providing the average of meta-reward behaviors by the CGM system.

However, our model does not consider the reference cost. We need future works while cooperation emerges if a CGM manager makes one's effort of that the reference behaviors has either little or no cost.

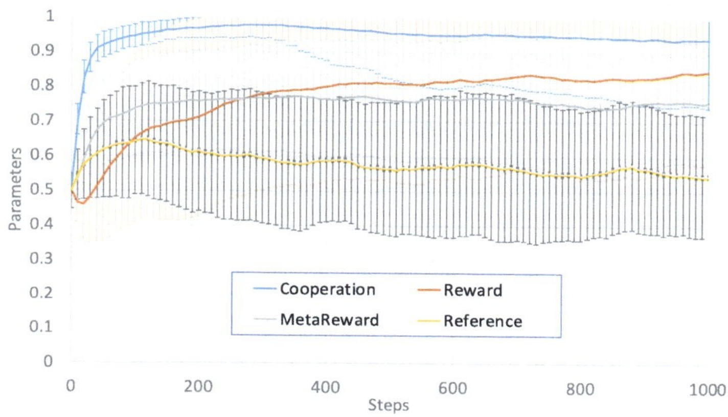

Fig. 5. Cases that agents learn their reference strategies

4 Conclusion

In this paper, we analyzed conditions to promote cooperation with no linkage between the first-order sanctions and the second-order ones by extending a model of [30]. Our results show that it can be if at least half agents observe the other agents' second-order sanctions. We also show that the rate of the observable agents for maintaining cooperation depends on what the other agents refer to. It is more likely to cooperation when the other agents observe the average of the meta-reward rates than when they believe that the meta-rewards are absolutely given to. Moreover, our results show that more than 60% agents should refer individual meta-reward rates to keep cooperation when they can learn their reference strategies.

We suggest the following story if a CGM provides a history of users' meta-reward actions. First, many users refer such a meta-rewards information and they selectively give rewards to users who tend to give meta-rewards to them only. Second, an information-provider with high rates that they give meta-rewards can receive rewards. The users learn this point, and they raise the meta-reward rates up. Third, they gain many rewards and they are motivated to provide information more. Finally, the CGM can be activated.

Our results have some challenges. We do not consider the costs to observe the other users' actions in this paper. We also do not consider the network structure among users. Those are future extensions.

Acknowledgement. This study is supported by Japan Society for the Promotion of Science (Grant nos. 16H03120, 17H02044, 17KK0055, 18H03498, 19K21570).

References

1. Toubia, O., Stephen, A.T.: Mark. Sci. **32**(3), 368 (2013)
2. Kim, A.J., Johnson, K.K.: Comput. Hum. Behav. **58**, 98 (2016)
3. Sung, Y., Lee, J.A., Kim, E., Choi, S.M.: Pers. Individ. Differ. **97**, 260 (2016)
4. Nakamura, H., Gao, Y., Gao, H., Zhang, H., Kiyohiro, A., Mine, T.: 3rd International Conference on Advanced Applied Informatics (2014)
5. Fulk, J., Flanagin, A.J., Kalman, M.E., Monge, P.R., Ryan, T.: Commun. Theory **6**(1), 60 (1996)
6. Fulk, J., Heino, R., Flanagin, A.J., Monge, P.R., Bar, F.: Organ. Sci. **15**(5), 569 (2004)
7. Leonardi, P.M., Huysman, M., Steinfield, C.: J. Comput.-Mediat. Commun. **19**(1), 1 (2013)
8. Hirahara, Y., Toriumi, F., Sugawara, T.: International Conference on Social Informatics, pp. 112–120. Springer (2014)
9. Osaka, K., Toriumi, F., Sugawara, T.: Comput. Soc. Netw. **4**(1), 2 (2017). https://doi.org/10.1186/s40649-017-0038-2
10. Miura, Y., Toriumi, F., Sugawara, T.: Companion Proceedings of the Web Conference 2018, International World Wide Web Conferences Steering Committee, Republic and Canton of Geneva, Switzerland, WWW 2018, pp. 1323–1329 (2018). https://doi.org/10.1145/3184558.3191573
11. Yamamoto, H., Ishida, K., Ohta, T.: Comput. Math. Organ. Theory **10**(2), 165 (2004)
12. Diekmann, A., Jann, B., Przepiorka, W., Wehrli, S.: Am. Sociol. Rev. **79**(1), 65 (2014). https://doi.org/10.1177/0003122413512316
13. Nowak, M.A.: Science **314**(5805), 1560 (2006)
14. McCusker, C., Carnevale, P.J.: Organ. Behav. Hum. Decis. Process. **61**(2), 190 (1995)
15. Fehr, E., Gächter, S.: Nature **415**(6868), 137 (2002). https://doi.org/10.1038/415137a
16. Sefton, M., Shupp, R., Walker, J.M.: Econ. Inq. **45**(4), 671 (2007)
17. Ule, A., Schram, A., Riedl, A., Cason, T.N.: Science **326**(5960), 1701 (2009). https://doi.org/10.1126/science.1178883
18. Balliet, D., Mulder, L.B., Van Lange, P.A.M.: Psychol. Bull. **137**(4), 594 (2011). https://doi.org/10.1037/a0023489
19. Okada, I., Yamamoto, H., Toriumi, F., Sasaki, T.: PLoS Comput. Biol. **11**(5), e1004232 (2015). https://doi.org/10.1371/journal.pcbi.1004232
20. Axelrod, R.: Am. Polit. Sci. Rev. **80**(4), 1095 (1986)
21. Galán, J.M., Izquierdo, L.R.: J. Artif. Soc. Soc. Simul. **8**(3), 2 (2005)
22. Yamamoto, H., Okada, I., Ogawa, Y.: J. Adv. Comput. Intell. Intell. Inf. **18**(4), 624 (2014). https://doi.org/10.20965/jaciii.2014.p0624

23. Yamamoto, H., Okada, I.: Phys. A **443**, 526 (2016). https://doi.org/10.1016/j.physa.2015.08.053
24. Toriumi, F., Yamamoto, H., Okada, I.: J. Artif. Soc. Soc. Simul. **19**, 4 (2016)
25. Horne, C.: Soc. Psychol. Q. **64**, 253–266 (2001)
26. Horne, C.: Rat. Soc. **19**(2), 139 (2007)
27. Yamagishi, T., Horita, Y., Mifune, N., Hashimoto, H., Li, Y., Shinada, M., Miura, A., Inukai, K., Takagishi, H., Simunovic, D.: Proc. Nat. Acad. Sci. **109**(50), 20364 (2012)
28. Egloff, B., Richter, D., Schmukle, S.C.: Proc. Nat. Acad. Sci. **110**(9), E786 (2013)
29. Kiyonari, T., Barclay, P.: J. Pers. Soc. Psychol. **95**(4), 826 (2008). https://doi.org/10.1037/a0011381
30. Toriumi, F., Yamamoto, H., Okada, I.: A belief in rewards accelerates cooperation on social media. J. Comput. Soc. Sci. **2019**(7), 1–13 (2019). https://doi.org/10.1007/s42001-019-00049-5
31. Toriumi, F., Yamamoto, H., Okada, I.: Proceedings of the The 2012 IEEE/WIC/ACM International Joint Conferences on Web Intelligence and Intelligent Agent Technology, vol. 02, pp. 43–50 (2012)
32. Axelrod, R., et al.: The dynamics of norms, pp. 1–16 (1987)

An Automated Negotiating Agent that Searches the Bids Around Nash Bargaining Solution to Obtain High Joint Utilities

Shan Liu[1(✉)], Shun Okuhara[2,3], Ahmed Moustafa[2], and Takayuki Ito[2]

[1] The University of Sheffield International College, 3 Solly Street, Sheffield S1 4DE, UK
liu.shan@itolab.nitech.ac.jp

[2] Nagoya Institute of Technology, 705 Building No. 4, Gokiso-cho, Showa-ku, Nagoya, Aichi, Japan
okuhara@itolab.nitech.ac.jp, {ahmed,ito.takayuki}@nitech.ac.jp

[3] Fujita Health University, Room 507, Building No. 6, 1-98 Dengakugakubo, Kutsukake-cho, Toyoake, Aichi, Japan
sokuhara@fujita-hu.ac.jp

Abstract. This is an extension from a selected paper from JSAI2019. Many researchers have focused on agent strategies in auto-negotiations. In this paper, we present a negotiating agent that can search for suitable bids that obtain high joint utility values near a Nash bargaining solution by using a novel bid searching strategy. The proposed agent finished in second place in the social welfare category in International Automated Negotiating Agents Competition (ANAC) 2018 and first place in the social welfare category in Pacific Rim International Automated Negotiation Agents Competition (PRIANAC) 2018.

Keywords: Automated negotiation · Multi-agent · Bid searching

1 Introduction

In this paper, we present a novel bid searching strategy for an automated negotiating agent that participated in International Automated Negotiating Agents Competition (ANAC) Repeated Multilateral Negotiation League (RMNL) 2018 and Pacific Rim International Automated Negotiation Agents Competition (PRIANAC) 2018 [1]. As a competitive challenge in ANAC 2018 [2], RMNL requires the participants to design and implement an automated negotiating agent that is able to negotiate with two opponent agents and learn from its previous negotiation experiences. Similarly, PRIANAC 2018 requires participants to design an automated negotiating agent that is able to negotiate only with one opponent agent. In addition, PRIANAC 2018 permits the usage of a local file to save the bids offered by the opponent agent and a machine learning library. To

Y. Ohsawa et al. (Eds.): JSAI 2019, AISC 1128, pp. 66–76, 2020.
https://doi.org/10.1007/978-3-030-39878-1_7

compete in these competitions, we propose a novel negotiation strategy that can find the preference issues and values of each agent and then combine them to generate the bids that provide high joint utility values near the Nash bargaining solution. This paper is organized as follows. Section 2 describes the negotiation environment in ANAC 2018 and PRIANAC 2018. Section 3 introduces our agent's negotiation strategy. Section 4 presents the ANAC 2018 and PRIANAC 2018 results. Section 5 analyzes the proposed negotiating strategy and discusses its evaluation results. Section 6 concludes the paper and mentioned future work.

2 Negotiation Environment

2.1 Negotiation Competition

RMNL in ANAC 2018 is a repeated multi-party closed negotiation competition among three agents. RMNL assumes no previous knowledge of the preferences and strategies of the opponent agents, wherein the negotiating agents use the Stacked Alternating Offers Protocol (SAOP) [3]. In each negotiation round, each agent is given three minutes to deliberate. Each negotiation round is repeated five times. Also, the utility functions are linear and the participant agents are able to negotiate over a large set of previously unknown preferences. Similarly, PRIANAC 2018 is a bilateral negotiation competition that assumes no previous knowledge of the preferences of the opponent agents, wherein the negotiating agents use the Alternative Offer Protocol (AOP) [3]. Each negotiation round lasts 10 s, and the number of negotiation rounds on the same configuration is set to 100. Both competitions use the GENIUS [4] platform shown in Fig. 1.

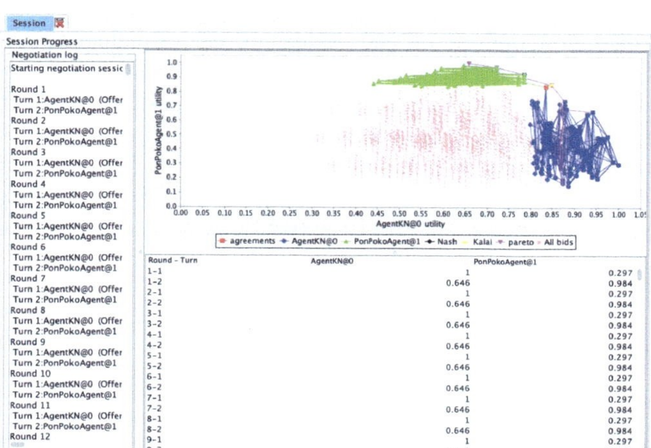

Fig. 1. GENIUS platform

2.2 Preferences of the Negotiating Agents

The preferences for each agent in all negotiation domains are represented by a weighted sum utility function. In this regard, each agent has its own utility function, which is expressed as follows:

$$u_a(b^t) = \sum_{j \in I} V_a(b_j^t) \cdot w_{a,j} \tag{1}$$

According to Eq. (1), each negotiation issue $j \in I$ can take a value v_j from a predefined set of valid values for that issue which is denoted by D_j (i.e., $v_j \in D_j$), where each agent can access this domain information. In addition, a bid $b = (b_1,...,b_{|I|})$ is an assignment of values to all issues where $b_1 \in D_1$. $V_a(v_j)$ denotes Agent a valuation of the value of issue j.

3 Proposed Approach

3.1 Bid Searching Strategy

The proposed bid searching strategy aims to find the bids around the Nash bargaining solution because these bids are expected to possess higher values of the joint utility function. To achieve this goal, the proposed method aims to find the priority issues for each opponent agent after analyzing the series of bids offered by this agent.

The main idea is dividing these bids into negotiation issue units, selecting the priority issue and value for each opponent agent, and then generating new bids that not only have high utilities but also include those prior issues and values. Therefore, the bid searching space becomes unrestricted to the already offered bids, so new bids that include all the preference issues from different agents are expected. In addition, in PRIANAC 2018, we save and analyze all the bids offered by the opponent agents, in all negotiation rounds of the same negotiation domain, to a local file. As the number of bids increases, the accuracy of determining the preference for the opponent agent also increases. This bid searching strategy that was used in both ANAC 2018 and PRIANAC 2018 includes five steps.

The bid searching strategy is represented by Algorithm 1 as follows:

Algorithm 1. Bid Searching Strategy

```
1: Issues: given by a scenario, such as: a, b, c, d, e.
2: Values: given for each issue by a scenario, such as: a1, a2, a3, a4, a5 for issue
   a.
3: for all Opponent participant agents do
4:     for all Issues: a, b, c, d, e do
5:         Calculate the average of all the values in all offered bids
6:         Calculate the standard deviation of all the values (VSD)
7:     end for
8:     calculate the standard deviation of all the issues' VSD (ISD)
9:     if ISD > 0.0 then
10:        The agent has preference
11:        Compare the preference of issues
12:        Choose the prior issue and the prior value
13:    else
14:        Choose the prior issue and the prior value randomly
15:    end if
16: end for
17: while True do
18:    Generate the bid with the prior issues and values of opponent participant agents
19:    if The utility of the generated bid > threshold then
20:        Offer the bid
21:    end if
22: end while
23: if The offered bid is accepted by all opponent agents then
24:    Negotiation succeeds
25:    Save the list of offered bids to a local file
26: else
27:    Update offered bids list and go to step 3
28: end if
```

In addition, a sample demonstration is according to the following steps. First, the standard deviation is calculated for each value in each issue, which is denoted as the value standard deviation. Second, the standard deviation is calculated for value standard deviation of each issue, which is denoted as the issue standard deviation. Third, issues with high value standard deviations are set as prior issues, and values with high frequencies are set as prior values. Fourth, the utility of the bids with prior issues and prior values are searched for randomly. If the utilities of these bids are higher than a predefined threshold value, an offer is made. Fifth, if the negotiation succeeds, the list of offered bids is saved to a local file. Otherwise, the list of offered bids is updated, and the process returns to the first step.

e.g.: A scenario is given that has five issues: a, b, c, d, and e. Agent 1 has the sample scenario's flow in Table 1.

Each issue has five values (a1, a2, a3, a4, and a5) assuming that Agent 1 is the opponent agent and provides bids as follows:

Table 1. The sample scenario's flow

Agent1
a1, b1, c2, d5,e1
a1, b1, c1, d5, e1
a1, b1, c3, d5, e1
a1, b1, c1, d2, e3
a1, b2, c2, d2, e3
a1, b2, c3, d2, e2
a1, b1, c1, d3, e2
a1, b2, c4, d3, e2
a1, b1, c4, d3, e2
a1, b1, c5, d4, e1

Step1

The standard deviation is calculated for a1, a2, a3, a4, and a5 frequency in an issue as follows:
Number: a1 = 10, a2 = 0, a4 = 0, a5 = 0.

$$Average(a) = \frac{10+0+0+0+0}{5} = 2.$$

Value Standard Deviation:

$$V(a) = \frac{\sqrt{(10-2)^2+(0-2)^2+(0-2)^2+(0-2)^2+(0-2)^2}}{5} = 1.79.$$

It is the same situation for issues b, c, d, e with their value standard deviations.
Value Standard Deviation (VSD): a: 1.79, b: 1.23, c: 0.28, d: 0.57, e: 0.8.

Step2

The standard deviation is calculated for issue a, b, c, d, e with their Value Standard Deviations.

$$Average\ of\ VSD = \frac{1.79+1.23+0.28+0.57+0.8}{5} = 0.94.$$

Issue Standard Deviation:

$$V(a) = \frac{(1.79-0.94)^2+(1.23-0.94)^2+(0.28-0.94)^2+(0.57-0.94)^2+(0.8-0.94)^2}{5}$$
$$= 0.24 > 0.$$

Agent1 has preference.

Step3

Preference of issues: a >b >e >d >c.
Prior issues of Agent1: a, b.
Prior values of Agent1: a1, b1.

Step4

Is utility greater than the threshold?
Yes, Offer: a1, b1, c1, d1, e1
No, Revise: a1, b1, c2, d1, e2
No, Revise: a1, b1, c1, d4, e1
Yes, Offer: a1, b1, c5, d1, e3
... : ...

Step5

Accepted offer
Offer: a1, b1, c1, d1, e1 → Accepted!
Updating offered bids list → A local file.

Rejected offer
Offer: a1, b1, c1, d1, e1 → Rejected!
Updating offered bids list.
Going back to step1.

3.2 Bid Acceptance Strategy

A compromise function is used in order to judge whether or not to accept an offered bid. If the utility of a certain bid is greater than a preset threshold value, then the proposed agent accepts this bid. This threshold value decreases as time passes. In this context, this threshold value is calculated using the following equation.

$$Threshold = \max\{(1-(1-df)\cdot\log(e-1.9+(e-1)^{\alpha})\cdot t), emax\} \quad (2)$$

In this equation, *df* represents a discount factor. α is a parameter that we set as 4.5. t represents the current time. *emax* represents the estimated maximum utility, which is calculated using the following equation.

$$emax(t) = \mu(t) + (1-\mu(t))\cdot d(t) \quad (3)$$

where, μ(t) is the mean utility of the opponent offers in the utility space of a certain agent. *d(t)* [8] is a function for estimating the utility width of the opponent offers in the utility space of this agent. This utility width is given by Farma Agent in the ANAC 2016 competition as follow:

$$d(t) = \frac{\sqrt{3}\sigma(t)}{\sqrt{\mu(t)(1-\mu(t))}} \tag{4}$$

where σ is the standard deviation.

4 Results and Evaluation

Both ANAC 2018 and PRIANAC 2018 have two categories: (1) the individual category, in which the participant agents are ranked in accordance with the individual utility they have obtained; and (2) the social welfare category, in which the participant agents are ranked in accordance with their social utility. This social utility is the sum of the individual utilities of all agents. In ANAC 2018, the 21 competing agents were divided into three groups randomly.

Pool-A Meng wan, AgentHerb, IQSun2018, PonPokoRampage, FullAgent, Seto, Lancelot

Pool-B Beta One, Yeela, SMAC Agent, AgreeableAgent2018, ConDAgent, Shiboy, Libra

Pool-C AgentNP1, GroupY, ATeamAgent, Sontag, Agent33, Agent Hama, Exp-Rubick

Each pool's performing top three agents proceeded to the final. Their pools' results are shown in Tables 2, 3, and 4. Our proposed agent (agent33) is in Pool-C. The purple and orange cells in Tables 2, 3, and 4 are the finalists for individual utility and social welfare, respectively.

Table 2. Results of qualification round for Pool-A

Agent name	Individual utility	Social welfare
meng wan	0.587	1.632
AgentHerb	0.492	1.854
IQSun2018	0.583	1.759
PonPokoRampage	0.567	1.547
FullAgent	0.549	1.767
Seto	0.5518	1.449
Lancelot	0.5178	1.534

In ANAC 2018, the final round involved the nine finalists in each category with four selected scenarios submitted by the participants. For each scenario,

Table 3. Results of qualification round for Pool-B

Agent name	Individual utility	Social welfare
Beta one	0.483	1.404
Yeela	0.399	1.423
SMAC_Agent	0.465	1.325
AgreeableAgent2018	0.506	1.505
ConDagent	0.471	1.519
Shiboy	0.503	1.411
Libra	0.446	1.228

Table 4. Results of qualification round for Pool-C

Agent name	Individual utility	Social welfare
AgentNP1	0512	1.530
GroupY	0.485	1.339
ATeamAgent	0.345	0.981
Sontag	0.485	1.515
Agent33	0.429	1.419
Agent_Hama	0.470	1.393

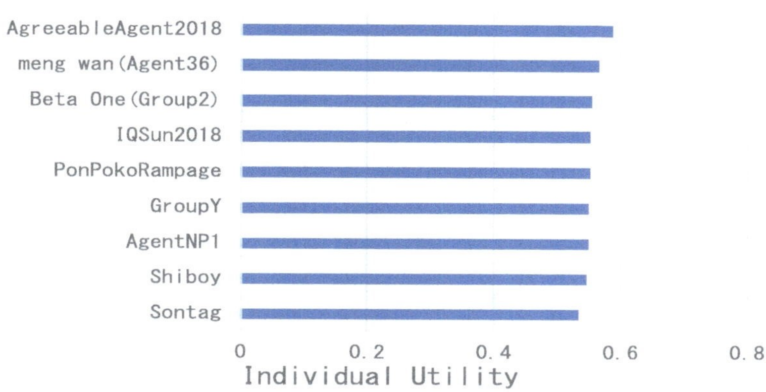

Fig. 2. Results of the individual utility in ANAC 2018

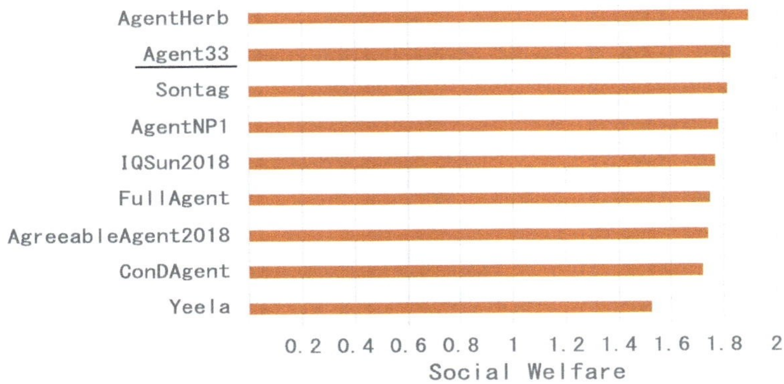

Fig. 3. Results of the social welfare in ANAC 2018

2520 negotiations were run. The results for the final round of the individual utility and social welfare categories are presented in Figs. 2 and 3, respectively. In Fig. 3, our proposed agent33 is underlined.

PRIANAC 2018 This competition involved 6 agents with 120 negotiation scenarios generated by the organizers. In each category, 12,000 sessions were run. The results for the final round of the individual utility and social welfare categories are presented in Figs. 4 and 5, respectively. In both figures, our proposed agent33 is underlined.

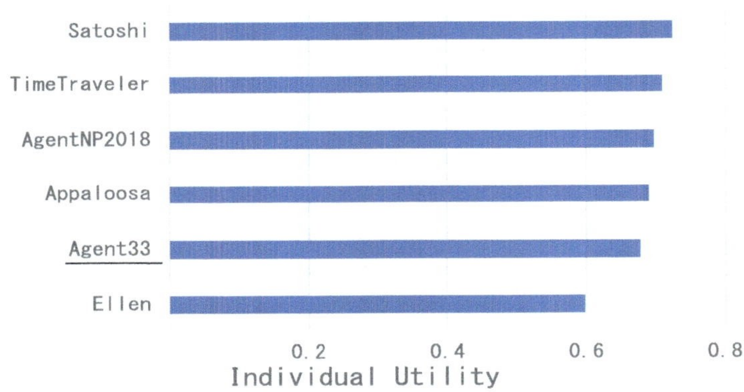

Fig. 4. Results in PRIANAC 2018

The proposed agent won second place in the final round of the social welfare category in ANAC 2018 and the first place of the social welfare category in PRIANAC 2018. These results demonstrate the efficiency of the proposed agent and its ability to find high joint utility solutions.

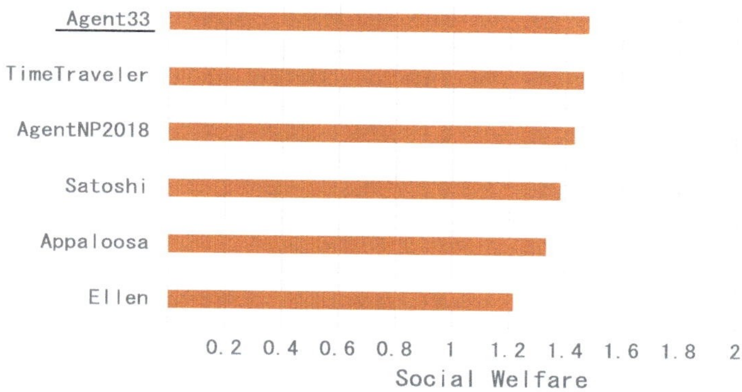

Fig. 5. Results in PRIANAC 2018

5 Related Works

In ANAC 2014 [5], Agent M [6] was the champion in both the individual and the social welfare categories. It used a bid combination strategy that offered a combined bid consisting of three bids: Best Bid, Best Offered Bid, and the Frequency Bid. This agent used the Best Bid as a base bid and replaced two issues' values of the Best Bid randomly with random issue values of both Best Offered Bid and Frequency Bid. To generate the Frequency Bid, Agent M counted up the number of all issue values and chose the most frequent value for each issue. In ANAC 2016 [7], Agent Farma and Agent Terra used the same consideration, but none of these three agents considered the importance level of each issue. In Agent Farma17 [8], the Agreement and Reject (A/R) idea is applied to offer a more acceptable bid, as this agent focuses on the frequency of bids offered by the opponents and calculates the cumulative probability. Agent Farma17 also sets up a Reject Rate and an Agreement Rate that include part of the bids accepted by the opponents. If the cumulative probability of the accepted bid is lower than the Reject Rate, the values of certain issues will be randomly changed into those values of a bid that has higher cumulative probability than the Agreement Rate. Just because a bid has a higher cumulative probability, however, does not mean that all of its issue values are acceptable. Instead of focusing on the Reject Rate and the Agreement Rate of the bids, an Agreement Rate of all negotiation issues should be considered.

6 Conclusion and Future Work

This paper proposed a negotiating agent that implements a novel bid reconstruction method to search for suitable bids that obtain high joint utility values. Towards this end, the bid reconstruction method utilizes the previously offered bids by the opponent agents to construct successful bids around the Nash bargaining solution. The final round results of ANAC RMNL 2018 and the results of PRIANAC 2018 demonstrate that the proposed agent is able to search the scope of suitable bids around the Nash bargaining solution and succeeds in leading other participant agents where the negotiating agent achieves higher values of joint utilities. Future work is to study the necessary improvements to achieve a high individual utility value and to investigate the usage of machine learning to help reduce the scope of bid searching.

References

1. Pacific Rim International Automated Negotiation Agents Competition, PRIANAC (2018). http://web.tuat.ac.jp/~katfuji/PRIANAC2018/
2. Ninth International Automated Negotiating Agents Competition, ANAC (2018). http://web.tuat.ac.jp/~katfuji/ANAC2018/
3. Aydoğan, R., Festen, D., Hindriks, K.V., Jonker, C.M.: Alternating offers protocols for multilateral negotiation. In: Modern Approaches to Agent-based Complex Automated Negotiation, pp. 153–167. Springer, Cham (2017)
4. Lin, R., Kraus, S., Baarslag, T., Tykhonov, D., Hindriks, K., Jonker, C.M.: Genius: an integrated environment for supporting the design of generic automated negotiators. Comput. Intell. **30**(1), 48–70 (2014)
5. Ninth International Automated Negotiating Agents Competition, ANAC (2014). http://www.itolab.nitech.ac.jp/ANAC2014/
6. Niimi, M., Ito, T.: Recent Advances in Agent-based Complex Automated Negotiation, pp. 235–240 (2016)
7. Ninth International Automated Negotiating Agents Competition, ANAC (2016). http://web.tuat.ac.jp/~katfuji/ANAC2016/
8. Toyama T., Takayuki, I.: Concession based on maximum utilities estimated by a divided uniform distribution (2016)

ANAC 2018: Repeated Multilateral Negotiation League

Reyhan Aydoğan[1(✉)], Katsuhide Fujita[2], Tim Baarslag[3,4], Catholijn M. Jonker[5,6], and Takayuki Ito[7]

[1] Department of Computer Science, Özyeğin University, Istanbul, Turkey
reyhan.aydogan@ozyegin.edu.tr
[2] Institute of Engineering, Tokyo University of Agriculture and Technology, Tokyo, Japan
katfuji@cc.tuat.ac.jp
[3] Centrum Wiskunde & Informatica (CWI), Amsterdam, The Netherlands
t.baarslag@cwi.nl
[4] Department of Computer Science, Utrecht University, Utrecht, The Netherlands
[5] Interactive Intelligence Group, Delft University of Technology, Delft, The Netherlands
c.m.jonker@tudelft.nl
[6] LIACS, Leiden University, Leiden, The Netherlands
[7] Nagoya Institute of Technology, Nagoya, Japan

Abstract. This is an extension from a selected paper from JSAI2019. There are a number of research challenges in the field of Automated Negotiation. The Ninth International Automated Negotiating Agent Competition encourages participants to develop effective negotiating agents, which can negotiate with multiple opponents more than once. This paper discusses research challenges for such negotiations as well as presenting the competition set-up and results. The results show that winner agents mostly adopt hybrid bidding strategies that take their opponents' preferences as well as their strategy into account.

Keywords: Multi-agent system · Negotiation · Competition

1 Introduction

In multi-agent systems, agents interact and collaborate with each other to achieve their goals; however, their interests and preferences may sometimes conflict. In such situations, agents can resolve their conflict and reach a consensus through negotiation. Therefore, it is important to design and develop efficient negotiation strategies for autonomous agents [7, 10, 16]. In order to facilitate the research on negotiation in multi-agent systems and to provide unique benchmarks for evaluation of the developed negotiation strategies, an international competition on automated negotiation (ANAC [18]) has been organized for several years. As the main organizers of this competition, we aim to address a

Y. Ohsawa et al. (Eds.): JSAI 2019, AISC 1128, pp. 77–89, 2020.
https://doi.org/10.1007/978-3-030-39878-1_8

variety of research challenges (e.g. uncertainty about opponent, reasoning on complex preferences, negotiating with multiple opponents). In ANAC 2018, we introduced three leagues: Human-agent negotiation [21], Negotiation Strategies for Diplomacy Game [17] and Repeated Multilateral Negotiation. This paper presents the competition setup and evaluation results of the repeated multilateral negotiation as well as briefly pointing out the main challenges of 2018.

Since in multilateral negotiation, there are more than two agents searching for a consensus, it is more complicated than bilateral negotiation [2,13,14]. When those agents have a long term relation, they may need to negotiate with each other more than once. In such a situation, it is important for agents to understand their opponents' needs and strategy well and adjust their strategy accordingly to find a better deal. It is also essential they realize that they are going to negotiate with the same opponents again. If their strategy is based on completely exploiting the other sides, their negotiation may end up in failure in the long run. In ANAC 2018, we encouraged the participants to design effective negotiation strategies for such repeated multilateral negotiations.

The competition results showed that successful agents mostly employ a hybrid bidding strategy in order to avoid being exploited and consider their opponents' best offer in their previous negotiation. Furthermore, they usually adopt a frequency based opponent modeling. Some agents model the likelihood of an offer to be accepted by other agents by analyzing the bids in their past negotiations while others use opponents' best offers in their bidding strategy. In the rest of this paper, some of the research challenges of 2018 are discussed, and competition setup and results are explained.

The rest of the paper is organized as: Sect. 2 shows ANAC2018 competition challenges. Section 3 explains negotiation setup while Sect. 4 presents the results of the competition in detailed. Related work are provided in Sect. 5. Lastly, we conclude the paper with the potential research challenges for upcoming competitions in Sect. 6.

2 ANAC 2018 Competition Challenges

Although agents negotiate with the same opponents several times, they do not know their exact preferences and strategies. The research challenge is to model their opponents' behavior or preferences based on their past negotiations and to incorporate those models into their bidding strategy so as to improve their negotiation outcome. However, this is not trivial since their opponents may change their behavior over time in spite of having the same preferences. Therefore, opponent models may mislead the agents. Furthermore, if an agent tries to exploit its opponents based on what it learned about them, they may reciprocate in a similar way next time, which would result in decreasing utility for all agents. Moreover, when agents negotiate with the same opponents, they should establish a good relationship while still aiming at maximizing their own utility. There is thus a trade-off acting nicely and considering your own benefit.

3 Competition Setup

In the competition, three agents negotiate on multiple issues to reach a consensus by following the Stacked Alternative Offers Protocol (SAOP) [1]. According to this protocol, one of the agents starts the negotiation with an offer. Agents can take their action in a turn-taking fashion. When an offer is made by any agent, the next agent in line can make a counter offer (overriding the previous offer), accept the current offer, or walk away (e.g. ending the negotiation without any agreement).

This process continues in a turn taking fashion until agents reach an agreement or the given deadline, or one of the agents walks away. It is worth noting that an agreement is reached if and only if all agents accept the agreed offer. If negotiation fails, agents receive their reservation utility (i.e., BATNA). The utility of agreement for each agent is calculated with respect to their own preferences. Note that in the competition preferences of each agent are represented by means of additive utility function as shown in Eq. 1 where $V_n(v_j)$ denotes agent n's valuation of the value for the issue j in the given bid and $w_{n,j}$ denotes the weights of that issue. In other words, agents sum up their weighted valuation of each issue value to calculate the overall utility. It is worth noting that each agent can only access their own preferences during the negotiation; that is, they do not know each others' truth preferences.

$$u_n(b^t) = \sum_{j \in I} V_n(b_j^t) \cdot w_{n,j} \tag{1}$$

In the competition, the deadline is set to three minutes and each negotiation session is repeated five times. Agents are allowed to access provided historical data from their past negotiations. The historical data involves the utility distribution of the exchanged offers in previous negotiation sessions according to agent's own utility space and previous agreements. In addition, agents are allowed to model their opponents' preferences by examining the bids exchanged during their negotiation and to store the learnt model where they can access in their further negotiations.

Genius 8.0.4 [20] was used to run negotiation tournaments in the competition. Agents were evaluated in four different negotiation scenarios described in Table 1. Note that all participants submitted a negotiation scenario consisting of three conflicting preference profiles and four of them were chosen for the tournaments based on varying size of their outcome space.

Table 1. Negotiation scenarios

Name	# of issue	# of values	# of outcomes
Meng wan	6	4,4,4,4,3,4	3072
BetaOne	3	4,4,4	64
IQSon	7	7,6,5,3,4,4,4	40320
Hamada	4	5,5,5,5	625

In order to complete such an extensive set of tournaments within a limited time frame, we used some high-spec computers, made available by Tokyo University of Agriculture and Technology, Japan. Specifically, each of these machines contained an *Intel Core i7* CPU, at least 64 GB of DDR3 memory, and a hard drive with at least 2 TB of capacity.

4 Result of the Competition

We have received 21 submissions from 10 institutions in eight different countries. The performance of the agents were evaluated according to their average *individual utility* and average sum of utilities (*social welfare*). The competition consists of two stages. In the *qualification round*, finalist agents are determined while the winners for each category (*individual* and *social welfare*) are determined in the *final round*. In the following sections, results for each round are presented.

- Bar Ilan University - Israel
- TU Delft - NL
- Tianjin University - China
- Nagoya Institute of Technology - Japan
- Özyegin University - Turkey
- University of Tehran - Iran
- Tokyo University of Agriculture and Technology - Japan
- Technical University Of Crete - Greece
- University of Southampton - UK
- Privateer - Japan

4.1 Qualification Results

Running the whole tournament involving 21 agents in four domains with 5 repetitions were not feasible within the given time. Therefore, three agent pools were generated randomly as follows:

- **Pool-A:** Meng wan, AgentHerb, IQSun2018, PonPokoRampage, FullAgent, Seto, Lancelot
- **Pool-B:** Beta One, Yeela, SMAC_Agent, AgreeableAgent2018, ConDAgent, Shiboy, Libra
- **Pool-C:** AgentNP1, GroupY, ATeamAgent, Sontag, Agent33, Agent_Hama, Exp-Rubick

The top 3 performing agents in each pool proceeded to the final. Therefore, there were nine finalists for each category after the qualification round.

According to the results of the pool-A, the *Mengwan*, *IQSun2018*, and *PonPokoRampage* qualified for the final round in the individual utility category. As far as the average sum of the utilities are concerned, *AgentHerb*, *IQSun2018*, and *FullAgent* are qualified for the final round (Fig. 1).

According to the results of the pool-B, the *BetaOne*, *AgreeableAgent2018*, and *Shiboy* qualified for the final round in the individual utility category. As far as the average sum of the utilities are concerned, *Yeela*, *AgreeableAgent2018*, and *ConDAgent* are qualified for the final round (Fig. 2).

Agent Name	Individual Utility	Social Welfare
meng wan	0.586923478	1.632116151
AgentHerb	0.49193542	1.854380384
IQSun2018	0.583185484	1.759158992
PonPokoRampage	0.567184966	1.547263306
FullAgent	0.549039084	1.766976108
Seto	0.551805443	1.449438396
Lancelot	0.517828051	1.53437244

Fig. 1. Results of qualification round for Pool-A

Agent Name	Individual Utility	Social Welfare
Beta One	0.482887311	1.404794645
Yeela	0.399430418	1.422654384
SMAC_Agent	0.464635439	1.325276881
AgreeableAgent2018	0.505584897	1.505251211
ConDAgent	0.471002153	1.518896956
Shiboy	0.502585371	1.411124211
Libra	0.445786573	1.227738202

Fig. 2. Results of qualification round for Pool-B

According to the results of the pool-C, the *AgentNP1*, *GroupY*, and *Sontag* qualified for the final round in the individual utility category. As far as the average sum of the utilities are concerned, *AgentNP1*, *Sontag*, and *Agent33* are qualified for the final round (Fig. 3).

4.2 Final Results: Individual Utility Category

In this category, we ran a tournament with the following qualified agents.

- meng wan, University of Southampton, UK
- IQSun2018, University Of Tehran, Iran
- PonPokoRampage, Tokyo University of Agriculture and Technology, Japan
- AgreeableAgent2018, University of Tehran, Iran
- Shiboy, Nagoya Institute of Technology, Japan
- Beta One, Özyeğin University, Turkey
- AgentNP1, privateer, Japan

Agent Name	Individual Utility	Social Welfare
AgentNP1	0.512177707	1.52976242
GroupY	0.485323479	1.33874982
ATeamAgent	0.345297603	0.980979189
Sontag	0.484535782	1.515047114
Agent33	0.428666974	1.419233621
Agent_Hama	0.469592183	1.393009022

Fig. 3. Results of qualification round for Pool-C

Fig. 4. Overall ranking w.r.t. individual utility

- GroupY, Özyeğin University, Turkey
- Sontag, Tokyo University of Agriculture and Technology, Japan

Figure 4 shows the average individual utility gained by each finalist over 2520 negotiations. Recall that each negotiating agent negotiates with all other agents five times for each negotiation scenario in Table 1. *AgreeableAgent2018* gained 0.59 on average and won the competition. *Meng wan* and *Beta One* agents were awarded second and third place respectively. The detailed description of the winner strategies according to the individual utility, are given.

- ***AgreeableAgent2018 by Sahar Mirzayi (University of Tehran, Iran):*** This agent tries to learn its opponents' preferences during the ongoing negotiations by using a frequency based modeling. It uses a time-based bidding strategy, which takes its opponents' preferences into account. Basically, it generates all candidate bids above estimated target utility and sorts those candidates according to opponent models. It makes the bid selected by

using Roulette Wheel Selection. It employs ACNext acceptance strategy. It also accepts it opponents' bid, which is higher than reservation utility if it is almost deadline.
This agent combines different bidding strategies for increasing social welfare and individual utility. initially, it extracts three different utilities using three different ways; and then according to the size of the domain, it calculates the target utility based on those three utilities. As a statement somewhat in detail for the way of extracting those three utilities, it can be said that initially, the agent calculates a utility based on time. Secondly a utility based on an idea from previous agents which also is somehow time based, but in a completely different way from the first one; and finally a utility according to the average of best five bids that were received by the agent and also best five bids that were given by the agent itself.
In their submission: The University of Tehran's agent (AgreeableAgent2018) is trying to improve the negotiation outcome by using opponent models. This agent learns the opponent model based on the history of bids during one session, then uses models to concede more effectively during the negotiation process. Agreeable agent also uses a time-based strategy, but does not necessarily offer a bid near target utility. This agent extracts some bids in a neighborhood threshold around the target's utility, sorts them using opponent models and selects the next bid using Roulette Wheel Selection. Moreover the neighborhood threshold increases during session to improve social welfare.

- ***Meng wan by Meng Wan, Hui Cui (University of Southampton, UK):*** It employs a hybrid bidding strategy composed of three bidding strategy. The first strategy generates a random bid above an estimated target utility (time-based) while the second strategy chooses a random bid among opponents' best offers so far. The third bidding strategy makes a bid whose utility is higher than target utility, which also maximizes its opponents' utility based on estimated frequency-based opponent models. It employs a time-based acceptance strategy, which is not inclined to accept any offer at the beginning. By the end of negotiation, this agent tends to ask opponents' best offers so far.
In order to ensure that an agreement can be reached at the end, its acceptance strategy is based on time change. Its bidding strategy is analyzing the rival agents' model and then throwing the offer based on their frequency.
- ***Beta One by Alper Sekerci, Abdulkadir Nurkalem (Özyeğin University):*** This agent has a tendency to be more stubborn against stubborn opponents while more generous against generous opponents. For the opponent modelling, the agent uses statistical analysis to decide whether it should concede or not. The personality of the agent is parameterized. By analyzing the history, the agent is able to tune its parameters (e.g. selfish ratio) to try to perform better in the upcoming negotiations.
Each domain presents different circumstances and the main concern of the agent is to determine the range where the offers are selfish yet logical and acceptable.

4.3 Social Welfare Category

In this category, we ran a tournament with the following qualified agents.

- AgentHerb, Bar Ilan University, Israel
- FullAgent, Bar Ilan University, Israel
- IQSun2018, University Of Tehran, Iran
- ConDAgent, Technical University Of Crete, Greece
- AgreeableAgent2018, University of Tehran, Iran
- Yeela, Bar-Ilan University, Israel
- AgentNP1, privateer, Japan
- Sontag, Tokyo University of Agriculture and Technology, Japan
- Agent33, Nagoya Institute of Technology, Japan

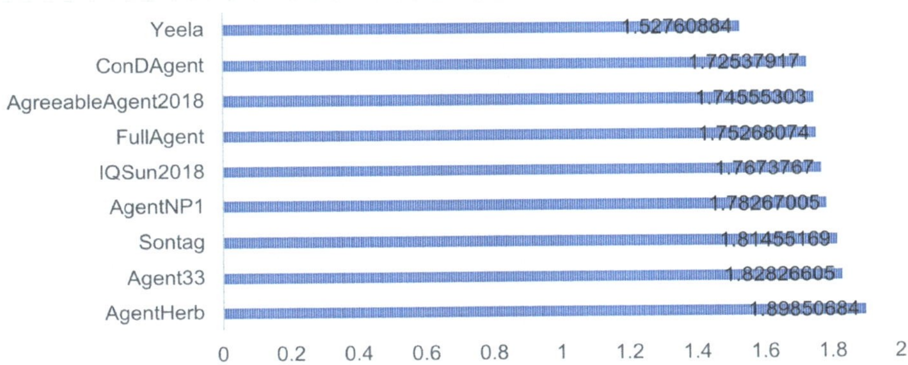

Fig. 5. Overall ranking wrt. social welfare

Figure 5 shows the average sum of utilities of the agreements reached by each agent. *AgentHerb* gained 1.89 on average and won the competition where *Agent33* and *Sontag* agents took the second and third place respectively. The description of the winner strategies are given below.

Table 2. Detailed results of winners tournaments (individual category)

Agreeable	MengWan	BetaOne
0.769856823	**0.805217872**	0.787584175
Social welfare	**Dist. to pareto**	**Dist. to nash**
2.362659083	0.019035083	0.223667667

Table 3. Detailed results of winners tournaments (social category)

Agreeable	MengWan	BetaOne
0.634797437	0.688591936	**0.899989331**
Social welfare	**Dist. to pareto**	**Dist. to nash**
2.223378333	0.027012083	0.32230525

- ***AgentHerb by Alon Stern, Amit Moryossef, Yehudit Reyzer, Karin Dahan (Bar Ilan University, Israel):*** This agent records the history of bids for each opponent and whether they were accepted or rejected by the opponents. Accordingly, Agent Herb uses a logistic regression model to predict the likelihood of acceptance of a bid by the opponents. It makes its bids based on its chances of acceptance by all the opponents while taking its own utility into account. The agent employs ACNext acceptance strategy with a discount factor.
- ***Agent33 by Liu Shan (Nagoya Institute of Technology, Japan):*** The bidding Strategy aims to propose bids around Nash bargaining solution (NBS). Agent33 uses a novel heuristic method in order to find the promising bids around the NBS. The proposed heuristic method aims to construct a list of the opponents prior issues, which is incrementally updated throughout the negotiation process by calculating the standard deviation of each issue value's frequency and the standard deviation of each issue's value.
  The process to find the opponents prior issues consists of two steps. The first step calculates the standard deviation of each value's frequency in each issue and names the result as Value Standard Deviation. While the second step calculates the standard deviation of each issue's Value Standard Deviation and names the result as Issue Standard Deviation. In this context, if a certain negotiator has large Issue Standard Deviation, it means this negotiator has an obvious preference of the prior issues. The issues with high Value Standard Deviation are the prior issues of this negotiator and the values with high frequency are the prior values of these prior issues. These prior issues and their prior values are set in a pair that is called Prior Pair.
- ***Sontag by Ryohei Kawata (Tokyo University of Agriculture and Technology, Japan):*** Sontag makes concessions in early to increase social welfare. Sontag does not model opponents and no learning from negotiation history. In other words, it expects that the opponents will make a concession finally. The proposal and acceptance of the bids are based on the following equation: $f(t) = t/2.5 - \log(t/2 + 0.1)$.

4.4 Further Analysis of ANAC 2018 Winners

After competition, we ran another tournament among the winners per each category. The settings of tournaments are same as ANAC 2018; however, the agents in the tournament consists of only the winners of each category (Individual

Utility Category: AgreeableAgent2018, MengWan, BetaOne; Social Welfare Category: AgentHerb, Agent33, Sontag). According to the results listed in Table 2, MengWan received the highest average utility although this agent took the second place in the competition. Similarly, when only winners of the social welfare category negotiate with each other, Songtag outperforms other winners according to the results listed in Table 3. The result should not surprise us because negotiation outcome highly depends on whom we are negotiating with. Similarly, the winners in individual category received higher average social welfare than the winners of social category. In addition, it can be seen that winners made agreements that are closer to the Pareto Frontier, received higher social welfare. Therefore, one of the important points of the competition is to predict the Pareto frontier and Nash bargaining solution to make high quality agreements socially.

5 Related Work

The automated negotiating agent competition follows in the footsteps of a series of successful competitions that aim to advance the state-of-the-art in artificial intelligence (other examples include the Annual Computer Poker Competition and the various Trading Agent Competitions, the Alimama International Advertising Algorithm Competition, and Angry Birds Competition). Over the years the competition has addressed various topics: varying the number of negotiators, the complexity of the negotiation domains, repeated negotiations with the same set of opponents, negotiations in special domains, and negotiating against humans.

Every year, new features have been incorporated into the competition environment to increase realism and to encourage the development of flexible and practical negotiation agents [3]. The first installment of ANAC was in 2010 and started as a jointly organized competition between Delft University of Technology and Bar-Ilan University [8]. Since the first edition, the local organization has rotated between various international institutions. In 2011, ANAC was organized by the 2010 winner, the Nagoya Institute of Technology [6]. Next editions were organized by University of Southampton (2012) [22], Ben Gurion University of the Negev (2013) [15]. From 2014 to now, Tokyo University of Agriculture and Technology, Özyeğin University, Nagoya Institute of Technology and Delft University of Technology [4] have been organized. Every installment has seen a growing number of participants: ANAC 2010 started with 7 teams from 5 different institutes, while in 2018, it featured 21 teams from 10 institutes.

By analyzing the results of ANAC, the stream of the strategies of ANAC and important factors for developing the competition have been shown. Baarslag et al. [5] presented an in-depth analysis of ANAC 2011. This paper mainly analyzes the different strategies using classifications of agents with respect to their concession behavior against a set of standard benchmark strategies and empirical game theory (EGT) to investigate the robustness of the strategies. It also shows that the most adaptive negotiation strategies, while robust across different opponents, are not necessarily the ones that win the competition. Furthermore, our EGT analysis highlights the importance of considering metrics.

Kawaguchi et al. [19] proposed a strategy for compromising the estimated maximum value based on estimated maximum utility. These papers have been important contributions for bilateral multi-issue closed negotiation; however, they don't deal with multi-time negotiation with learning and reusing the past negotiation sessions. This agent is the winner of ANAC2010. After that, Fujita [11,12] proposed the compromising strategy with adjusting the speed of making agreements using the Conflict Mode, and focused on multi-time negotiations. However, these strategies only focused on the linear utility functions.

In addition, some studies have focused on the divided parts of negotiating strategies in the alternating offering protocol: proposals, responses, and opponent modeling. Effective strategies can be achieved by combinations of top agents' strategies in the competitions depending on the opponent's strategies and negotiation environments. Many of the sophisticated agent strategies that currently exist are comprised of a fixed set of modules. Therefore, the studies for proposing the negotiation strategies focusing on the modules are important and influential [9,23].

6 Conclusion

This paper discusses the main ANAC 2018 competition setup and results as well as explaining briefly the research challenges in repeated multilateral negotiations. As a future work, it would be interesting to analyze the performance of agents in a more extensive setup. Therefore, we are planning to run additional tournaments including ANAC 2017 winner agents and compare their performance.

Acknowledgement. This work is part of the Veni research programme with project number 639.021.751, which is financed by the The Dutch Research Council (NWO). We are thankful for the sponsorship of the ANAC competition by the Artificial Intelligence Journal.

References

1. Aydoğan, R., Festen, D., Hindriks, K.V., Jonker, C.M.: Alternating offers protocols for multilateral negotiation. In: Fujita, K., Bai, Q., Ito, T., Zhang, M., Ren, F., Aydoğan, R., Hadfi, R. (eds.) Modern Approaches to Agent-based Complex Automated Negotiation, pp. 153–167. Springer (2017)
2. Aydoğan, R., Hindriks, K.V., Jonker, C.M.: Multilateral mediated negotiation protocols with feedback. In: Novel Insights in Agent-based Complex Automated Negotiation, pp. 43–59. Springer (2014)
3. Baarslag, T.: What to bid and when to stop. Dissertation, Delft University of Technology (2014). https://doi.org/10.4233/uuid:3df6e234-a7c1-4dbe-9eb9-baadabc04bca
4. Baarslag, T., Aydoğan, R., Hindriks, K.V., Fuijita, K., Ito, T., Jonker, C.M.: The automated negotiating agents competition, 2010–2015. AI Mag. **36**(4), 115–118 (2015). http://www.aaai.org/ojs/index.php/aimagazine/article/view/2609

5. Baarslag, T., Fujita, K., Gerding, E., Hindriks, K., Ito, T., Jennings, N.R., Jonker, C., Kraus, S., Lin, R., Robu, V., Williams, C.: Evaluating practical negotiating agents: results and analysis of the 2011 international competition. Artif. Intell. J. (AIJ) **198**, 73–103 (2013)
6. Baarslag, T., Fujita, K., Gerding, E.H., Hindriks, K., Ito, T., Jennings, N.R., Jonker, C., Kraus, S., Lin, R., Robu, V., Williams, C.R.: Evaluating practical negotiating agents: results and analysis of the 2011 international competition. Artif. Intell. **198**, 73–103 (2013). https://doi.org/10.1016/j.artint.2012.09.004
7. Baarslag, T., Gerding, E.H., Aydoğan, R., Schraefel, M.: Optimal negotiation decision functions in time-sensitive domains. In: IEEE International Conference on Web Intelligence and Intelligent Agent Technology, vol. 2, pp. 190–197. IEEE (2015)
8. Baarslag, T., Hindriks, K., Jonker, C.M., Kraus, S., Lin, R.: The first automated negotiating agents competition (ANAC 2010). In: Ito, T., Zhang, M., Robu, V., Fatima, S., Matsuo, T. (eds.) New Trends in Agent-based Complex Automated Negotiations. Series of Studies in Computational Intelligence, pp. 113–135. Springer, Heidelberg (2012). https://doi.org/10.1007/978-3-642-24696-8_7
9. Baarslag, T., Hindriks, K.V.: Accepting optimally in automated negotiation with incomplete information. In: Proceedings of the 12th International Conference on Autonomous Agents and Multi-agent Systems (AAMAS 2013), pp. 715–722 (2013)
10. Fatima, S., Kraus, S., Wooldridge, M.: Principles of Automated Negotiation. Cambridge University Press, New York (2014)
11. Fujita, K.: Automated strategy adaptation for multi-times bilateral closed negotiations. In: Proceedings of the 13th International Conference on Autonomous Agents and Multiagent Systems (AAMAS 2014), pp. 1509–1510 (2014)
12. Fujita, K.: Compromising adjustment strategy based on TKI conflict mode for multi-times bilateral closed negotiations. Comput. Intell. (2017). https://doi.org/10.1111/coin.12107
13. Fujita, K., Ito, T., Klein, M.: A secure and fair protocol that addresses weaknesses of the nash bargaining solution in nonlinear negotiation. Group Decis. Negot. **21**(1), 29–47 (2012)
14. Fujita, K., Ito, T., Klein, M.: Efficient issue-grouping approach for multiple interdependent issues negotiation between exaggerator agents. Decis. Support Syst. **60**, 10–17 (2014)
15. Gal, K., Ilany, L.: The fourth automated negotiation competition, pp. 129–136. Springer, Tokyo (2015). https://doi.org/10.1007/978-4-431-55525-4_8
16. Jennings, N.R., Faratin, P., Lomuscio, A., Parsons, S., Sierra, C., Wooldridge, M.: Automated negotiation: prospects, methods and challenges (1998)
17. de Jonge, D., Baarslag, T., Aydoğan, R., Jonker, C.M., Fujita, K., Ito, T.: The challenge of negotiation in the game of diplomacy. In: Proceedings of the 6th Conference on Agreement Technologies (2019)
18. Jonker, C.M., Aydoğan, R., Baarslag, T., Fujita, K., Ito, T., Hindriks, K.V.: Automated negotiating agents competition (ANAC). In: AAAI, pp. 5070–5072 (2017)
19. Kawaguchi, S., Fujita, K., Ito, T.: Compromising strategy based on estimated maximum utility for automated negotiation agents competition (ANAC-10). In: 24th International Conference on Industrial Engineering and Other Applications of Applied Intelligent Systems (IEA/AIE d2011), pp. 501–510 (2011)
20. Lin, R., Kraus, S., Baarslag, T., Tykhonov, D., Hindriks, K., Jonker, C.M.: Genius: an integrated environment for supporting the design of generic automated negotiators. Comput. Intell. **30**(1), 48–70 (2014). https://doi.org/10.1111/j.1467-8640.2012.00463.x

21. Mell, J., Gratch, J., Baarslag, T., Aydoğan, R., Jonker, C.M.: Results of the first annual human-agent league of the automated negotiating agents competition. In: Proceedings of the 18th International Conference on Intelligent Virtual Agents, IVA 2018, pp. 23–28. ACM, New York (2018). https://doi.org/10.1145/3267851.3267907
22. Williams, C., Robu, V., Gerding, E., Jennings, N.: An overview of the results and insights from the third automated negotiating agents competition (ANAC 2012). In: Marsa-Maestre, I., Lopez-Carmona, M., Ito, T., Zhang, M., Bai, Q., Fujita, K. (eds.) Novel Insights in Agent-based Complex Automated Negotiation. Studies in Computational Intelligence, vol. 535, pp. 151–162. Springer, Japan (2014)
23. Zafari, F., Mofakham, F.N.: POPPONENT: highly accurate, individually and socially efficient opponent preference model in bilateral multi issue negotiations. In: Proceedings of the Twenty-Sixth International Joint Conference on Artificial Intelligence (IJCAI 2017), pp. 5100–5104 (2017)

Flexibility of Emulation Learning from Pioneers in Nonstationary Environments

Moto Shinriki, Hiroaki Wakabayashi, Yu Kono, and Tatsuji Takahashi(✉)

Tokyo Denki University, Ishizaka, Hatoyama, Hiki, Saitama 350-0394, Japan
tatsujit@mail.dendai.ac.jp
https://sites.google.com/site/tatsujitakahashi/

Abstract. This is an extension from a selected paper from JSAI2019. Social learning is crucial in acquisition of the intelligent behaviors of humans and many kinds of animals, as it makes behavior learning far more efficient than pure trial-and-error. In imitation learning, a representative form of social learning, the agent observes specific action-state pair sequences produced by another agent (*expert*) and reflect them into its own action. One of its implementations in reinforcement learning is the inverse reinforcement learning. We propose another form of social learning, emulation learning, which requires much less information from another agent (*pioneer*). In emulation learning, the agent is given only a certain level of achievement by another agent, or a record. In this study, we implement emulation learning in the reinforcement learning setting by applying a model of satisficing action policy. We show that the emulation learning algorithm works well both in stationary and non-stationary reinforcement learning tasks, breaking the often observed trade-off like relationship between efficiency and flexibility.

Keywords: Reinforcement learning · Social learning · Imitation

1 Introduction

Humans usually begin learning with some kind of prior knowledge. If the knowledge is about the structure of the environment, the past experience, or other's action history, the learning will be somehow model-based, by transfer, or supervised, respectively. Imitation learning requires an *expert* who provides an exemplar behavior. However, the expert's action data may be very expensive or unavailable in general. On the other hand, there are cases where (only) the information of someone's achievement level is obtained. Our search is often accelerated by a peer's high achievement or record breaking, as in sports or in invention. As it is closely related to *end state emulation* in social learning [1], we call this form of learning *emulation learning*. In emulation, the outcome of an action sequence ("what") is socially transmitted and referred, while in imitation the process or the procedure ("how") is observed and assimilated.

Y. Ohsawa et al. (Eds.): JSAI 2019, AISC 1128, pp. 90–101, 2020.
https://doi.org/10.1007/978-3-030-39878-1_9

In this study, we test the performance of the three algorithms: Q-Learning, imitation learning with inverse reinforcement learning, and our emulation learning algorithm with satisficing reinforcement learning, from two aspects. The first aspect is the speed of learning. The second aspect is the flexibility in adaptation to environment. The learning tasks include nonstationary reinforcement learning tasks, and we evaluate how flexibly the algorithm can respond to the environmental changes.

2 Reinforcement Learning Algorithms

Reinforcement learning is a type of machine learning in which an agent learns an appropriate action sequence through trial-and-error type interaction with the environment. Recently, as researches in game AI and autonomous robot have been actively conducted, reinforcement learning is gaining more attention as a method for autonomous learning in unknown environments. The scheme of reinforcement learning is shown in Fig. 1.

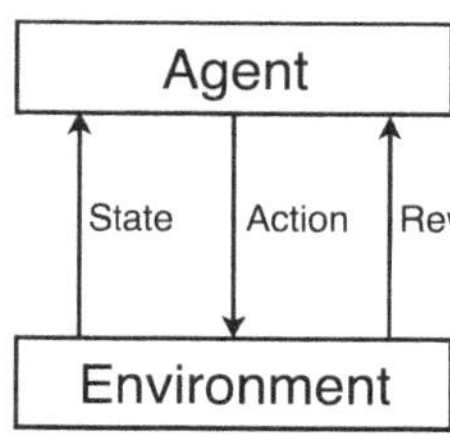

Fig. 1. A scheme of reinforcement learning

The learning subject of reinforcement learning is called an agent. State in the environment is perceived by the agent, and the state at step t is denoted by s_t. The agent chooses an action from the set of actions available at the current state. The action taken at step t is denoted by a_t. The (probabilistic) mapping from the states to the actions is called a policy, denoted by π. When the state observed at a certain time t is s_t, the agent selects the action at based on a certain behavior policy $\pi(s_t)$. Reward is evaluation of the agent's action at the state by the environment. As a result of the action, the environment transitions from state s_t to s_{t+1}. The agent obtains a reward according to the reward function $R(s_t, a_t)$.

2.1 Q-Learning

Q-Learning is the representative algorithm for reinforcement learning. The action value $Q(s_t, a_t)$ of Q-Learning is updated according to

$$Q(s_t, a_t) \leftarrow Q(s_t, a_t) + \alpha\Big(r_t + \gamma \max_a Q(s_{t+1}, a) - Q(s_t, a_t)\Big), \tag{1}$$

where α is the learning rate and γ is the discount rate.

2.2 Inverse Reinforcement Learning as Imitation

Inverse reinforcement learning (IRL) is a form of learning of which the goal is to infer a reward function $R(s)$ from the expert's behavior trajectory T. A reward function $R(s)$ is usually a function which returns a reward calculated by multiplying the parameter θ and the one-hot-vector of the state. The behavior trajectory is a pair of expert's states and actions, $(s_0, a_0, s_1, a_1, ...)$. In this study, we adopt the maximum entropy IRL (MaxEntIRL) as the implementation [4].

2.3 Emulation by Risk-Sensitive Satisficing (RS)

Humans tend not to exhaustively search for optimization. Rather, we satisfice. That is, we confront a task with certain reference level (aspiration, $\aleph$ (aleph)) and finish searching when we find a satisfactory action better than the reference [6]. When the aspiration level is given socially, satisficing means emulation. Satisficing policy will converge the actions to take, when the aspiration is satisfied, after the limited search. We implement emulation with the risk-sensitive satisficing (RS) model, a cognitive satisficing value function with reflective risk attitudes as in the prospect theory in behavioral economics [2,3]. The RS value function is defined as follows:

$$\begin{aligned} RS(s_t, a_t) &= \tau(s_t, a_t)\delta(s_t, a_t) && (2)\\ &= \tau(s_t, a_t)\big(Q(s_t, a_t) - \aleph(s_t)\big), && (3) \end{aligned}$$

and the action to be selected is determined as:

$$a_t^{\mathrm{sel}} = \arg\max_a RS(s_t, a). \qquad (4)$$

RS's valuation qualitatively changes according to the sign of δ, the difference between the Q and $\aleph$. It leads to optimistic exploration (uncertainty seeking) when the sign of all δ's for the actions is negative (no satisfactory action is found). When a satisfactory action is found (some action has positive δ), it pessimistically exploits the knowledge (uncertainty aversion). It considers the reliability of the Q-value with τ that approximates how many times the action has been chosen. In the multi-armed bandit problems, which form the most basic class of reinforcement learning problems, it is proven and empirically shown that RS realizes very efficient satisficing [2,3]. τ is defined as follows, where γ_τ is the discount rate and α_τ is the learning rate.

$$\begin{aligned} \tau(s_t, a_t) &= \tau_{\mathrm{curr}}(s_t, a_t) + \tau_{\mathrm{post}}(s_t, a_t), && (5)\\ \tau_{\mathrm{curr}}(s_t, a_t) &\leftarrow \tau_{\mathrm{curr}}(s_t, a_t) + 1, && (6)\\ \tau_{\mathrm{post}}(s_t, a_t) &\leftarrow (1 - \alpha_\tau)\tau_{\mathrm{post}}(s_t, a_t)\\ &\quad + \alpha_\tau \gamma_\tau \tau(s_{t+1}, a_{t+1}^{\mathrm{sel}}). && (7) \end{aligned}$$

Global Reference Conversion (GRC). While RS works as intended in the multi-armed bandit problems that may be considered as a single-state reinforcement learning task, it is generally difficult to assign the optimal aspiration to multiple states, when only the global aspiration (i.e., for the entire episode), $\aleph_G$ is available from a pioneer. It is $\aleph_G$ that works as a social goal-setting trigger, such as someone's high performance or record breaking. The global reference conversion (GRC) algorithm allocates the global aspiration to each state by defining the global return E_G, the level of achievement. E_G is defined using the temporary return (E_{tmp}) which is updated at the end of each episode.

$$E_G \leftarrow \frac{E_{\mathrm{tmp}} + \gamma_G N_G E_G}{1 + \gamma_G N_G} \tag{8}$$

$$N_G \leftarrow 1 + \gamma_G N_G \tag{9}$$

Using E_G, the global difference δ_G and the local aspiration at each state $\aleph(s)$ is defined as follows. δ_G is defined to be non-positive so that optimistic exploration is to be realized.

$$\delta_G = \min(E_G - \aleph_G, 0), \tag{10}$$

$$\aleph(s) = \max_a Q(s, a) - \zeta(s)\delta_G \tag{11}$$

The parameter $\zeta(s)$ is introduced to adjust the proportion of the global aspiration and the Q-value.

3 First Experiment

3.1 Task: SwitchWorld

The first task is SwitchWorld where there are multiple, numbered switches in a lattice space. A state is a location in the space, and an action is a move to a neighboring state (the upper, lower, left, or right adjacent cell). The state space of the task is shown in Fig. 2. The red cells are where switches are placed, and the green cell is where the agent is placed at the initial step of an episode. When the agent passes through one of the switch cells, the agent is notified of it (i.e., the states form an augmented state space [5]). When the agent has acted for 99 times, each episode ends. In order for the agent to gain a reward, the switches must be checked in the correct order: switch 1, 2, and then 3. When the agent presses the last switch, a reward 1 is given, and the state of the switches resets. The agent observes that a switch is checked when having passed one. There are 1,000 episodes and the result is the average of 100 simulations. An optimal trajectory going through switches 1, 2, and 3 takes 16 steps, as shown by the arrows in Fig. 2. If the agent keeps repeating the trajectory, the optimal reward 6 is achieved.

3.2 Task: UnsteadySwitchWorld

To test the speed and flexibility of the learning algorithms, we conduct UnsteadySwitchWorld, which is based on SwitchWorld, but the switches are randomly shuffled. The agent ran 10,000 episodes and the result is the average of 1,000 simulations. The switches change its location randomly every 1,000 episodes, under the constraint that the new switch configuration is not the same at the previous one. The start state is moved to a central cell so that the maximum reward in an episode stays 6, as in Fig. 3.

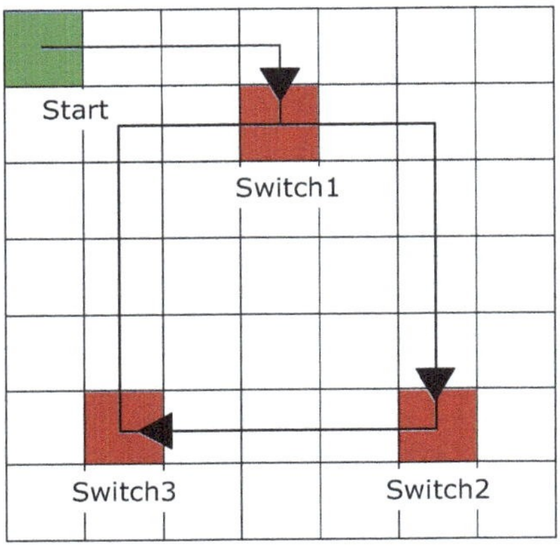

Fig. 2. SwitchWorld task

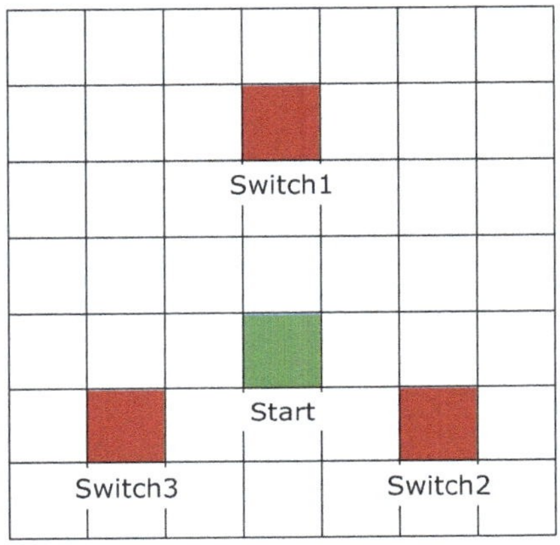

Fig. 3. UnsteadySwitchWorld task

3.3 Simulation and Result

Q-Learning and MaxEntIRL are operated under the ϵ-greedy policy, in which the agent selects an action at random at probability ϵ and selects the greedy action a with the highest action value $Q(s_t, a_t)$ at probability $1-\epsilon$. ϵ starts from 1.0 and then decreases by 0.005 per episode. In the steady task (SwitchWorld), ϵ reaches 0.0 at episode 200. In the unsteady task (UnsteadySwitchWorld), ϵ reaches 0.025 and thereafter kept constant so that the two algorithms can have the adaptability to environmental changes. The global aspiration for entire episodes was $\aleph_G = 0.06$, which represents the pioneer's record of reward 6 achieved during 100 steps. γ_G was set to 0.9. The scaling parameter $\zeta(s) = 0.05$ for all s [7]. Learning rate α was 0.1 in all methods, and the discount rate γ was set to 0.9 for all the algorithms.

For MaxEntIRL, the learning rate β was 0.01 for reward function estimation and the epoch was 20. For generating the expert's trajectories, we used the Q-values of the Q-Learning agent after the learning. Expert route information was extracted from the result of Q-Learning for each divided tasks, which are: Start to Switch 1, Switch 1 to Switch 2, and Switch 2 to Switch 3. For each of the three subtasks, 33 trajectories with random initial state were generated. Hence the sample size of expert's trajectory for MaxEntIRL was 99 in total.

SwitchWorld. Figure 4 shows the time development of the obtained reward for each episode. RS+GRC reaches the optimal reward 6 after 100 episodes. MaxEntIRL reaches there after 200 episodes. The average accumulated reward for Q-Learning did not reach the optimum.

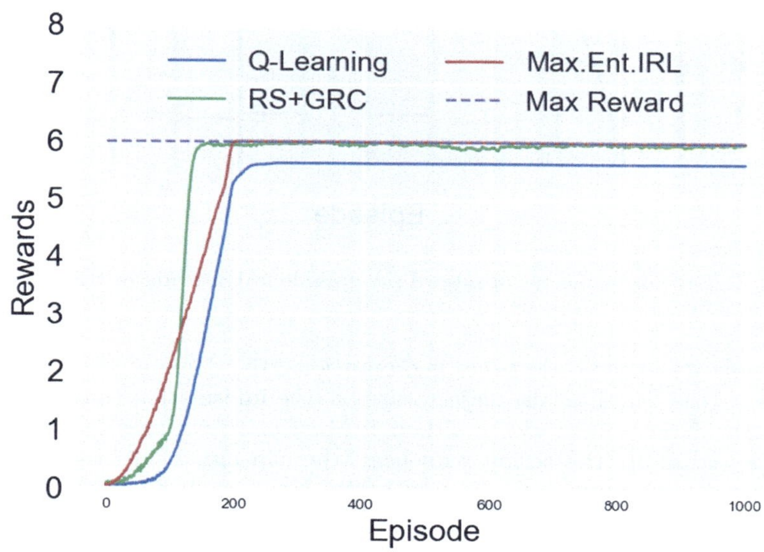

Fig. 4. Time development of reward per episode in SwitchWorld

UnsteadySwitchWorld. Figure 5 shows the time development of the obtained reward for each episode. RS+GRC was overall capable to obtain the rewards, while MaxEntIRL failed to obtain the reward stably. Q-Learning can gradually adapt to the environmental change, but slower than RS+GRC.

3.4 Discussion

From the result of SwitchWorld, we see that RS+GRC and MaxEntIRL can reach the optimal reward 6. Let us focus on the difference in the amount of prior information for the two social learning algorithms. The amount of prior information on RS+GRC is just the global aspiration that is only a numerical value (at most between 2 to 8 bytes), $\aleph_G = 0.06$. On the other hand, MaxEntIRL needs action trajectories as prior information.

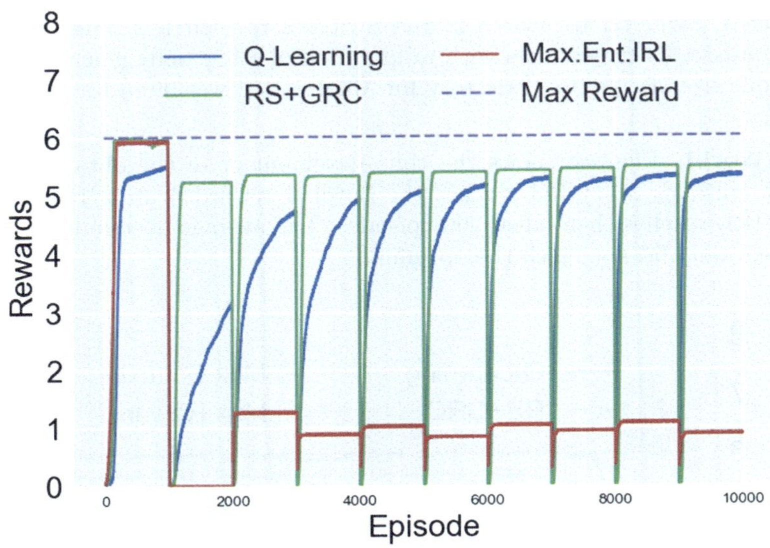

Fig. 5. Time development of reward per episode in UnsteadySwitchWorld

The capacity of the trajectories is proportional to (the number of trajectory samples) × (the length of the trajectories) × (the information to represent the unit length of the trajectories). Although MaxEntIRL utilizes only the state sequences and omits the action sequences, the capacity is at least 100 times larger than the prior information for RS+GRC. Despite that RS+GRC requires much smaller amount of prior information, it can learn faster than MaxEntIRL.

From the results of UnsteadySwitchWorld, we see that RS+GRC can learn faster than Q-Learning and MaxEntIRL. The latter did not cope well with non-stationary environment. Considering the learning speed of RS+GRC, it conducted exploration with an optimistic directionality based on the satisficing policy and efficient samplings, rather than ϵ-greedy applied to Q-Learning, which is a random search with no directionality. As for the difference in the behavior of RS+GRC and MaxEntIRL in the non-stationary environment, RS+GRC judges the superiority or inferiority of the action sequence only with its result, E_G, compared to $\aleph_G$. MaxEntIRL compares the whole action sequence with the trajectories of the expert to judge its superiority or inferiority. RS+GRC was able to cope with the non-stationary environment because it can truncate the existing action sequence and search for a new action sequence if the result is lower than the reference level.

However, there is a property of this environment that makes its nonstationary version slightly trivial. In this environment, whatever the configuration of the switches is, the optimal action sequence is to pass through the switches either clockwise or counter-clockwise. There are more than one possible shortest paths between any two switches. If the agent learns different paths for clockwise and counter-clockwise travels, then the change (reconfiguration of switches) can be dealt with just by changing the actions to take just at the switches. That way, the paths can be easily reserved. The Q-values at the switches would converge to similar values that lead to quick switching enabled.

4 Second Experiment

In the first experiment, there were multiple possible paths between the switches that may lead to the adaptation to the environmental changes much easier. For this reason, we perform another experiment where the agent cannot cope with the environmental changes unless the actions to take on the single path must be overwritten in terms of the Q-value. The tasks are based on SwitchWorld, and the difference is that there is only one (shortest) path between switches.

The steady version of the task may be easier than SwitchWorld in that the former requires narrower exploration due to the walls around the paths.

4.1 Task: LoopSwitchWorld and LoopUnsteadySwitchWorld

The task, LoopSwitchWorld, is shown in Fig. 6 in which there is walls around shortest paths between the switches. The agent is initially located at Start, but once it gets in to the rectangular loop area, it cannot come out of it. The action to get out of the loop area places the agent at the same location. There are steady and unsteady versions, as in the first experiment.

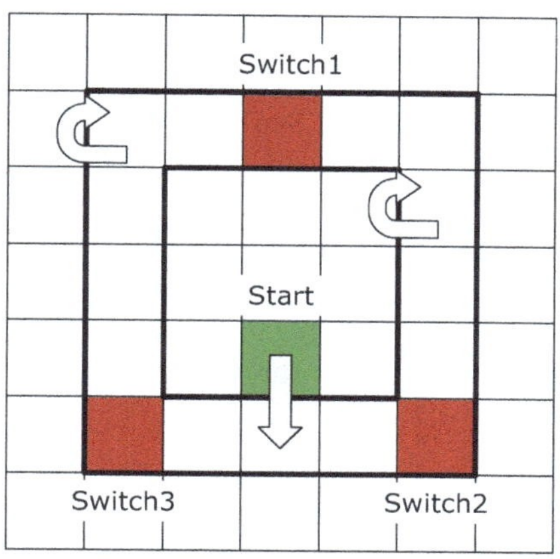

Fig. 6. LoopSwitchWorld

4.2 Simulation and Result

The experiment conformed to the same parameter setting in the first experiment. The result of LoopSwitchWorld and LoopUnsteadySwitchWorld are shown in Figs. 7 and 8, respectively.

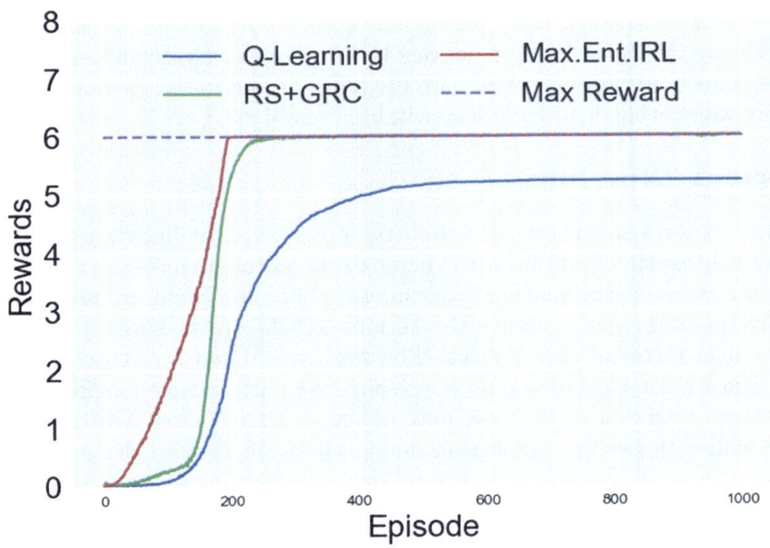

Fig. 7. Time development of reward per episode in LoopSwitchWorld

4.3 Discussion

From Figs. 7 and 8, we see that Q-Learning earns lower rewards compared to SwitchWorld and UnsteadySwitchWorld, as in Figs. 4 and 5. It is because of the random action selection of Q-Learning that leads to hitting a wall (and stays to the same location) which does not lead to any reward. Regarding LoopUnsteadySwitchWorld, it seems that Q-Learning, which was able to find side paths in the first experiment in Sect. 3, can no longer escape from overwriting the Q-values of all the states in the paths, which lead to slower and lesser increase in reward after the switch shuffle. RS+GRC and MaxEntIRL behaved similarly as in the last experiment. The next chapter explains why RS+GRC can cope with the environmental changes even when the Q-values of many actions have to be overwritten.

5 General Discussion

Two experiments showed that RS+GRC as an emulation algorithm works well in both stationary and non-stationary environments, with the same parameter setting. We focus on whether the algorithm utilizes some specific mechanisms in dealing with non-stationary environments, or it requires much larger amount of calculation. In the following, we will discuss whether RS+GRC is fair compared to Q-Learning in terms of computational complexity and data volume, and explain how the dynamic exploration is realized over and over in non-stationary environments.

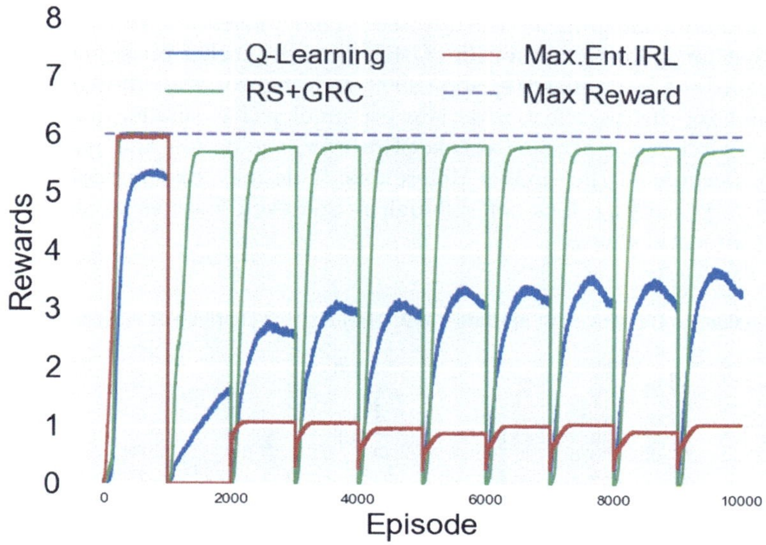

Fig. 8. Time development of reward per episode in LoopUnsteadySwitchWorld

5.1 RS+GRC in Terms of Spatial and Temporal Efficienty

Compared to Q-Learning, RS+GRC needs the τ value for each state and some numbers such as E_{tmp}, N_G, and E_G. At each step, while (1-step backup) Q-Learning updates only the Q-value of the current state-action pair, RS+GRC updates three values, Q, τ, and E_{tmp}. Hence the computation takes a constant multiple of Q-Learning's. In action selection, the RS+GRC values as in (2) are calculated for the possible actions at the current state. Only a simple product-sum operation is needed.

5.2 Adaptation to Non-stationary Environments

The result of Q-Learning becomes higher and higher in the non-stationary environments, as shown in Figs. 5 and 8. The reason for the former may be the multiple possible paths, so we discuss only the latter. In the latter task, LoopUnsteadySwitchWorld, either clockwise or counter-clockwise run leads to reward 1, so one of them is the optimal policy. At an environmental change, the rotative direction may change. As the agent acquires rewards in both directions, the Q-values of the moves along both directions get higher and converge to similar values. Then, when the Q-value of a move gets lower, the Q-value of the other move may be higher, leading to a different action selection. Then, what is required to respond to the environmental change is to switch the magnitude relations of the Q-values of the moves. Figure 9 shows the Q-values in LoopUnsteadySwitchWorld. There are four bar plots. Each shows the Q-values of two effective actions in three states. The states are in between Switch3 and Switch2.

The red and blue bars indicate the Q-values of moving left and right, respectively. We see that for Q-Learning and RS+GRC, from the earlier phase (episode 1999, left figure) to the later phase (episode 9999, right figure), the differences between the Q-values of the two effective actions get smaller. The smaller differences lead to easier switching of actions to take. However, the action and policy become unstable, because of the smaller differences. This may be the reason why the reward for Q-Learning does not get higher, and even lower after the rise, after the environmental change.

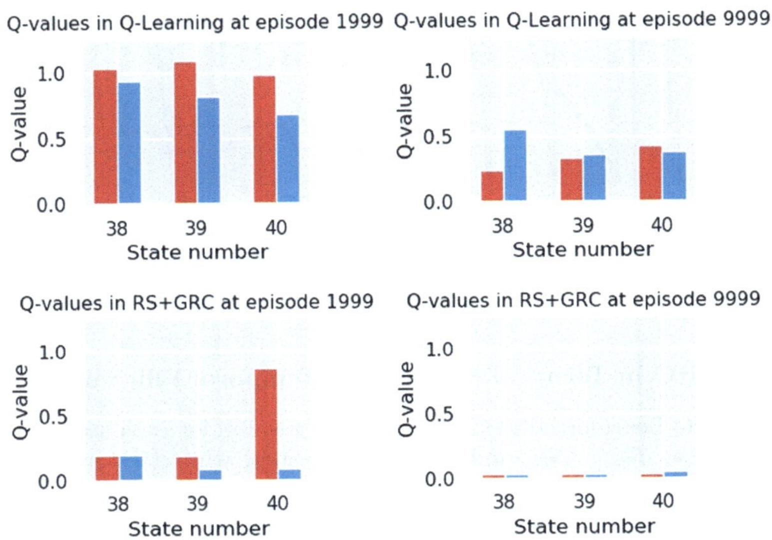

Fig. 9. Comparison of Q-values in LoopUnsteadySwitchWorld

RS, on the other hand, responds to non-stationary environments utilizing the satisficing policy that according to the difference, especially its sign, between the aspiration and currently obtained profit. If it is negative, RS starts exploration. The larger the difference is, the further it explores. The decision-making process is very different from changes in behavioral choice depending merely on the magnitude relationship of Q-values. Specifically, RS will change how it distributes the τ value to each action, according to how large the difference is [8]. Since RS makes an action selection that satisfies the target τ value, the search will start before the magnitude relationship of the Q-value changes. That is why RS can handle non-stationary environments, without much more computation and memory. In addition, since the Q-value update itself is the same for both Q-Learning and RS+GRC, it can be seen that RS+GRC becomes more responsive to environmental transitions as the number of environmental mutations increases.

6 Conclusion

In this study, we showed that emulation implemented in RS can learn faster than Q-Learning, and has more flexible search capability than imitation implemented as MaxEntIRL. One of the future tasks is to test our emulation algorithm compared with imitation in continuous environments and to clarify the functional roles of imitation and emulation learning in a broader perspective such as general machine intelligence.

Acknowledgments. This work was supported by JSPS KAKENHI Grant Number 17H04696.

References

1. Whiten, A., McGuigan, N., Marshall-Pescini, S., Hopper, L.M.: Emulation, imitation, over-imitation and the scope of culture for child and chimpanzee. Philos. Trans. R. Soc. B: Biol. Sci. **364**(1528), 2417–2428 (2009). https://doi.org/10.1098/rstb.2009.0069
2. Takahashi, T., Kohno, Y., Uragami, D.: Cognitive satisficing: bounded rationality in reinforcement learning. Trans. Jpn. Soc. Artif. Intell. **31**(6), AI30-M_1–11 (2016). (in Japanese)
3. Tamatsukuri, A., Takahashi, T.: Guaranteed satisficing and finite regret: analysis of a cognitive satisficing value function. BioSystems **180**, 46–53 (2019)
4. Andrew Maas, J., Bagnell, A., Dey, A.K., Ziebart, B.D.: Maximum entropy inverse reinforcement learning. In: AAAI 2008 (2008)
5. Levy, K.Y., Shimkin, N.: Unified inter and intra options learning using policy gradient methods. In: EWRL, pp. 153–164 (2011)
6. Simon, H.A.: Rational choice and the structure of the environment. Psychol. Rev. **63**(2), 129–138 (1956)
7. Ushida, U., Kono, Y., Takahashi, T.: Satisficing reinforcement learning for survival. In: Proceedings of JSAI 2017, 4C2-2in2 (2017). (in Japanese)
8. Kono, Y., Takahashi, T.: Autonomous optimal exploration through satisficing. In: Proceedings of JSAI 2018, 1Z3-04 (2018). (in Japanese)

Education and Culture

Computerized Adaptive Testing Method Using Integer Programming to Minimize Item Exposure

Yoshimitsu Miyazawa[1](✉) and Maomi Ueno[2]

[1] The National Center for University Entrance Examinations, Tokyo, Japan
miyazawa@rd.dnc.ac.jp

[2] The University of Electro-Communications, Tokyo, Japan

Abstract. This is an extension from a selected paper from JSAI2019. Computerized adaptive testing (CAT) estimates an examinee's ability sequentially and selects test items that have the highest accuracy for estimating the ability. However, conventional CAT selects the same items for examinees who have equivalent ability. As described herein, we propose CAT that minimizes item exposure and which adaptively selects different items for examinees of equal ability, while retaining accuracy. This paper presents the method's effectiveness using simulation data and actual test data.

Keywords: Computerized adaptive testing · e-Testing · Uniform test form assembly · Item response theory

1 Introduction

Computerized adaptive testing (CAT) estimates an examinee's ability after every answer and selects an item with the highest accuracy of estimating the ability [14].

This item selection presents an important benefit: the number of items to be selected and the testing time can be decreased for an examinee without reducing the accuracy. Furthermore, the abilities of all examinees are measured using the same degree of accuracy. Nevertheless, conventional CAT has the following two shortcomings.

1. Conventional CAT is inapplicable for conditions in which the same examinee takes a test multiple times because the same group of items tends to be selected when the same examinee takes a test multiple times.
2. Conventional CAT provides the same items to examinees of equivalent ability. Therefore, not all items in an item pool can be used effectively. Excessive item exposure engenders disclosure of the item contents to examinees, and might therefore degrade reliability [17].

Y. Ohsawa et al. (Eds.): JSAI 2019, AISC 1128, pp. 105–113, 2020.
https://doi.org/10.1007/978-3-030-39878-1_10

Earlier studies [5] demonstrated that frequently exposed items deteriorate rapidly. To resolve this difficulty, Kingsbury and Zara [5] proposed partitioning of an item pool into several groups of items and then selected the optimal item that maximizes Fisher information from the group minimizing item exposure [5]. Furthermore, van der Linden and et al. [3,15,16] proposed a shadow-test approach that maximizes Fisher information under several constraints (e.g., test area and item exposure frequency) using integer programming [3,15,16]. However, deviation of accuracy in the selected items produces great differences among examinees in terms of the test length and accuracy.

To resolve that shortcoming, we proposed a framework that can control the balance between item exposure and measurement accuracy [12]. More specifically, we use a uniform test assembly technique [4] to divide an item pool into several equivalent groups of items and thereby adjust the degree of item exposure. However, they encountered the difficulty that increases calculation cost of assembling equivalent groups.

We propose a CAT to resolve this difficulty. This study proposes CAT item selection using integer programming. It can select a different item while maintaining equal accuracy, even for an examinee with the same ability. The approach of the proposed method is described as presented below. (1) A group that satisfies test constraints as Fisher information is assembled using integer programming for minimizing item exposure. (2) An item that has the highest information is selected from the group. The proposed method uses integer programming for minimizing item exposure and satisfying upper and lower bounds of information. Accordingly, a different item set can be selected, even for an examinee with equal ability. Moreover, the improved diversity of item selection is expected to encourage thorough use of items in an item pool and to mitigate deviation in exposure. This paper demonstrates the effectiveness of the proposed method through simulation and actual data.

2 Item Response Theory

In CAT, an examinee's ability parameter is estimated based on Item Response Theory (IRT) [6] to select the optimum item with the highest information. In the two-parameter logistic model (2PLM), the most popular IRT model, the probability of a correct answer to item i by examinee j with ability $\theta \in (-\infty, \infty)$ is assumed as

$$p(u_i = 1|\theta) = \frac{1}{1 + exp[-1.7a_i(\theta - b_i)]}. \tag{1}$$

Therein, u_i is 1 when an examinee answers item i correctly, and 0 otherwise. Furthermore, $a_i \in [0, \infty)$ and $b_i \in (\infty, \infty)$ respectively denote the discrimination parameter of item i and the difficulty parameter of item i.

The asymptotic variance of estimated ability based on the item response theory is known to approach the inverse of Fisher information [6]. Accordingly, item response theory usually employs Fisher information as an index representing the accuracy. In 2PLM, the Fisher information is defined when item i provides an examinee's ability θ using the following equations.

$$I_i(\theta) = \frac{[p'(u_i = 1|\theta)]^2}{p(u_i = 1|\theta)[1 - p(u_i = 1|\theta)]} \tag{2}$$

where

$$p'(u_i = 1|\theta) = \frac{\partial}{\partial \theta} p(u_i = 1|\theta).$$

Results imply that the examinee's ability can be discriminated using an item with high Fisher information $I_i(\theta)$. Accordingly, that ability estimation can be expected to be implemented by selecting items with the highest amount of Fisher information given an examinee's ability estimate $\hat{\theta}$. The test information function $\mathrm{I}_T(\theta)$ of a test form T is defined as $\mathrm{I}_T(\theta) = \sum_{i \in T} \mathrm{I}_i(\theta)$. The asymptotic error of ability estimate $\hat{\theta}$: $\mathrm{SE}(\hat{\theta})$ can be obtained as the inverse of square root of the test information function at a given ability estimate $\hat{\theta}$ as $\mathrm{SE}_T(\theta) = \frac{1}{\sqrt{\mathrm{I}_T(\theta)}}$.

3 Computerized Adaptive Testing

In conventional CAT, adaptive items are selected from an item pool using the following procedures.

1. An examinee's ability is initialized to $\hat{\theta} = 0$.
2. An item maximizing Fisher information for given ability is selected from the item pool. It is presented to the examinee.
3. The examinee's ability estimate is updated from the correct/wrong response data to the item.
4. Procedures 2 and 3 are repeated until the update difference of the examinee's ability estimate decreases to a constant value ϵ or less.

Consequently, CAT can reduce the number of items examined so that it does not reduce the test accuracy in comparison to that of the same fixed test.

3.1 Constrained CAT with Item Exposure Control

In CAT, it is highly likely that the same set of items will be presented to examinees exhibiting similar abilities. Therefore, conventional CAT cannot be used practically in situations where the same examinee can take a test multiple times. Furthermore, because the ability variable follows the standard normal distribution, items with higher information around $\theta = 0$ tend to be exposed frequently. Therefore, bias of item exposure frequency occurs in an item pool. An earlier report [5] described that frequently exposed items tend to deteriorate rapidly. To resolve this difficulty, Kingsbury and Zara [5] proposed the partitioning of an item pool into several groups of items and then selected the optimal item maximizing Fisher information from the group minimizing item exposure [3,5,15]. The item-pool partitioning procedure is the following.

1. An item pool is partitioned into several groups of items.
2. The estimated ability of an examinee is initialized to $\hat{\theta} = 0$.
3. The group minimizing the number of exposure items is selected from an item pool.
4. The item maximizing Fischer information is selected from the group and is presented to the examinee.
5. After each response, the examinee's ability estimate is updated.
6. Procedures 2, 3, and 4 are repeated until the update difference of the estimated ability decreases to ϵ or less. ϵ is set to 0.01, which is used conventionally for actual computerized adaptive testing.

The number of groups was ascertained by comparing the respective performances of several numbers of groups. Actually, van der Linden and et al. [3,15,16] proposed a shadow-test approach that maximizes Fisher information under several constraints (e.g., test area and item exposure frequency) using integer programming [3,15,16]. The procedure used for constrained computerized adaptive testing is the following.

1. The estimated ability of an examinee is initialized to $\hat{\theta} = 0$.
2. The item set (shadow test with I items) maximizing Fischer information is then assembled using the integer programming shown below.

$$\text{maximize} \sum_{i=1}^{I} I_i(\theta) x_i \tag{3}$$

$$\text{subject to}$$

$$\sum_{i=1}^{I} x_i = n; \text{(test length)},$$

$$x_i = 1 \text{ if item } i \text{ is included in the shadow test, } x_i = 0 \text{ otherwise.}$$

$$\text{If the exposure count of item } i \text{ is greater than } R\text{, then } x_i = 0,$$

$$\text{where } R \text{ is the upper bound of the exposure count by the user.}$$

3. The item maximizing Fischer information is selected from the shadow test and is presented to an examinee.
4. After each response, the examinee's ability estimate is updated.
5. Procedures 2, 3, and 4 are repeated until the update difference of the estimated ability decreases to $\epsilon = 0.01$ or less.

Earlier methods mitigated the bias of item exposure frequency in an item pool. However, they led to the important difficulty of increased bias of measurement accuracies (errors) for examinees. Furthermore, this difficulty necessarily engenders a bias of examinees' required test lengths in CAT. In fact, a tradeoff exists between minimizing item exposure and maximizing the measurement accuracy (test information). Nevertheless, earlier methods do not adjust the tradeoff.

4 Proposed Method

The concept of the proposed method is to assemble a group with test constraints as Fisher information, but with different items using integer programming, to select an item with higher information, and to minimize item exposure from the group. Details of the proposed method are described below.

1. The estimated ability of an examinee is initialized.
2. The group maximizing Fischer information is assembled using the integer programming presented below.

$$\text{Minimize } y = \sum_{i=1}^{I} e_i x_i \tag{4}$$

subject to

$$\sum_{i=1}^{I} I(\theta_k) x_i \geq r_k, k = 1, 2, \cdots, K \tag{5}$$

$$\sum_{i=1}^{I} I(\theta_k) x_i \leq s_k, k = 1, 2, \cdots, K \tag{6}$$

$$\sum_{i=1}^{I} x_i = n(\text{Test length}) \tag{7}$$

To expand the method proposed by Adema (1989) [1] in which test forms are assembled using integer programming, we propose an optimization problem in which we embed an objective function minimizing item exposure. The lower boundaries and the upper boundaries of test information function at a set of the examinee's ability, $\Theta = \{\theta_1, \ldots, \theta_K\}$, is r_k and s_k. Also, $I(\theta_k)$ denotes the test information function at the examinee's ability θ_k. The exposure count of item i is e_i. If item i is selected into the group, then $x_i = 1$; otherwise $x_i = 0$.

3. An item is selected from a group. Then response data are generated with the given true ability and item parameters.
4. Ability $\hat{\theta}$ is estimated by expected a posteriori (EAP) [2].
5. Procedures 2–4 are repeated until the update difference of the estimated ability decreases to ϵ or less.

The proposed method selects an item from a different group for each examinee. Accordingly, a different item is expected to be selected, even to an examinee of equivalent ability. Furthermore, improved diversity of item selection is anticipated to encourage thorough use of items in an item pool and to mitigate deviation in exposure.

5 Numerical Evaluation

This section presents a comparison of the performance of the proposed method (designated as Proposal) to those of other computerized adaptive testing methods (conventional adaptive testing in 2.3 (designated as CAT), Kingsbury and

Zara [5] CAT in 2.4 (designated as KZ), and van der Linden's CAT in 2.4 (designated as LAT). We conducted the following two experiments using simulation data and actual data.

5.1 Simulation Experiment

We conducted a simulation experiment as described hereinafter.

1. Item pools with 500, 1000, 2000 and 3000 items are generated. The true parameters of each item are generated from $a_i \sim U(0,1)$ and $b_i \sim N(0,1)$.
2. The true abilities of examinees are sampled from $\theta \sim N(0,1)$.
3. The estimated ability of an examinee is initialized to $\hat{\theta} = 0$.
4. Each adaptive testing method is conducted using each item pool. The examinees' response data are generated from $p(u_i \mid \widehat{\theta})$. Correct response data are generated if $p(u_i \mid \widehat{\theta}) > 0.5$. Incorrect response data are generated otherwise. The convergence (Stopping) criterion is $\epsilon = 0.01$, which is used conventionally for actual computerized adaptive testing [14].
5. Procedures 2–4 are repeated 1000 times.

Table 1 presents the results. "Avg. test length" represents the average test length which reflects the measurement accuracy of the test (the standard error of test lengths in parenthesis), and "S.D. estimates error" expresses the standard deviation of asymptotic errors of estimates $\widehat{\theta}$. When the value of "S.D. estimates error" approaches to zero, the tests presented to the different examinees by the CAT have the same estimation accuracy. Also, "Max. No. exposure item" shows the maximum number of exposure items. "Avg. exposure item" expresses the average exposure count of an item (the standard error of numbers of exposure items in parenthesis).

Table 1 presents the results. Results confirmed that the proposed method yielded the shortest test length under all conditions. The proposed method constrains the number of items with item exposure control. Therefore, it has a property of not selecting items with an extremely large amount of information only to a certain estimated value. This property shortens the test length, thereby reducing the exposure of items. The proposed method also indicates the smallest standard deviation of test length under all conditions. The proposed method selects items from an item set of uniform information. Therefore, it can render the number of items to take for convergence of the estimated ability uniform. "S. D. estimates error" using conventional CAT was lowest. It repeatedly selects some item sets with much information. However, the proposed method controls exposure and selects different items for each examinee. "S. D. estimates error" was low because of the deviation of measurement accuracy in the selected items. The maximum exposure was the least when the proposed method was used. Results demonstrate that the proposed method can control maximum exposure directly in the same manner as the fraction of different items. This result is interpreted as attributable to the exposure setting to use as many items in an item pool as possible in this experiment. The averages of exposure obtained using the proposed method were the lowest.

Table 1. Results obtained using simulation data

Item pool size	Methods	Avg. test length	S. D. estimates error	Max. No. exposure item	Avg. exposure item
500	CAT	65.41 (8.00)	0.02	1000	130.81 (228.74)
	LAT	73.98 (31.21)	37854.16	150	147.96 (16.63)
	KZ	40.47 (4.57)	0.04	1000	80.94 (252.83)
	Proposal	21.16 (2.56)	0.07	43	42.15 (1.33)
1000	CAT	73.19 (7.24)	0.02	1000	73.19 (156.89)
	LAT	79.18 (45.64)	7390254.11	80	79.18 (7.98)
	KZ	43.70 (5.22)	0.04	1000	43.70 (192.97)
	Proposal	17.86 (2.39)	0.07	18	17.08 (0.95)
2000	CAT	80.60 (7.41)	0.02	1000	40.30 (112.06)
	LAT	98.88 (61.30)	5628898.42	50	49.44 (5.21)
	KZ	42.55 (11.58)	0.04	1000	21.28 (130.02)
	Proposal	19.42 (2.39)	0.07	10	9.57 (0.51)
3000	CAT	86.54 (7.62)	0.01	1000	28.85 (88.94)
	LAT	118.50 (65.53)	1239701.12	40	39.50 (4.42)
	KZ	38.17 (6.46)	0.04	1000	12.72 (101.38)
	Proposal	18.18 (2.55)	0.07	7	6.03 (0.22)

Table 2. Results obtained using actual data

Item pool size	Methods	Avg. test length	S. D. estimates error	Max. No. exposure item	Avg. exposure item
978	CAT	65.22 (10.54)	0.02	1000	66.69 (165.95)
	LAT	98.64 (23.42)	0.04	90	88.05 (10.08)
	KZ	54.67 (6.06)	0.04	1000	55.90 (216.23)
	Proposal	21.48 (3.03)	0.06	1	0.55 (0.5)

5.2 Experiment Conducted Using Actual Data

This section presents evaluation of the effectiveness of the proposed method using actual data. An experiment was conducted using the item pool of real data, with 978 items, and a test constraint used in the synthetic personality inventory (SPI) examination (actual CAT style), which is a popular aptitude test in Japan [7]. Table 2 suggests the following characteristics: (1) The test length produced using the proposed method is the shortest. The standard deviation of test length produced using the proposed method is the smallest, which suggests that the test length selected to examinees has little dispersion. (2) "S. D. estimates error" by the proposed method is the third lowest to CAT, so the same accuracy was maintained using the proposed method. (3) The maximum exposure was smallest by the proposed method. The average tended to be small when obtained using the proposed methods.

6 Conclusions

This study has examined a proposed CAT implementation in which different items can be selected, even by an examinee with equivalent ability, while maintaining equal accuracy. Specifically, we proposed a method by which a group is assembled using integer programming: an item is selected from the group. Using simulation and real data, the effectiveness of the proposed method was demonstrated. For this study, we used Fischer information measure as an item selection criterion. Although the Fischer information measure becomes accurate for late stage of CAT because it is an asymptotic approximation, recent studies have proposed more accurate information measures and the item selection algorithms [8,9,13]. We expect to apply the proposed uniform partition of item pool technique to the information measure and its applications [10,11] in future studies.

References

1. Adema, J.J.: Implementations of the branch-and-bound method for test construction problems. Methodika **6**(2), 99–117 (1992)
2. Baker, F.B., Kim, S.H. (eds.): Item Response Theory: Parameter Estimation Techniques. CRC Press, Boca Raton (2004)
3. Choi, S.W., Moellering, K.T., Li, J., van der Linden, W.J.: Optimal reassembly of shadow tests in CAT. Appl. Psychol. Measur. **40**(7), 469–485 (2016)
4. Ishii, T., Songmuang, P., Ueno, M.: Maximum clique algorithm and its approximation for uniform test form assembly. IEEE Trans. Learn. Technol. **7**(1), 83–95 (2014)
5. Kingsbury, G.G., Zara, A.R.: Procedures for selecting items for computerized adaptive tests. Appl. Measur. Educ. **2**(4), 359–375 (1989)
6. Lord, F., Novick, M.R.: Statistical Theories of Mental Test Scores. Addison-Wesley, Boston (1968)
7. Recruit: Synthetic personality inventory (SPI). http://www.spi.recruit.co.jp/
8. Ueno, M.: An extension of the IRT to a network model. Behaviormetrika **29**(1), 59–79 (2002)
9. Ueno, M.: Adaptive testing based on Bayesian decision theory. In: The Proceedings of the 16th International Conference on Artificial Intelligence in Education, pp. 712–716 (2013)
10. Ueno, M., Miyazawa, Y.: Pobability based scaffolding system with fading. In: The Proceedings of the 16th International Conference on Artificial Intelligence in Education, pp. 492–503 (2015)
11. Ueno, M., Miyazawa, Y.: IRT-based adaptive hints to scaffold learning in programming. IEEE Trans. Learn. Technol. **11**(4), 415–428 (2018)
12. Ueno, M., Miyazawa, Y.: Uniform adaptive testing using maximum clique algorithm. In: The Proceedings of the 20th International Conference on Artificial Intelligence in Education, pp. 482–493 (2019)
13. Ueno, M., Songmuang, P.: Computerized adaptive testing based on decision tree. In: 2010 10th IEEE International Conference on Advanced Learning Technologies, pp. 191–193. IEEE (2010)

14. van der Linden, W.J., Glas, C.A.W. (eds.): Elements of Adaptive Testing. Springer, Heidelberg (2010)
15. van der Linden, W.J., Reese, L.M.: A model for optimal constrained adaptive testing. Appl. Psychol. Measur. **22**(3), 259–270 (1998)
16. van der Linden, W.J., Veldkamp, B.P.: Constraining item exposure in computerized adaptive testing with shadow tests. J. Educ. Behav. Stat. **29**(3), 273–291 (2004)
17. Way, W.D.: Protecting the integrity of computerized testing item pools. Educ. Measur. Issues Pract. **17**, 17–27 (1998)

Proposal of Context-Aware Music Recommender System Using Negative Sampling

Jin-cheng Zhang(✉) and Yasufumi Takama

Tokyo Metropolitan University, 6-6 Asahigaoka, Hino, Tokyo 191-0065, Japan
aiba_hz@hotmail.com, ytakama@tmu.ac.jp

Abstract. This is an extension from a selected paper from JSAI2019. This paper proposes a method for recommending music items considering listeners' context information. Recently, users can enjoy music easily regardless of time and a place due to evolution of online music services such as Spotify. However, it is difficult for us to find appropriate music items from enormous resources. On the other hand, because of listening style and characteristic of music items, music items do not usually have explicit rating. Therefore, implicit feedback such as playing count has been widely used to construct recommender systems. As additional information, this paper considers listeners' context. The proposed method employs FMs (Factorization Machines), in which the context information is treated as factors. Negative sampling is applied to reduce the number of negative samples (music items a user has yet to be listened). The effectiveness of the proposed method and the effect of negative sampling are evaluated with an offline experiment. The experimental result on nowplaying-rs dataset shows that the proposed method outperforms wALS (weighted Alternating Least Squares) method. Furthermore, different negative sampling methods such as popularity-based one and sampling with different time window size are also investigated.

Keywords: Music recommendation · Context-aware recommendation · Factorization Machines · Negative sampling

1 Introduction

Recently, with the development of communication technology and portable electronic devices, the way people consume music has changed from hard devices to online music streaming services. Specifically, users need not buy albums from traditional brick-and-mortar shops, and play it by CD players or extract song from CD to digital devices. Instead, they can enjoy music easily regardless of time and place by accessing online music streaming services such as Spotify, Apple Music and Amazon Music. These up-to-date music streaming services offer thousands of tracks in wide variety of music genres via Internet-accessible

Y. Ohsawa et al. (Eds.): JSAI 2019, AISC 1128, pp. 114–125, 2020.
https://doi.org/10.1007/978-3-030-39878-1_11

smartphones or tablets. The library size of Apple Music, Spotify, and Amazon Music have exceeded 45, 35, 16 millions tracks on 2018, respectively.[1]

As a result, a gap of users' ability to access music items and the volume of accessible items was large. To help user discover their favorite music items, music recommendation systems have been studied [9,11,19,20]. As in the case with other domains such as movies, e-commerce and documents, Collaborative Filtering that finds people having similar interests with a target user based on the interaction of users and items to predict ratings [1] and Content-based Filtering that is based on a description of items and a profile of the user's preferences [2] are the two most common approaches. Moreover, previous research [10] has shown that context information such as location, time, weather, demographic information and emotional state of the user plays an important role in the music recommendation domain. On the other hand, music domain has different challenges from movie and e-commerce domain: explicit feedback such as rating is not usually available due to the way of listening to music and characteristic of music items. Instead, the interaction between users and items is recorded as a listening event (LE), from which user feedback is usually described implicitly such as total playing count. Therefore, in addition to cold-start and data sparsity problems which are common regardless of domains, the above-mentioned two challenges should also be tackled in music domain.

Regarding how to use limited implicit feedback for recommending music items more effectively, playing count has been used by existing music recommender systems [4]. Regarding how to utilize the context for music recommendation, as mentioned above the influence of users' context while listening to music on their preference is supposed to be higher than other domains. By extending the conventional matrix factorization approach, a tensor factorization method has been studied: it can consider context axis in addition to user and item axes [9]. However, it is difficult to optimize the use of contexts due to high time complexity. Therefore, this paper employs Factorization Machines (FMs) [8], in which listeners' contexts are treated as factors. It is shown that FMs can flexibly consider the interaction of users, items, and context with relatively low time complexity. In other words, this advantage of FMs can also help to discover useful context effectively.

As FMs needs both positive and negative samples, a straightforward approach is to use LEs as positive samples, and music items a user has yet to be listened are used as negative samples. The problem is the number of items regarded as negative samples are much larger than that of positive ones. To reduce the number of negative samples, this paper employs negative sampling. The proposed method randomly selects music items which are not played by a user under the same context as actual LEs as negative samples. Furthermore, popularity of music items is also considered when determining negative samples. This paper investigates two approaches: reducing the probability of being selected as negative samples linearly related to music item's popularity, and reducing the probability of selecting a part of popular items as negative samples.

[1] https://www.macobserver.com/news/songs-in-streaming-music-service-libraries/.

Effectiveness of the proposed method is evaluated by comparing it with weighted Alternating Least Squares (wALS) using n-fold real-life split strategy [23] cross-validation. The effect of negative sampling is also investigated. A dataset used in the experiment is nowplaying-rs that is generated from all tweets with hashtag #nowplaying posted in 2014. The experimental result shows that the proposed method outperforms wALS in term of prediction accuracy. It is also confirmed that prediction accuracy could be increased further by considering popularity of music items when selecting negative samples. The influence of the size of time windows when selecting negative samples is also investigated.

The rest of this paper is organized as follows. The next section introduces some related work about music recommendation before describing the detail of the proposed method in Sect. 3. Experimental results are shown in Sect. 4.

2 Related Work

This section introduces some related works of music recommendation, including implicit feedback recommendation, and recommendation methods leveraging additional information such as context, social information and audio signals of music.

2.1 Implicit Feedback

How to use implicit feedback effectively have been a crucial theme for recommendation regardless of domain [3]. Koren et al. proposed a collaborative filtering method called wALS for implicit feedback datasets [4]. As an implicit feedback such as play count is not as reliable as explicit rating, it is difficult to make recommendation by solving the same loss function as methods based on Matrix Factorization. They introduced a confidence computation to a loss function based on playing count and considered all missing LEs as negative samples. For reducing the cost of computation, Pan et al. [5] proposed sampling-based ALS for implicit feedback recommendation. Instead of using all missing LEs as negative samples, this method just considers a part of those based on negative example sampling. This sampling-based method approximates the exact solution with much lower computational costs for large scale sparse datasets than original wALS. Furthermore, a method that dynamically chooses negative samples from the ranked list of Top-N recommendation produced has been proposed [6]. As another approach, Steffen et al. [7] have introduced a generic optimization criterion Bayesian Personalized Ranking optimization (BPR-OPT), which is derived from maximum estimator for optimal personalized ranking.

2.2 Context Information

Kaminskas et al. [10] have classified context information on music domain into user-related ones (emotional state, demographic information and activity of user), environment-related ones (location, current time, weather and noise level) and multi-media (image of album, lyrics and review of a song).

Deng et al. [11] have explored user's emotion in microblogs service for music recommendation. In accordance with the hypothesis that users have higher similarity when they played similar music in the same emotional state, this method utilizes the text of microblogs service to extract users' emotion at different granularity levels and during different size of time window. Wang et al. [12] have enhanced emotion-aware approach by modeling the relations among user, music, and emotion as an Emotion Aware Graph. Kaminskas et al. [13] recommended music suited for places of interest based on the auto-tagging and the knowledge of the semantic relations among items.

Regarding the utilization of multi-media information, Pichl et al. [14] proposed a pre-filtering method, which divides users into clusters by the name of playlists given by users, and recommend music items for each user cluster with kNN-based CF. To improve the performance of this approach, they have used the cluster label as one of the features of FMs to overcome the drawbacks of dividing user groups in advance by pre-filtering. This method does not sample any negative LE: instead, they let a rating equal to 1 when a certain user listened to a certain track in a certain cluster, equal to -1 otherwise.

Schedl and Schnitzer [16] have introduced multiple types of context information into hybrid music recommendation. The process of similarity computation considers genres, styles, instruments, moods from "virtual artist document," rhythmic features of music item, and user context such as timestamp and location.

Regarding different information from context, [17] and [18] proposed a high-order SVD using social tagging. Schedl et al. [21] introduced the concept of user mainstreamness to judge whether users prefer music items which are currently popular or ignore such popular items. When using audio signals for recommendation, Convolutional Neural Network has been used to extract characteristics affecting user preference directly from audio signals [19]. This approach tries to reduce the semantic gap [20] between low-level descriptors and the concepts music listeners have.

3 Proposed Method

The details of the proposed context-aware music recommendation method are described in this section. First, the reason why this paper employs FMs as well as how to use it are described. Negative sampling methods are also introduced in this section. Figure 1 shows the flowchart of the proposed method.

3.1 Context-Aware Recommendation

This paper employs FMs [8] to recommend music items, in which users, music items and users' contexts are used as feature variables of FMs. FMs can not only learn the latent factors of users and items like Matrix Factorization methods,

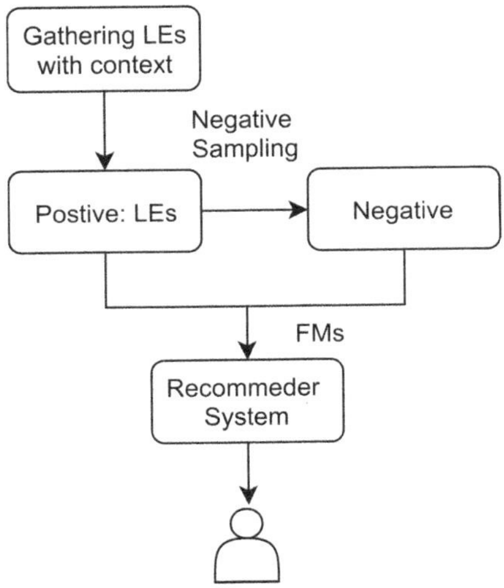

Fig. 1. Flowchart of proposed method

but also flexibly consider the interaction of users, items, and any other features with relatively low time complexity. All LEs are described by a design matrix $X \in \{0,1\}^{s \times n}$, where s, n are the number of LEs and feature variables of each LE, respectively. Given $\boldsymbol{x} \in \{0,1\}^n$ as a LE, the model equation for a 2-way FMs (degree $d = 2$) is defined as:

$$\hat{r}(\boldsymbol{x}) := \omega_0 + \sum_{i=1}^{n} \omega_i x_i + \sum_{i=1}^{n-1} \sum_{j=i+1}^{n} \hat{\omega}_{i,j} x_i x_j, \tag{1}$$

$$\hat{\omega}_{i,j} := \sum_{f=1}^{k} v_{i,f} v_{j,f} \tag{2}$$

where x_i is a feature variable of LE $\boldsymbol{x}$ (shown in Fig. 2), $\hat{r}(\boldsymbol{x})$ is the prediction value for $\boldsymbol{x}$. ω_0 is the global bias, ω_i is the weight of x_i. The $\hat{\omega}_{i,j}$ shows the interaction of x_i and x_j with k factors. It can capture all single and pairwise interactions between these variables.

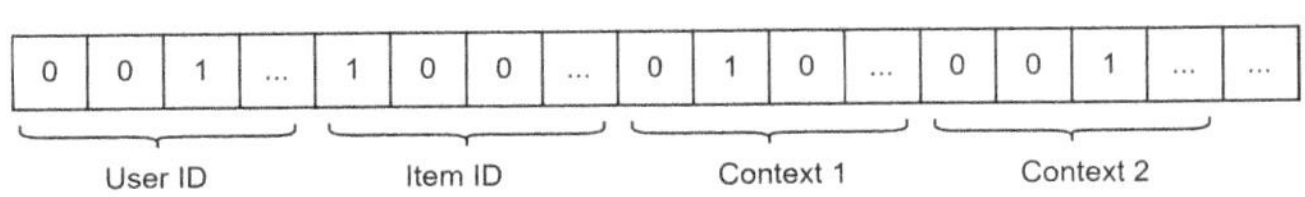

Fig. 2. a LE x

This paper considers season, time zone, time stamp, and mother language when a user listened music item as context information, which are determined on the basis of the result of preliminary experiment.

Obviously, FMs needs both positive and negative samples in order to obtain the latent factors of users, items and context. However, only positive sample (LE) could be obtained when users play music items. Therefore, this paper considers to utilize negative sampling to find negative samples for solving this problem.

3.2 User-Oriented Negative Sampling

Regarding all music items a user has yet to be listened as negative is a naive method. Although it could be effective in Matrix Factorization method such as wALS [4], the number of negative samples becomes huge, which causes a problem of time complexity especially when training a model by FMs. To solve this problem, this paper selects a part of music items that a user has yet to be listened as negative samples. This paper proposes two kinds of negative sampling methods, which are compared in the experiment.

Method 1. This method selects negative samples in according with the hypothesis that **if a music item is not played by a user in some context, s/he would not be interested in it in that context.** When training FMs model, for each LE, this method selects music item(s) on random from music items which are not played in this LE and uses it (those) as negative sample(s). The same values of context information as the LE is added to the context information of the negative sample(s). Because the more active users will obtain the more negative samples, Method 1 is also called user-oriented negative sampling.

Method 2. Although selecting negative samples on random is a comparatively fair method, it is supposed that the popularity of music item could influence the result of learning FMs. Therefore, in addition to the hypothesis of Method 1, this paper adopts another hypothesis that **if a music item is popular, it would be a positive sample for most users**. Therefore, this paper investigates two approaches: reducing the probability of being selected as negative samples linearly related to music item's popularity, and reducing the probability of selecting a part of popular items as negative samples. After selecting negative sample(s) for each LE in accordance with the popularity, the context information is given in the same manner as Method 1.

We set rating $r(x_i) = 1$ for positive samples. For negative samples, this paper investigates the influence of changing the value, which will be described in Sect. 4. The ratio of negative samples are determined by parameter $m = \frac{\#Negative\ samples}{\#Positive\ samples}$. In other words, the proposed method selects m negative sample(s) for each LE. On the other hand, it is supposed that the attention to a music item would decrease since its release date. Based on this idea, this paper changes the size of time windows to reduce the sample space of negative samples.

Furthermore, this paper considers both top-N recommendation and rating prediction task. In top-N recommendation task, a top-N recommendation list of music items is generated for each LE. The purpose of this task is to recommend appropriate music items to users. On the other hand, rating prediction task predicts whether or not a specific music item is played by a user in specific context.

4 Experiment

4.1 Dataset

This paper uses nowplaying-RS dataset[2] that is extracted from all tweets with hashtag #nowplaying posted in 2014. Poddar et al. [22] extracted the name of music items, artists, and the context of a user when playing music items from tweets. The number of user, music items, and LEs are 138,781, 346,273, and 11,639,541, respectively. For avoiding the influence of tweet bots and extremely not active users, we removed the users who listened less than 10 and larger than 5000 music items from the dataset. After pre-processing, the number of users, music items, and LEs become to 18,946, 22,023, and 1,835,993, respectively.

4.2 Experimental Setup

This paper compares the proposed method to baseline method wALS [4]. Equation (2) shows the loss function of wALS,

$$Ł(x_u, y_i) = \sum_{u,i} c_{u,i}(p_{u,i} - x_u^T y_i)^2 + \lambda(\sum_u ||x_u||^2 + \sum_i ||y_i||^2), \tag{3}$$

where x_u, y_i is the latent factors of users and items, $p_{u,i}$ is the element of user-item binary matrix representing whether or not a user u played a music item i. $c_{u,i}$ is the element of confidence matrix: $c_{u,i} = 1 + \alpha r_{u,i}$, where α is a hyper-parameter and $r_{u,i}$ is the play count of i by u.

This paper uses 5-fold real-life split strategy [23] based cross-validation. Experiments are repeated 5 times, and t-test is applied to confirm whether or not significant difference to wALS is observed. Used parameters are: $\alpha = 250$, $\#latent\,factor = 15$, $\lambda = 0.05$ for wALS by hyper-parameter tuning.

For the proposed method, ratio of negative samples is set to $m = 3$. Because negative samples are a kind of virtual LE, they are not used in evaluation but only used in the process of training models for ensuring the fairness of evaluation.

4.3 Evaluation Metric

This paper employs MPR (Mean Percentage Ranking) (Eq. (4)) to evaluate the performance of each method for top-N recommendation task. $rank_{u,i}$ represents the percentile-ranking of a music item i in the recommend list for a user u, and $r_{u,i}$ corresponds to $r(x_i)$ in Eq. (1): $r_{u,1} = 1$ if a user u actually listened a music item i, otherwise 0. Smaller MPR indicates better performance of Top-N recommendation.

$$MPR = \frac{\sum_{u,i} r_{u,i} \times rank_{u,i}}{\sum_{u,i} r_{u,i}} \tag{4}$$

[2] http://dbis-nowplaying.uibk.ac.at/#nowplayingrs.

For rating prediction task, this paper computes $RMSE$ (Root Mean Squared Error) (Eq. (5)) between prediction value $\hat{r_i}$ and actual value r_i. As described in Sect. 3, we set rating $r_i = 1$ for positive samples, and test set includes positive samples only. Therefore, all rating of actual values are equal to 1. Smaller $RMSE$ indicates more accurate rating prediction.

$$RMSE = \sqrt{\frac{\sum_{i \in TestSet} (r_i - \hat{r_i})^2}{|TestSet|}} \tag{5}$$

4.4 Result and Discussion

Comparing with wALS. Table 1 shows MPR with the p-value among wALS, UO (user-oriented negative sample)+FMs and UO+FMs+Contexts. The result shows that UO+FMs is more effective than wALS, and its performance is improved by introducing context information.

Table 1. MPR and p-values for comparison of proposed and baseline methods

	wALS	UO+FMs	UO+FMs+Contexts
wALS	-	2.363e−05	5.368e−07
UO+FMs	2.363e−05	-	0.0003
UO+FMs+Contexts	5.368e−07	0.0003	-
MPR	0.1092	**0.1041**	**0.1017**

Influence of m. Figure 3 shows the MPR against the ratio m of negative samples to positive samples. It is observed that setting m larger than 1 can obtain more effective result than wALS. Although MPR decreases as m increases, the effect is gradually reduced. The computation time when $m = 2$ and 6 are 688(s) and 2142(s) per fold in cross-validation, respectively. This means that m should be set by considering the balance between time complexity and accuracy of recommendation.

Rating of Negative Samples. We changed the rating of all negative samples equally from 0 to 1 inspired by [5]. The result is shown in Fig. 4, in which the shadow indicates standard deviation. The best performance on MPR (p-value = 0.000862) could be obtained when rating of negative samples was set to 0.7 and considered context information. Although we should verify it with other datasets, this result indicates that rating of negative samples should be set to appropriate value when sampling.

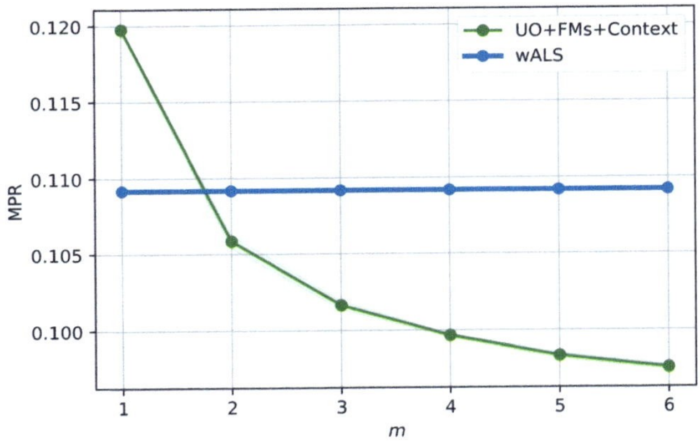

Fig. 3. Effect m on MPR

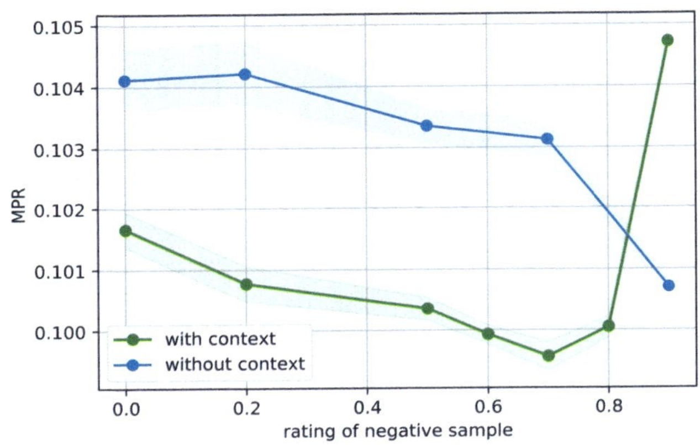

Fig. 4. Effect rating of negative on MPR

Sampling Method. Table 2 shows the comparison of Method 1 (user-oriented negative sample) and Method 2 (popularity-based sampling) when utilizing same context information. Pop (linear) is the approach that reduces the probability of being selected as negative samples linearly related to music item's popularity. Pop (top 25%) reduces the probability of selecting top 25% popular items as negative samples. It shows that all of popularity-based approaches could not improve the performance on MPR significantly. However, RMSE could be reduced while retaining MPR in the case of using Pop (linear) and Pop (top 25%). This means that considering the popularity of music items when selecting negative samples could improve the performance of rating prediction.

Table 2. Comparison among sampling methods

Sampling approach	MPR	p-value to MPR	RMSE	p-value to RMSE
UO	0.102339	-	0.128363	-
Pop (linear)	0.107094	0.000251	**0.121178**	**0.000164**
Pop (top 25%)	0.102536	0.746596	**0.125259**	**0.003541**
Pop (top 12.5%)	0.101941	0.528584	0.126204	0.040512

Time Window Size. Table 3 shows the influence of time window size. The values in the parentheses show the size of time window. It is shown that both MPR and RMSE could not be reduced.

Table 3. Influence of time window size

	MPR	p-value to MPR	RMSE	p-value to RMSE
UO	0.102339	-	0.128363	-
UO (tw = 15 days)	0.104043	**0.026417**	0.132658	**0.012262**
UO (tw = 30 days)	0.102462	0.834677	0.131792	**0.016182**
UO (tw = 60 days)	0.102658	0.405607	0.129667	0.453952
UO (tw = 90 days)	0.102746	0.613417	0.130347	0.196908

5 Conclusion and Future Work

This paper proposes a music recommendation method using FMs considering listeners' context information. To generate negative samples, two types of sampling methods are also proposed. Experimental results show the proposed method is more effective than wALS. It is also shown that its performance can be improved by popularity-based sampling method on rating prediction task. On the other hand, the rating values of negative samples should be set to appropriate value when sampling. Furthermore, it was found that considering the size of time window when selecting negative samples did not improve the result. In future work, we will investigate other negative sampling methods to improve performance, and leverage the other additional features such as user genre profiling and acoustic features of music.

References

1. Resnick, P., Iacovou, N., Suchak, M., Bergstrom, P., Riedl, J.: An open architecture for collaborative filtering of netnews. In: ACM Conference on Computer Supported Cooperative Work, pp. 175-186 (1994)

2. Aggarwal, C.C.: Recommender Systems, pp. 1–28. Springer, Heidelberg (2016)
3. Oard, D.W., Kim, J.: Implicit feedback for recommender systems. In: AAAI Workshop on Recommender Systems, vol. 83, pp. 81–83 (1998)
4. Hu, Y., Koren, Y., Volinsky, C.: Collaborative filtering for implicit feedback datasets. In: 8th IEEE International Conference on Data Mining, pp. 263–272 (2008)
5. Pan, R., Zhou, Y., Cao, B., Liu, N.N., Lukose, R., Scholz, M., Yang, Q.: One-class collaborative filtering. In: 8th IEEE International Conference on Data Mining, pp. 502–511 (2008)
6. Zhang, W., Chen, T., Wang, J., Yu, Y.: Optimizing top-N collaborative filtering via dynamic negative item sampling. In: 36th International ACM SIGIR Conference on Research and Development in Information Retrieval, pp. 785–788 (2013)
7. Rendle, S., Freudenthaler, C., Gantner, Z., Schmidt-Thieme, L.: BPR: Bayesian personalized ranking from implicit feedback. In: 25th Conference on Uncertainty in Artificial Intelligence, pp. 452–461 (2009)
8. Rendle, S.: Factorization machines with libFM. ACM Trans. Intell. Syst. Technol. **3**(3), 1–22 (2012). Article no. 57
9. Adomavicius, G., Tuzhilin, A.: Context-aware recommender systems. In: Recommender Systems Handbook, pp. 217–253. Springer, Heidelberg (2011)
10. Kaminskas, M., Ricci, F.: Contextual music information retrieval and recommendation: state of the art and challenges. Comput. Sci. Rev. **6**, 89–119 (2012)
11. Deng, S., Wang, D., Li, X., Xu, G.: Exploring user emotion in microblogs for music recommendation. Expert Syst. Appl. **42**(23), 9284–9293 (2015)
12. Wang, D., Deng, S., Xu, G.: GEMRec: a graph-based emotion-aware music recommendation approach. In: International Conference on Web Information Systems Engineering, pp. 92–106 (2016)
13. Kaminskas, M., Ricci, F., Schedl, M.: Location-aware music recommendation using auto-tagging and hybrid matching. In: 7th ACM Conference on Recommender Systems, pp. 17–24 (2013)
14. Pichl, M., Zangerle, E., Specht, G.: Towards a context-aware music recommendation approach: what is hidden in the playlist name. In: IEEE International Conference on Data Mining Workshop, pp. 1360–1365 (2015)
15. Pichl, M., Zangerle, E., Specht, G.: Improving context-aware music recommender systems: beyond the pre-filtering approach. In: 2017 ACM on International Conference on Multimedia Retrieval, pp. 201–208 (2017)
16. Schedl, M., Schnitzer, D.: Hybrid retrieval approaches to geospatial music recommendation. In: 36th International ACM SIGIR Conference on Research and Development in Information Retrieval, pp. 793–796 (2013)
17. Symeonidis, P., Ruxanda, M., Nanopoulos, A., Manolopoulos, Y.: Ternary semantic analysis of social tags for personalized music recommendation. Int. Soc. Music Inf. Retr. **8**, 219–224 (2008)
18. Nanopoulos, A., Rafailidis, D., Symeonidis, P., Manolopoulos, Y.: MusicBox: personalized music recommendation based on cubic analysis of social tags. IEEE Trans. Audio Speech Lang. Process. **18**(2), 407–412 (2009)
19. Van den Oord, A., Dieleman, S., Schrauwen, B.: Deep content-based music recommendation. In: Advances in Neural Information Processing Systems, pp. 2643–2651 (2013)
20. Celma, O.: Music recommendation. In: Music Recommendation and Discovery, pp. 43–85. Springer (2010)

21. Schedl, M., Bauer, C.: Distance-and rank-based music mainstreaminess measurement. In: 25th Conference on User Modeling, Adaptation and Personalization, pp. 364–367 (2017)
22. Poddar, A., Zangerle, E., Yang, Y.: #nowplaying-RS: a new benchmark dataset for building context-aware music recommender systems. In: 15th Sound and Music Computing Conference (2018)
23. Wang, C.-S., Lin, S.-L., Yang, H.-L.: Evaluating music recommendation in a real-world setting: on data splitting and evaluation metrics. In: IEEE International Conference on Multimedia and Expo, pp. 1–6 (2015)

The Influence of Story Creating Activities While Appreciating Abstract Artworks

Kotone Tadaki(✉) and Akinori Abe

Graduate School of Humanities and Public Affairs, Chiba University, Chiba, Japan
kotonetadaki@yahoo.co.jp, ave@ultimaVI.arc.net.my

Abstract. This is an extension from a selected paper from JSAI2019. In this paper, the authors will introduce our experiment to determine the influence of story creation on the abstract artwork appreciation. From previous researches, it was known that novice viewers tend to dislike abstract artworks more than representative artworks, because abstract artworks lack of the forms. The authors tried to relax this trend by making viewers create a story about the artwork while their art appreciation. To answer our research question which is "will individual viewers create similar stories for the same artworks?," the authors conducted the experiment. The participants answer questions in the worksheet while appreciating abstract artworks and representational artworks. As results, we found out that individual viewers create the similar story to the same artwork from the corresponding analysis and bold lines on the artworks were the important factor to imagine a story. Our result can suggest the possible advantages and disadvantages of using similar activities and how to create effective captions for the artworks.

Keywords: Art appreciation · Story creation · Reality constraint

1 Introduction

1.1 Abstract and Representational Paintings

The mechanism of art appreciation has been a major issue for years. Schmidt et al. focused on a novice who have not performed art appreciations (visual analysis) nor created works as a professional artist and monitored their verbal protocols during art appreciations [8]. By using these data, they tried to clarify the novices' strategies for understanding paintings to build a computer-based education program. Based on these researches, they showed that novices tend to use more semantic features or contents (e.g. theme, symbolism) more than formal elements (e.g. lines, shape, colour) when appreciating representational paintings. However, when they appreciate abstract paintings, the result was completely opposite, which means novices frequently use more formal elements than semantic features. Since instructors in schools and museums tend to use formal elements when they teach a visual analysis to novices, this result suggested that

Y. Ohsawa et al. (Eds.): JSAI 2019, AISC 1128, pp. 126–135, 2020.
https://doi.org/10.1007/978-3-030-39878-1_12

an art education might be unsuitable for a novice. In addition, O'hare pointed out that novices tend to dislike abstract artworks [6]. Okada and Inoue described that these differences in art appreciation occur because abstract paintings lack of the specific form [7]. They conducted an experiment to determine how novice would evaluate abstract and representational artworks. They concluded that the artistic evaluation of abstract artworks is highly affected by the softness and the preference whereas the artistic evaluation of representational artworks is highly affected by the interestingness and the originality of the artwork. Ishibashi and Okada call this phenomenon the reality constraint and pointed out that nowadays an art education (usually in an elementary and a secondary school) tends to focus on representational artworks, but not abstract paintings [5]. They showed that the reality constraint can be relaxed by copying the artwork by experiment. From this result, they suggested by copying the artwork, novice played the artist's role and acquired the artist's point of view. Tanaka and Matsumoto tried to relax the reality constraint by using a commentary which is a short text explaining the artwork [11]. The result suggested that reading a commentary with technical information (style, colour, technique and various technical aspects of the artwork) relaxed the reality constraint. The participants' descriptions about the artwork showed that when reading these commentaries, novice tried to understand beyond what was drawn on the painting.

1.2 Creating Story About Abstract Paintings

There are two possible solutions to relax the reality constraint. Two research groups, Ishibashi and Okada, and Tanaka and Matsumoto tried to relax the reality constraint by obtaining the artist's point of view [5,11]. Viewers in their experiment played a role (copying the work) and knew how they created the artwork technically. However, they did not try to overcome to understand the lack of forms. In this paper, we will discuss whether novice can overcome to understand lack of forms by creating a story about abstract artworks. When creating a story, we need a character, stage, situation and many other elements. When we try to create a story form an abstract artwork which lacks of a specific form, we need to imagine its form. In other words, a viewer may create their own forms in their minds for abstract artworks. In art museums, we can find several kinds of tools and activities which aim to make visitors create a story about artworks. In this section, we will introduce such tools and activities in museums. The most typical tools are worksheets. When we visit art museums, we can obtain information from captions displayed next to the artwork. However, since the amount of information which can be written on captions are very limited, art museums usually provide a tool to help visitor's understanding. The most popular type of tools is worksheets with a short question (e.g. "What is this girl on the artwork doing?") and empty space to write the answer to the question. In The Metropolitan Museum of Art's family guide "Shall We Dance?", they focused on three artworks related to dancing and provided detailed facts and activities to help visitors understanding [12]. The worksheet instructed to create own story and suggested to eight sample sentences such as "Once upon a time there was a young girl and boy. Their names were ... and ...", "They went to a costume party to celebrate the ..." or "The party was held at ...". This worksheet can be downloaded from The Metropolitan Museum's website ([12]). The National Museum of Modern Art, Tokyo (MOMAT)'s MOMAT Collection Children's Self Guide also has questions to encourage visitors to create a story. On the Ryusei Kishida's artwork, there is a brown-coloured road crossing the canvas from bottom to up. And in the background, we can see blue sky with white clouds. The question on the sheet is "if you walk up the road, what will you see?" and the smaller text on the same page says "You are climbing up the hill road step by step. You can see blue sky in front of you." There is a blank box below the text to answer the question. In addition, art museums in Japan usually provide activities such as a guided tour and a workshop. Many of the tour and worksheets are organised based on the idea of Visual Thinking Strategies (VTS) [13]. In Japan, this methodology is widely accepted in art museums. VTS is a methodology of art appreciation which focuses on explaining and sharing viewers' impression between viewers. The first question in the VTS workshop is "what is going on in this picture?". Viewers answer, explain and share what they saw in the artwork.

1.3 Explaining Abstract Paintings

Our research question is that will individual viewers create similar stories for the same artworks? In this paper, the authors will introduce the experiment correspond to the research question and discuss the results of the experiment. We should note that the experiment in this paper was conducted in Japanese. All stimuli used in the experiment was written in Japanese. Therefore, this experiment's procedures and results may improve the display strategies in art museums in Japan. However, it might not fit non-Japanese museums or visitors.

2 Materials and Methods

2.1 Participants

Participants were 29 adults including university students and teachers. All the participants are asked to answer about their art educational background. Based on the answers in the worksheet, we chose participants who have not educated nor experienced to create their own artworks as a professional artist so that we could regard them as novices.

2.2 Stimuli

The experiment was conducted in a room in Chiba University on 23rd December 2017. We used 18 artworks created in from 1947 to 2017 as stimuli. Six of them were representational paintings and others were abstract paintings. Each artwork had numbers to distinguish them. All artworks were displayed with simple captions and labels. Information included in captions were chosen based on Tanaka and Matsumoto's experiment [11]. All participants answered the worksheet and questionnaire during appreciating the artworks. The questionnaire included questions about participants' experiences of art education and creation. We made two types of worksheets to apply to them according to conditions. The first type of worksheet was a worksheet with two questions which are (Q1) Write about your impression about the artwork and (Q2) Mark the point which fits best to your impression on the following list. Words pairs in Q2 were selected based on a semantic differential technique (SD) used in Okada and Inoue's experiment [7]. They were, for instance, bright–dark, stable–unstable, sharp–dull, like–dislike and artistic–not-artistic. The second type of worksheet included additional question. They were (Q1) Write about your impression about the artwork, (Q2) Create a story from the artwork and (Q3) Mark the point which fits best to your impression on the following list.

2.3 Conditions

We used the following three conditions for each participant (within-subjects factor). As they entered the room, they saw six abstract paintings without creating a story (abstract condition), six representational paintings without creating a story (representational condition) and six abstract paintings using a worksheet to create a story (story condition). All participants followed this order. If the story creating worksheet helped to image a form on the painting, in other words, if the following hypothesis was true, the result from story condition would have a similar trend compared to the representational condition and different trend compared with the abstract condition.

2.4 Hypothesis

Our hypothesis is that stories created for an artwork are different from stories created for other artworks. And all viewers create similar stories for the same artwork.

3 Results and Discussion

3.1 Similar Stories for an Artwork?

Our hypothesis is "Stories created for an artwork are different from stories created for other artworks." We used the data from "story condition" in this analysis. The result in Fig. 1 shows that there are three groups in stories and each group are isolated from other groups by correspondence analysis. We used KH coder [3,4] which was a free software for text mining and analysis. Since answers from participants were all written in Japanese, the result in Fig. 1 was plotted with English labels translated by the authors.

First, we focused on the words which participants frequently used. Numbers in square shape such as C13 are an artwork's number which mean C13 corresponds to the artwork No. 13. There are three groups on the top left (No. 15),

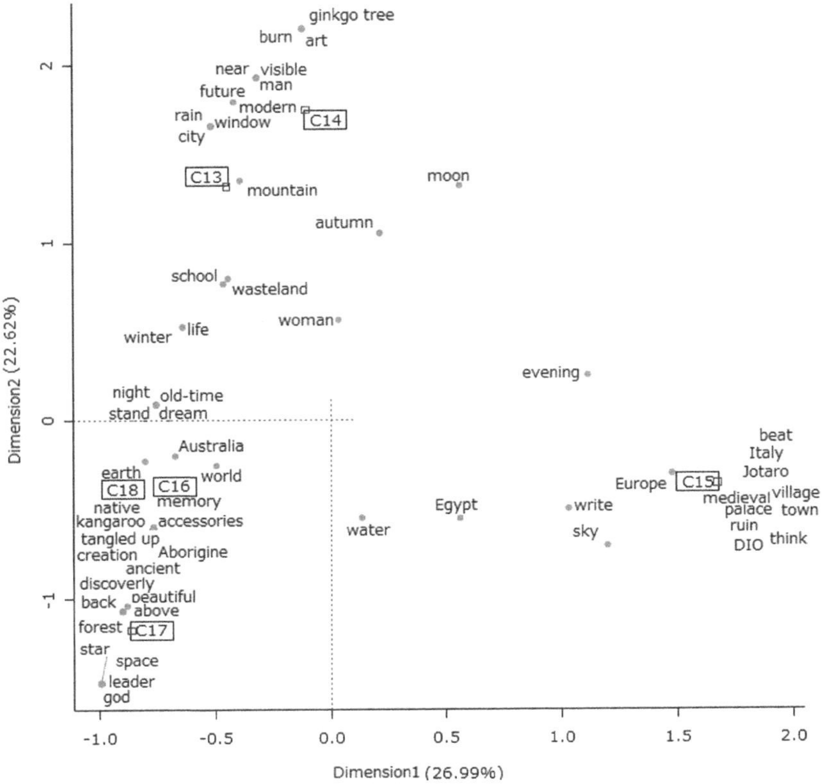

Fig. 1. Correspondence analysis result

bottom left (No. 16, No. 17, No. 18) and bottom right (No. 13, No. 14). In the dimension 2 (y-axis in Fig. 1), words such as “ginkgo tree”, “modern”, “art”, “‘window”, whereas words such as “space”, “ancient”, “god”, “native peoples”, “palace” and “ruin” gather on the negative side. We can estimate dimension 2 (y-axis) is a time axis.

In the dimension 1 (x-axis in Fig. 1), words such as “think”, “village”, “Italy” gather on the positive side of x-axis, whereas words such as “star”, “accessory”, “school”, “mountain” and “future” gather on the negative side. In the middle, there are words such as “moon”, “Egypt” and “water”. From this result, we could not estimate what this axis indicates.

No. 15 on the positive side of x-axis is an artwork painted with strong colours and compositions. The word group used in the story for the artwork No. 15 (C15) is isolated from the word groups used in the other stories. Accordingly, we can guess words used in the artwork No. 15 are quite different from words used in the other artworks. In fact, the stories for the artwork No. 15 are drawn in the context of “Europe.” For instance, the following stories were generated.

It was in 2010, in a certain town in the Europe, peoples saw a vividly red setting sun as if the sky have been burning.
In the evening, in the Europe, a boy traveling alone was rowing a boat with seeing a sunset glow in the sky. and

In fact, we can see in the artwork at least old style buildings in the Europe-like place are drawn. In addition, main colour of the painting is red. These stories are reasonable.

However, artworks No. 16, No. 17 and No. 18 have strong and bold lines in common. Of course, since these artworks are abstract paintings, we cannot clearly identify what is drawn. The artwork No. 16 does not seem a landscape painting, but we can see something are pile up. Accordingly participants generated their stories in the context of piling up something or group. Since in the artwork No. 17, somebody like alien is drawn, many participants generated stories in the context of alien. For the artwork No. 18, many participants generated stories in the aboriginal context after knowing that the paint was drawn by Aborigine. By the above analysis (Fig. 1), these paints gather in the same area. The common feature in the generated story is an extraordinariness. In addition, as shown above, the common feature of the three paintings is very clear lines, though it is rather difficult to determine what are drawn in the paintings. Okada and Inoue pointed out that the set of strong lines is important factor to recognise the existence of forms [7]. Thus we can guess the existence of the clear, strong and bold lines could influence the words in creating stories. The words are an extraordinariness and the universe.

This result suggests the importance of strong lines to show the object's form. Okada and Inoue suggested that abstract paintings are difficult to appreciate for a novice because they lack of forms [7]. As shown above, our result can support their suggestion that strong lines are the important factors to determine the existence of forms. Accordingly even novices can appreciate/understand certain abstract paintings with strong lines.

This result shows that there are three patterns of stories created for abstract artworks in this experiment. And when the groups of stories are compared with each other, we can suggest strong and bold line are the reason for the separation of these groups. This result means the bold lines are the important factors in appreciation. However, this result will not mean that the appreciation of abstract artworks becomes easier in some other artworks.

3.2 Settings: Place

In the following sections, we focus on settings of the viewers' story. We assume that stories will become more detailed and familiar when the appreciation is easier. The stories include information about the place where the story took place. For example, viewers wrote "under the water", "near the sun" and "a village in Europe" in their stories about artworks. We picked up the word about the place and added labels to each story based on their stories occurring place. For each answer, a label was chosen from three types of labels which were "common noun", "mixed" and "proper noun". Label "mixed" is for the word with a common noun characterized by the phrase including proper nouns such as "huge wasteland like we see in the USA or Australia". We chose these three labels because we tend to use proper nouns when we talk about the familiar or specific places.

Figure 2 shows the result of analysis of noun types in each artwork. There are significant differences between No. 15 and No. 13, and No. 15 and No. 17 by Tukey's HSD test when converting labels into scores. A more proper noun is used to explain No. 15 than No. 13 and No. 17. This result has a similar pattern to the result from the correspondence analysis shown on Fig. 1.

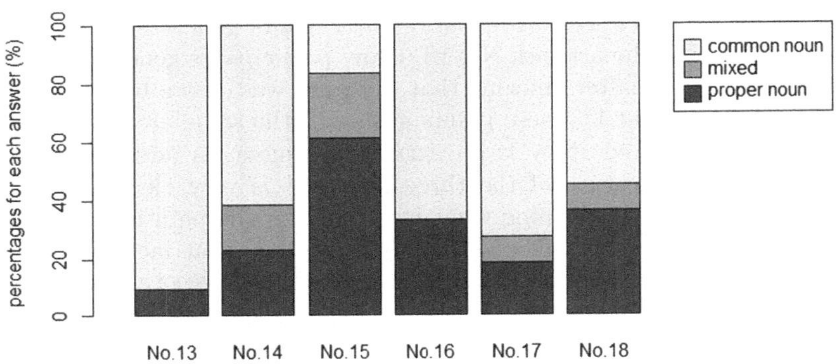

Fig. 2. Relationships between artworks and noun types in story

There are a place impossible to reach such as "another planet" and a place easy to reach such as "school". We added three labels which are "went", "possible" and "impossible". The place where viewer is considered to have visited such as school and their laboratory is labelled as "went". The result is shown in Fig. 3. However, no significant difference is found by Tukey's HSD test when we convert labels into scores.

3.3 Settings: Time

There is also information about the time when the story happens such as "yesterday" and "raining night".

We picked up the words about the time and added labels. For each answer, a label is chosen from six types of labels which are not-limited (eg. "future"), period (eg. "medieval"), year (eg. "1970"), season (eg. "before harvesting wheat"), day (eg. "December 20th 2015"), and hour (eg. "evening").

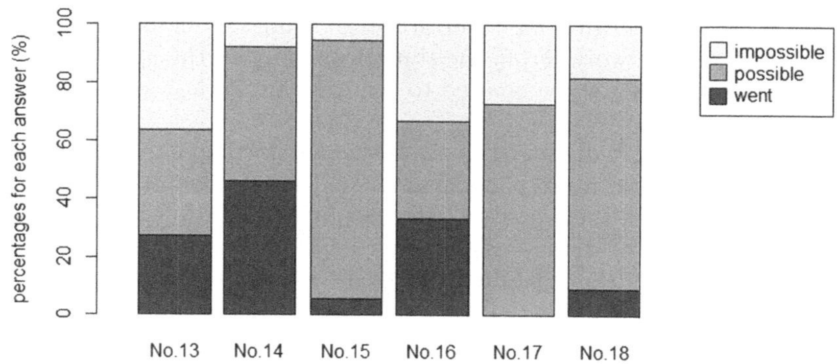

Fig. 3. Relationships between artworks and place types in story

Figure 4 shows the result which indicates word types about the time in each artwork. There are significant differences between No. 15 and No. 13, and No. 13 and No. 17 by Tukey's HSD test when we convert labels into scores. This result also had the similar pattern to the analysis shown in Figs. 2 and 4 which means three groups found in Fig. 2 were isolated from each other.

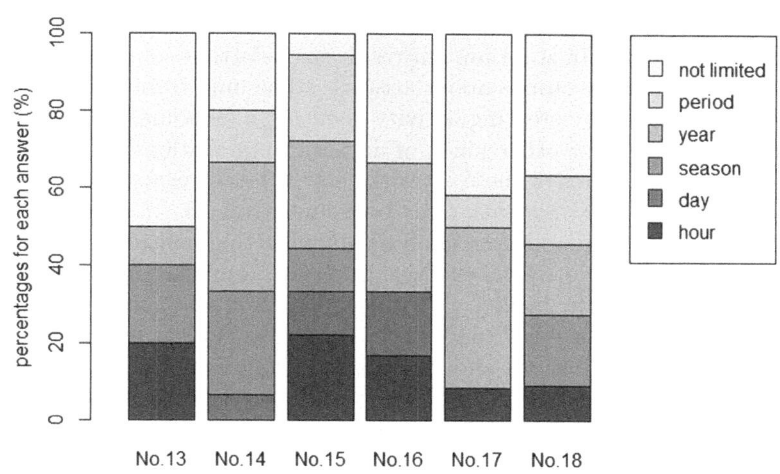

Fig. 4. Relationships between artworks and time setting of the story

3.4 Discussions

As discussed in this section, during the art appreciation, it is meaningful to make people create stories. Our hypothesis was that stories created for an artwork are different from stories created for other artworks. And all viewers create similar stories for the same artwork. From the statistical analysis, the hypothesis seems true. In addition, stories to be created to the same artwork is almost the same or similar.

In the paper, we only discussed the words which participants used. From the words we can guess the image participants could obtain from the painting. Of course from the stories they created, we can obtain more information. In this paper, we dealt with the data statistically. It is also important to deal with the data qualitatively. For the qualitative analysis, in [1], we discussed the effect of information offering during the art appreciation. We found that if the new information was added, new thinking or discovery would become possible. For instance, when we offered the phrase "Ivana Živić was born in 1979 in Sarajevo," participants become to appreciate the painting from the context of the Yugoslavia conflict. This interpretation might be wrong, but by the information, the other story can be created. The data should be dealt with both qualitatively and quantitatively.

4 Conclusions

We conducted the experiment to test the hypothesis whether stories created for an artwork are different from stories created for the other artworks. We found out that individual viewers create a similar story to the same artwork from the corresponding analysis. In addition, our result suggests that some patterns exists when participants create stories about artworks. The important suggestion from these results is that story creating activity is effective for some kinds of artwork appreciations but not for other kinds of artwork appreciations. We cannot not determine general features of these artworks now. The correspondence analysis's result suggests that artworks with clear bold lines (No. 16, 17, 18) are effective in an art appreciation. In addition, analysis about settings suggests artwork with coloured composition (No. 15) is effective in an art appreciation.

Since museums are the facility with the educational and conservational aim, museums must appeal the importance of conservation of cultural and natural heritage through their exhibitions and educational programs. This research aimed to help art museums to create the effective activities (educational programs) for visitors who are mainly novices. Therefore, we tried to relax the feelings of "dislike" to abstract artworks by converting abstract artworks to representational artworks in viewers' mind by creating a story. However, we cannot assume that the feeling of "like" is the main aim of an art appreciation. Some artworks even tried to raise an issue by inducing the negative feelings. We need further research and discussions about what is exactly the art appreciation and what is exactly the "fun" in an art appreciation. This viewpoint will not affect the result of our experiment in this paper. However, we must continue to discuss these topics when we try to somehow evaluate or verbalise the other people's art appreciation experience. For the relationship between an aesthetic value of senses and knowledge of art, we have started the research in [2].

References

1. Abe, A., Tadaki, K.: Captions with several levels of explanation. Procedia Comput. Sci. **159**, 2335–2344 (2019). Proceedings of KES2019
2. Fukushima, K., Abe, A., Kawada, R.: How will sense of values change during art appreciation? In: Proceedings of 2019 IEEE International Conference on Data Mining Workshops (CDEC 2019) (2019, to appear)
3. Higuchi, K.: A two-step approach to quantitative content analysis: KH coder tutorial using Anne of green gables (part I). Ritsumeikan Soc. Sci. Rev. **52**(3), 77–91 (2016). (in Japanese)
4. Higuchi, K.: A two-step approach to quantitative content analysis: KH coder tutorial using Anne of green gables (part II). Ritsumeikan Soc. Sci. Rev. **53**(1), 137–147 (2017). (in Japanese)
5. Ishibashi, K., Okada, T.: Facilitating creative drawing by copying art works by others. Cogn. Stud. Bull. Jpn. Cogn. Sci. Soc. **17**(1), 196–223 (2010). (in Japanese)
6. O'hare, D.: Individual differences in perceived similarity and preference for visual art: a multidimensional scaling analysis. Percept. Psychophys. **20**(6), 445–452 (1976)
7. Okada, M., Inoue, J.: A psychological analysis about the elements of artistic evaluation on viewing paintings. Educ. Sci. J. Yokohama Natl. Univ. **31**, 45–66 (1991). (in Japanese)
8. Schmidt, J.-A., McLaughlin, J.-P., Leighten, P.: Novice strategies for understanding paintings. Appl. Cogn. Psychol. **3**, 65–72 (1989)
9. Tadaki, K., Abe, A.: The influence of story writing worksheets on art appreciation. In: Proceedings of the 2018 International Conference on Artificial ALife and Robotics (ICAROB 2018), pp. 687–690 (2018)
10. Tadaki, K., Abe, A.: Effects of context creation work in impression of abstract paintings. In: Proceedings of the 32th Annual Conference of the Japanese Society for Artificial Intelligence (JSAI 2018), 3B1-OS-22a-03 (2018). (in Japanese)
11. Tanaka, Y., Matsumoto, S.: Cognitive constraints and its relaxation in appreciation of painting. Cogn. Stud. Bull. Jpn. Cogn. Sci. Soc. **20**(1), 130–151 (2013). (in Japanese)
12. The Metropolitan Museum of Art: Family Guides (2005). https://www.metmuseum.org/learn/kids-and-families/familyguides
13. Visual Thinking Strategies Site by Joel Smith Media: Visual Thinking Strategies (2018). https://vtshome.org

Natural Language Processing

Unsupervised Joint Learning for Headline Generation and Discourse Structure of Reviews

Masaru Isonuma[1(✉)], Junichiro Mori[1,2], and Ichiro Sakata[1]

[1] The University of Tokyo, 3-7-1 Hongo, Bunkyo, Tokyo, Japan
{isonuma,isakata}@ipr-ctr.t.u-tokyo.ac.jp
[2] RIKEN, 1-4-1 Nihonbashi, Chuo, Tokyo, Japan
mori@mi.u-tokyo.ac.jp

Abstract. This is an extension from a selected paper from JSAI2019. Recently, using a large number of reference summaries, supervised neural summarization models have achieved success. However, such data is rare, and trained models cannot be shared across domains. As a solution for such a problem, we propose the first unsupervised end-to-end headline generation model for a single review. We assume that a review can be described as a discourse tree in which the headline is the root and the child sentences elaborate on their parent. By estimating the parent from their children recursively, our model induces the tree and generates the headline that describes the entire review. Through the evaluation of the generated headline on actual reviews, our model achieved competitive performance with supervised models, especially on relatively long reviews. In induced trees, we confirmed that the child sentences explain the parent in detail and the generated headlines abstract for the entire review.

Keywords: Document summarization · Unsupervised learning · Discourse structure

1 Introduction

The need for automatic document summarization is widely increasing, with the recently growing vast amounts of online textual data such as reviews on E-commerce websites [17]. Under these circumstances, supervised neural network models have widely achieved success, using a large number of reference summaries [19,32]. However, the model trained from them cannot be adopted in other domains as salient phrases are not common across domains. Few or no examples of summaries exist for most documents, and preparing such large volumes of reference summaries is very expensive [12].

An unsupervised approach is a possible solution for such a problem. Traditionally, the unsupervised approach has been widely applied to sentence extraction [23,26]. The extractive approach can be effective for some types of documents, e.g. news articles, since the salient sentences should be the summary even

Y. Ohsawa et al. (Eds.): JSAI 2019, AISC 1128, pp. 139–149, 2020.
https://doi.org/10.1007/978-3-030-39878-1_13

though they describe only a part of the document. On the other hand, as for reviews, an appropriate summary is generally concise sentences that summarize the entire review. Therefore, the abstractive method is more effective for reviews because it condenses an entire review via paraphrasing and generalization [2,9]. Our work focuses on the headline generation of reviews, a kind of abstractive summarization tasks, without supervision.

Abstractive summarization techniques sometimes use discourse parsers [9, 11]; however, Ji and Smith indicates the limitations of using external discourse parsers [13]. In this context, a model that encodes a document while automatically inducing the discourse tree is proposed [18]. Inspired by the work, we aim to generate a headline while inducing the discourse tree.

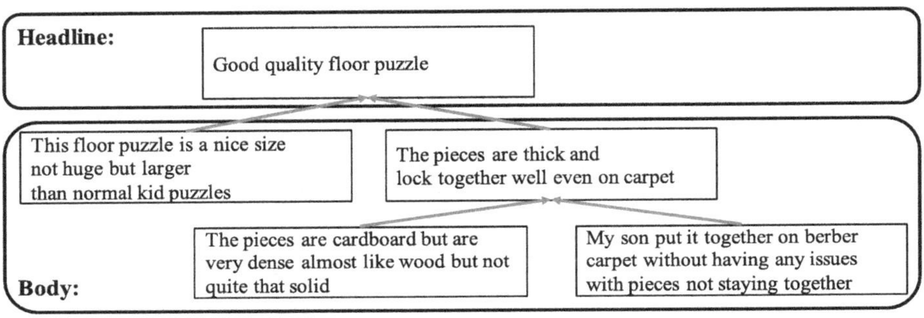

Fig. 1. Example of the discourse tree in a jigsaw puzzle review

Figure 1 shows the example of a jigsaw puzzle review and its dependency-based discourse tree [11]. The headline describes its quality, and the child sentences explain it in terms of the size and thickness. Each child sentence elaborates on its parent in detail.

Thus, we assume that reviews can generally be described as a multi-root non-projective discourse tree in which the headline is the root, and the body sentences are the nodes. In such trees, child sentences elaborate on or provide background to the parent sentence. By estimating the parent from their children recursively, our model learns such a discourse tree and generates a headline.

In this work, we propose the first end-to-end unsupervised model that generates the headline of a single review. Although there have been unsupervised summarization model using external parser [6,8], our work is, to the best of our knowledge, the first unsupervised headline generation model that requires no external parser. Through the evaluation of the headlines generated for actual online reviews and the induced discourse tree, we validate our assumption; child sentences elaborate on the parent sentences, and as a result, the generated headline summarizes the entire review.

2 Related Research

2.1 Supervised Review Summary Generation

Several previous studies have addressed abstractive summarization for product reviews [1,2,5,34]; however, their output summaries are not guaranteed to be grammatical [32]. Neural sequence-to-sequence models have improved the quality of abstractive summarization. Beginning with the adaptation to sentence summarization [3,27], several studies have tackled the generation of an abstractive summary of news articles [24,25,28,29]. With regard to product reviews, the neural sequence-to-sequence based model [32] and joint learning with sentiment classification [19,31] have improved the performance of one-sentence summarization. Our work is also based on the neural sequence-to-sequence model, while introducing the new concept of generating the summary by recursively reconstructing a parent sentence from its children.

2.2 Un-/Semi-supervised Summarization

Most of unsupervised summarization techniques have been focused on extractive approaches [7,23]. For example, Opinosis, a graph-based method is proposed for generating review summaries. On the other hand, with the recently growing neural summarization models, unsupervised or semi-supervised summary generation is being attempted. Miao and Blunsom used a variational auto-encoder for sentence compression [22]. They compressed sentences so that they do not lose essential meaning of the input. With respect to multi-review summarization, Chu and Liu use an auto-encoder to gain the summary representation [4]. They regarded the mean of the document embeddings as the summary, while learning a language model via the reconstruction of each review. However, their model cannot directly be applied to a single document. Against such a background, our model generates the headline of a single review without supervision.

2.3 Discourse Parsing and Its Application

Discourse parsing has been broadly studied and used for various applications. Discourse structure is generally parsed based on the corpus in which a rhetorical structure theory-based discourse tree (RST-DT [20]) is annotated. Some work transform the RST-DT into a dependency-based discourse tree for the summarization [11,15,33]. Ji and Smith also constructed a dependency-based discourse tree, and applied a recursive neural network model for document classification [13]. They indicated the limitations of using external corpus, showing the performance depends on the amount of the RST-DT and the domain of documents. Against such a background, Liu and Lapata proposed a model inducing a latent document structure without external corpus [18]. They reported that the child presents additional information regarding the parent in the induced document structure. While they induced a latent discourse tree via supervised document classification, we induce the tree and generate the headline via recursively reconstructing parent sentences from their children.

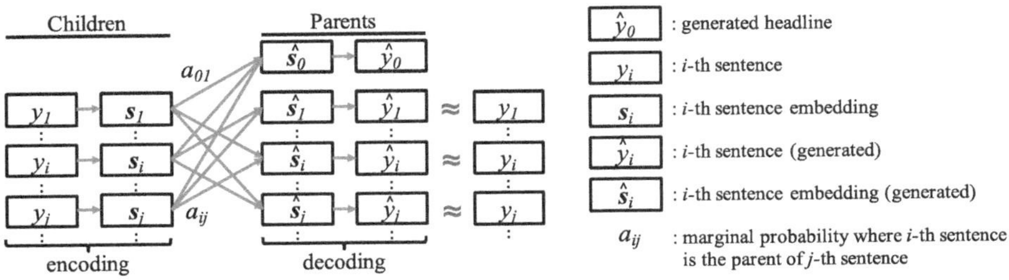

Fig. 2. Outline of the proposed model

3 Proposed Model

In this section, we present our headline generation model by inducing a discourse tree without external parsers. Figure 2 shows the outline of our proposed model. In the following, we explain the training method and computation of the marginal probability of the dependency edges.

3.1 Model Training

In Fig. 2, y_i and $\boldsymbol{s}_i \in \mathcal{R}^d$ indicate the i-th sentence and its embedding in a document $D = \{y_1, y_2, \ldots, y_n\}$ respectively. w_i^t is the t-th word in a sentence $y_i = \{w_i^1, w_i^2, \ldots, w_i^l\}$. $\boldsymbol{s}_i$ is computed via max-pooling operation across hidden states $\boldsymbol{h}_i^t \in \mathcal{R}^d$ of Bi-directional Gated Recurrent Units (Bi-GRU):

$$\overrightarrow{\boldsymbol{h}}_i^t = \overrightarrow{\mathrm{GRU}}(\overrightarrow{\boldsymbol{h}}_i^{t-1}, w_i^t) \tag{1}$$

$$\overleftarrow{\boldsymbol{h}}_i^t = \overleftarrow{\mathrm{GRU}}(\overleftarrow{\boldsymbol{h}}_i^{t+1}, w_i^t) \tag{2}$$

$$\boldsymbol{h}_i^t = [\overrightarrow{\boldsymbol{h}}_i^t, \overleftarrow{\boldsymbol{h}}_i^t] \tag{3}$$

$$\forall m \in \{1, \ldots, d\},\ \boldsymbol{s}_{i,m} = \max_t \boldsymbol{h}_{i,m}^t \tag{4}$$

Here, we assume that the document D involves a discourse tree in which the root is the headline, and all the sentences are the nodes. We denote a_{ij} and a_{0j} as the marginal probability, where sentence i and the root are the parent node of sentence j under the constraint $\sum_{i=0}^{n} a_{ij} = 1$ (see Fig. 2). From the sentence embeddings and a_{ij}, we compute the embedding of the parent sentence $\hat{s}_i$ and that of the headline $\hat{s}_0$. $\hat{s}_i$ ($i \in \{0, \ldots, n\}$) are defined with parameters $\boldsymbol{W}_s \in \mathcal{R}^{d*d}$ and $\boldsymbol{b}_s \in \mathcal{R}^d$ as shown below:

$$\hat{\boldsymbol{s}}_i = \tanh\Big\{\boldsymbol{W}_s(\sum_{j=1}^{n} a_{ij}\boldsymbol{s}_j) + \boldsymbol{b}_s\Big\} \tag{5}$$

The higher the marginal probability a_{ij} is, the more the information of $\boldsymbol{s}_j$ are input into $\hat{\boldsymbol{s}}_i$. From $\hat{\boldsymbol{s}}_i$, the GRU-decoder learns to reconstruct the sentence i, i.e., search the parameters $\boldsymbol{\theta}$ that maximize the following log likelihood:

$$\sum_{i=1}^{n}\sum_{t=1}^{l}\log P(w_i^t|w_i^{<t},\hat{\boldsymbol{s}}_i,\boldsymbol{\theta}) \tag{6}$$

3.2 Headline Generation

Here, we explain how training the model contributes to headline generation. By reconstructing the sentences in documents, the decoder learns a language model to generate grammatical sentences. Therefore, the model can decode the headline embedding $\hat{\boldsymbol{s}}_0$ as a fluent sentence.

Besides, the more the j-th sentence contributes to generating the i-th sentence, the higher a_{ij} can be. This mechanism models our assumption; child sentences can generate their parent, but not vice versa, because the children have more information about the parents. Based on this assumption, the most abstractive k-th sentences in the body make less contribution to reconstruction of any other sentences, i.e., a_{ik} get to be lower for $\forall i : i \neq 0$. From the constraint $\sum_{i=0}^{n} a_{ik} = 1$, a_{0k} is expected to be larger and contributes to the construction of the headline embedding $\hat{\boldsymbol{s}}_0$.

3.3 Marginal Probability of Dependency

We explain how to calculate the marginal probability a_{ij}. We first define the unnormalized weight f_{ij} of the edge between a parent node i and the child node j via a bilinear function and a linear function. For convenience, we assume the weighted adjacency matrix $\boldsymbol{F} = (f_{ij}) \in \mathcal{R}^{(n+1)*(n+1)}$. The index of the first column and row are 0, which denotes the root node. We assume that the discourse structure can be described as a multi-root non-projective tree. Therefore, f_{ij} is defined as:

$$f_{ij} = \begin{cases} \exp(\boldsymbol{w}_r^\top \boldsymbol{s}_j) & (i = 0 \wedge j \geq 1) \\ \exp(\boldsymbol{p}_i^\top \boldsymbol{W}_f \boldsymbol{c}_j) & (i \geq 1 \wedge j \geq 1 \wedge i \neq j) \\ 0 & (j = 0 \vee i = j) \end{cases} \tag{7}$$

$$\boldsymbol{p}_i = \tanh(\boldsymbol{W}_p \boldsymbol{s}_i + \boldsymbol{b}_p) \tag{8}$$

$$\boldsymbol{c}_j = \tanh(\boldsymbol{W}_c \boldsymbol{s}_j + \boldsymbol{b}_c) \tag{9}$$

where $\boldsymbol{W}_f \in \mathcal{R}^{d*d}$ and $\boldsymbol{w}_r \in \mathcal{R}^d$ are parameters for the transformation. $\boldsymbol{W}_p \in \mathcal{R}^{d*d}$ and $\boldsymbol{b}_p \in \mathcal{R}^d$ are the weights and the bias for constructing the representation of the parent nodes. $\boldsymbol{W}_c \in \mathcal{R}^{d*d}$ and $\boldsymbol{b}_c \in \mathcal{R}^d$ are those of the child nodes.

We normalize f_{ij} into a_{ij}, following the study about structure prediction [16]. a_{ij} corresponds the proportion of the total weight of all the spanning trees containing the edge (i, j):

$$\begin{aligned} a_{ij}(\boldsymbol{F}) &= \frac{\sum_{g \in G:(i,j) \in g} v(g|\boldsymbol{F})}{\sum_{g \in G} v(g|\boldsymbol{F})} \\ &= \frac{\partial \log Z(\boldsymbol{F})}{\partial f_{ij}} \end{aligned} \tag{10}$$

$$v(g|\boldsymbol{F}) = \prod_{(i,j)\in g} f_{ij} \tag{11}$$

$$Z(\boldsymbol{F}) = \sum_{g\in G} v(g|\boldsymbol{F}) \tag{12}$$

Here, G denotes the set of all the spanning trees in a document D. $v(g|\boldsymbol{F})$ is the weight of a tree $g \in G$, and $Z(\boldsymbol{F})$ denotes the sum of the weights of all the trees in G. From the Matrix-Tree Theorem [30], $Z(\boldsymbol{F})$ can be rephrased as:

$$Z(\boldsymbol{F}) = |L_0(\boldsymbol{F})| \tag{13}$$

where $L(\boldsymbol{F}) \in \mathcal{R}^{(n+1)*(n+1)}$ and $L_0(\boldsymbol{F}) \in \mathcal{R}^{n*n}$ are the Laplacian matrix of $\boldsymbol{F}$ and its principal submatrix formed by deleting row 0 and column 0, respectively. By solving Eq. 10, a_{ij} is given by:

$$a_{0j} = f_{0j}\left[L_0^{-1}(\boldsymbol{F})\right]_{jj} \tag{14}$$

$$a_{ij} = f_{ij}\left[L_0^{-1}(\boldsymbol{F})\right]_{jj} - f_{ij}\left[L_0^{-1}(\boldsymbol{F})\right]_{ji} \tag{15}$$

4 Experiments

In this section we present our experiments for evaluating the performance of headline generation. We compared the generated headlines on actual online reviews. The following explains the details of our experiments and the results.

4.1 Dataset

Our experiment uses Amazon product review data (Toys and Games) [10,21], which previous works used to evaluate the headline generation model of a review [19,31]. This dataset contains actual online reviews and their headlines.

Because our model generates a headline via learning the discourse tree, we assume that training will fail if the number of sentences in the review is too small. Therefore, we use reviews in which the number of sentence is in [10, 20) for training and [5, 20) for validation and evaluation. The number of reviews for training, validation, and evaluation are 21791, 416, and 464, respectively.

4.2 Experimental Details

For all the experiments, our model has 300-dimensional word embeddings and Bi-GRU with 256-dimensional hidden states. We initialize the word embeddings with pre-trained FastText vectors [14]. We train the model using Ada-grad with a learning rate of 10^{-1}, an initial accumulator value of 0.1, and a batch size of 16. At evaluation time, we use a beam search with a beam size of 10.

Similar to the previous work [19], our evaluation metric is the ROUGE-F1 score. We use ROUGE-1, ROUGE-2, and ROUGE-L, which measure the word-overlap, bigram-overlap, and longest common sequence between the reference and generated summaries, respectively.

4.3 Baseline

Following the previous work [19,31], our baseline models are the supervised vanilla sequence-to-sequence models for headline generation. We denote the sequence-to-sequence model as Seq-Seq and that with the attention mechanism as Seq-Seq-att. Implementation details are the same as above.

4.4 Results

Table 1 shows the ROUGE score of our model and the baseline models on the evaluation sets. Our model achieves a slightly low performance, compared to Seq-Seq.

In Fig. 3, we report the performance on the evaluation sets in which the number of the sentences are in [5, 10), [10, 15) and [15, 20), respectively. We compared to the supervised baseline model (Seq-Seq-Att). In the case of the dataset with under 10 sentences, the performance of our model is inferior to that of the baseline. On the other hand, on the one with 10 or more sentences, our model achieves a competitive performance as for ROUGE-1 and ROUGE-L. Because our model generates headlines via learning discourse tree, our model sometimes appears to fail at constructing a tree as for short documents. It results in a decline in the performance.

Table 1. ROUGE-F1 score on the evaluation set (%)

Models	ROUGE-1	ROUGE-2	ROUGE-L
Seq-Seq	12.8	1.8	10.2
Seq-Seq-att	13.8	2.5	10.9
Our model	11.4	1.6	9.1

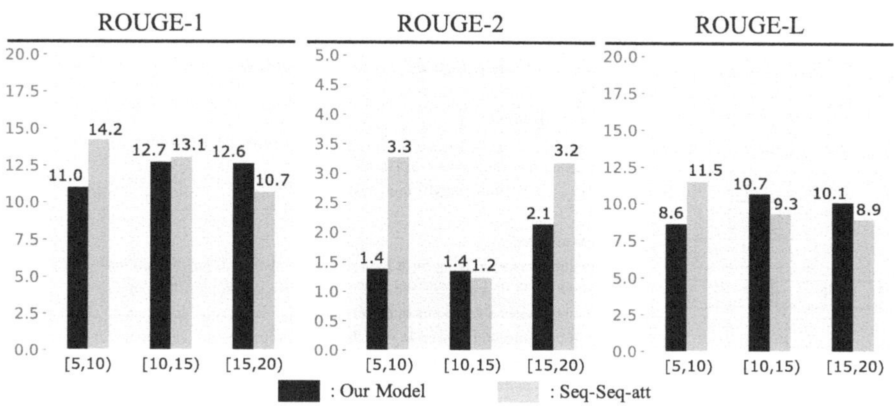

Fig. 3. ROUGE-F1 score on evaluation set with various number of sentences

5 Discussion

Figure 4 shows the generated headline and the discourse tree induced by our model. We obtain the maximum spanning tree from the probability distribution of dependency, using Chu–Liu–Edmonds algorithm.

In Fig. 4(a), the generated headline is "this is a great game for a young child", while the actual headline is "great game." On the discourse tree, the child nodes of the root are the 2nd and 5th sentence. Both of them elaborate on the generated

Induced Discourse Tree | **Generated Headline / Input Sentences**

(a)

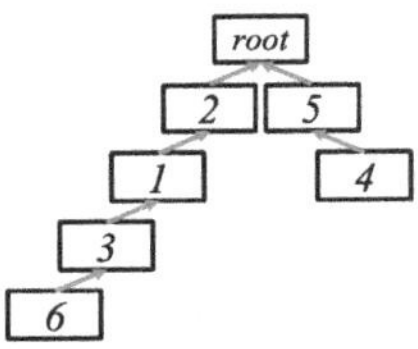

- **Generated Headline**
  - Reference: great game
  - StrSum (Unsupervised): this is a great game for a young child
  - Seq-Seq-att (Supervised):fun game
- **Input Sentences**
  1. this is a fun game and it scales really well from 2-4 players
  2. my friends and family have really enjoyed playing this
  3. the scoring rules are what really make this game great
  4. having the lowest number no matter what the color be the player that loses makes for some great take that blocking opportunities
  5. also for parents of children my 6 year old was able to play this with me with only minimal coaching
  6. also the box is designed well to house all the components which are very well made

(b)

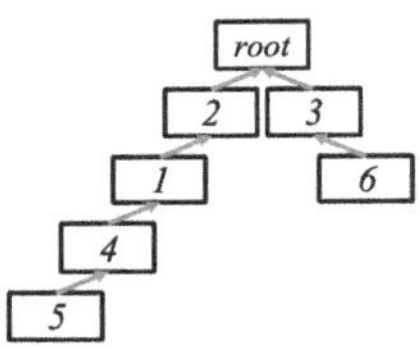

- **Generated Headline**
  - Reference: love this game
  - StrSum (Unsupervised): i love this game
  - Seq-Seq-att (Supervised):fun game
- **Input Sentences**
  1. I love this game
  2. It is so much fun
  3. I'm all about new and different games
  4. I love to play this with my brother because he is very bad at keeping score so I win most of the time and he loves to tell each characters story
  5. And he loves to tell each characters story and to tell why each person got what fate
  6. It's a must buy if you want a fun and fast card game

(c)

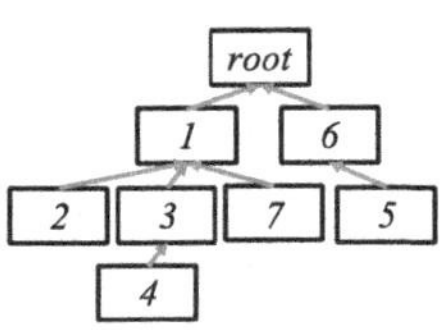

- **Generated Headline**
  - Reference: a good game but extremely time consuming and tedious
  - StrSum (Unsupervised): it 's ok
  - Seq-Seq-att (Supervised): fun game
- **Input Sentences**
  1. twilight has a lot going for it but i personally think it is just too top heavy for play
  2. you have to be a true fan and dedicate an entire day to play this game
  3. the box states a few hours but it takes an entire hour to set up the game alone
  4. and the concept is flawed as well though it has a very nice idea behind it
  5. my personal bone to pick would be the artwork
  6. i happen to think it is silly
  7. but that is me and not a fair thing to say other than i am the customer

Fig. 4. Examples of the generated headlines and induced discourse trees headline. The 3rd sentence explains the parent (the 1st sentence) by explaining the cause of fun.

Figure 4(b) shows that our model generates the headline "i love this game", against the reference, "love this game". In the induced tree, the 2nd sentence elaborates on the generated headline, while the 3rd sentence describes its background. The 4th and 5th sentences describe why the author loves the game, i.e., they explain the 1st sentence in detail.

In Fig. 4(c), we show the example of a *negative* review about a board game. The induced tree shows that the 1st and 6th sentences present additional information about the generated summary. While the 1st to 4th sentences indicate the *heaviness* of the game, the 5th and 6th ones criticize the *artwork*. The 2nd, 3rd, and 4th sentences present the additional information about the parent.

As shown above, we confirmed that the child sentences elaborate on the parent in the induced discourse tree, while the headline abstracts for the child sentences.

6 Conclusion

In this work, we proposed unsupervised end-to-end headline generation model of a single review. Our model induces the latent discourse tree and generate the headline as the root sentence without reference summaries or external parser.

We evaluate our model in comparison with supervised models on actual reviews. Our model achieves a competitive performance when the number of sentences is relatively large.

Furthermore, our model induced a discourse tree in which the child sentences elaborate on the parent. We also confirmed that the headline abstracts for the entire review.

Acknowledgements. This work was supported by CREST, JST, the New Energy and Industrial Technology Development Organization (NEDO) and Deloitte Tohmatsu Financial Advisory LLC.

References

1. Bing, L., Li, P., Liao, Y., Lam, W., Guo, W., Passonneau, R.: Abstractive multi-document summarization via phrase selection and merging. In: Proceedings of the 53rd Annual Meeting of the Association for Computational Linguistics and the 7th International Joint Conference on Natural Language Processing of the Asian Federation of Natural Language Processing, vol. 1, pp. 1587–1597 (2015)
2. Carenini, G., Cheung, J.C.K., Pauls, A.: Multi-document summarization of evaluative text. Comput. Intell. **29**(4), 545–576 (2013)
3. Chopra, S., Auli, M., Rush, A.M.: Abstractive sentence summarization with attentive recurrent neural networks. In: Proceedings of the 2016 Conference of the North American Chapter of the Association for Computational Linguistics: Human Language Technologies, pp. 93–98 (2016)

4. Chu, E., Liu, P.: MeanSum: a neural model for unsupervised multi-document abstractive summarization. In: Proceedings of the 36th International Conference on Machine Learning, vol. 97, pp. 1223–1232 (2019)
5. Di Fabbrizio, G., Stent, A., Gaizauskas, R.: A hybrid approach to multi-document summarization of opinions in reviews. In: Proceedings of the 8th International Natural Language Generation Conference, pp. 54–63 (2014)
6. Dohare, S., Gupta, V., Karnick, H.: Unsupervised semantic abstractive summarization. In: Proceedings of the 56th Annual Meeting of the Association for Computational Linguistics, Student Research Workshop, pp. 74–83 (2018)
7. Erkan, G., Radev, D.R.: Lexpagerank: prestige in multi-document text summarization. In: Proceedings of the 2004 Conference on Empirical Methods in Natural Language Processing, pp. 365–371 (2004)
8. Fang, Y., Zhu, H., Muszyńska, E., Kuhnle, A., Teufel, S.: A proposition-based abstractive summariser. In: Proceedings of COLING 2016, the 26th International Conference on Computational Linguistics, pp. 567–578 (2016)
9. Gerani, S., Mehdad, Y., Carenini, G., Ng, R.T., Nejat, B.: Abstractive summarization of product reviews using discourse structure. In: Proceedings of the 2014 Conference on Empirical Methods in Natural Language Processing, pp. 1602–1613 (2014)
10. He, R., McAuley, J.: Ups and downs: modeling the visual evolution of fashion trends with one-class collaborative filtering. In: Proceedings of the 25th International Conference on World Wide Web, pp. 507–517 (2016)
11. Hirao, T., Yoshida, Y., Nishino, M., Yasuda, N., Nagata, M.: Single-document summarization as a tree knapsack problem. In: Proceedings of the 2013 Conference on Empirical Methods in Natural Language Processing, pp. 1515–1520 (2013)
12. Isonuma, M., Fujino, T., Mori, J., Matsuo, Y., Sakata, I.: Extractive summarization using multi-task learning with document classification. In: Proceedings of the 2017 Conference on Empirical Methods in Natural Language Processing, pp. 2101–2110 (2017)
13. Ji, Y., Smith, N.A.: Neural discourse structure for text categorization. In: Proceedings of the 55th Annual Meeting of the Association for Computational Linguistics, vol. 1, pp. 996–1005 (2017)
14. Joulin, A., Grave, E., Mikolov, P.B.T.: Bag of tricks for efficient text classification. In: Proceedings of the 15th Conference of the European Chapter of the Association for Computational Linguistics, vol. 2, pp. 427–431 (2017)
15. Kikuchi, Y., Hirao, T., Takamura, H., Okumura, M., Nagata, M.: Single document summarization based on nested tree structure. In: Proceedings of the 52nd Annual Meeting of the Association for Computational Linguistics, vol. 2, pp. 315–320 (2014)
16. Koo, T., Globerson, A., Carreras, X., Collins, M.: Structured prediction models via the matrix-tree theorem. In: Proceedings of the 2007 Joint Conference on Empirical Methods in Natural Language Processing and Computational Natural Language Learning, pp. 141–150 (2007)
17. Liu, B., Zhang, L.: A survey of opinion mining and sentiment analysis. In: Aggarwal, C., Zhai, C. (eds.) Mining Text Data, pp. 415–463. Springer, Boston (2012)
18. Liu, Y., Lapata, M.: Learning structured text representations. Trans. Assoc. Comput. Linguist. **6**, 63–75 (2018)
19. Ma, S., Sun, X., Lin, J., Ren, X.: A hierarchical end-to-end model for jointly improving text summarization and sentiment classification. In: Proceedings of the 27th International Joint Conference on Artificial Intelligence, pp. 4251–4257 (2018)

20. Mann, W.C., Thompson, S.A.: Rhetorical structure theory: toward a functional theory of text organization. Text-Interdiscip. J. Study Discourse **8**(3), 243–281 (1988)
21. McAuley, J., Targett, C., Shi, Q., Van Den Hengel, A.: Image-based recommendations on styles and substitutes. In: Proceedings of the 38th International ACM SIGIR Conference on Research and Development in Information Retrieval, pp. 43–52 (2015)
22. Miao, Y., Blunsom, P.: Language as a latent variable: discrete generative models for sentence compression. In: Proceedings of the 2016 Conference on Empirical Methods in Natural Language Processing, pp. 319–328 (2016)
23. Mihalcea, R., Tarau, P.: Textrank: bringing order into texts. In: Proceedings of the 2004 Conference on Empirical Methods in Natural Language Processing, pp. 404–411 (2004)
24. Nallapati, R., Zhou, B., dos Santos, C., Gulcehre, C., Xiang, B.: Abstractive text summarization using sequence-to-sequence RNNs and beyond. In: Proceedings of the 20th SIGNLL Conference on Computational Natural Language Learning, pp. 280–290 (2016)
25. Paulus, R., Xiong, C., Socher, R.: A deep reinforced model for abstractive summarization. In: Proceedings of the 6th International Conference on Learning Representations (2018)
26. Radev, D.R., Jing, H., Styś, M., Tam, D.: Centroid-based summarization of multiple documents. Inf. Process. Manag. **40**(6), 919–938 (2004)
27. Rush, A.M., Chopra, S., Weston, J.: A neural attention model for abstractive sentence summarization. In: Proceedings of the 2015 Conference on Empirical Methods in Natural Language Processing, pp. 379–389 (2015)
28. See, A., Liu, P.J., Manning, C.D.: Get to the point: summarization with pointer-generator networks. In: Proceedings of the 55th Annual Meeting of the Association for Computational Linguistics, vol. 1, pp. 1073–1083 (2017)
29. Tan, J., Wan, X., Xiao, J.: Abstractive document summarization with a graph-based attentional neural model. In: Proceedings of the 55th Annual Meeting of the Association for Computational Linguistics, vol. 1, pp. 1171–1181 (2017)
30. Tutte, W.T.: Graph Theory, vol. 21. Addison-Wesley, Boston (1984)
31. Wang, H., Ren, J.: A self-attentive hierarchical model for jointly improving text summarization and sentiment classification. In: Proceedings of the 10th Asian Conference on Machine Learning, pp. 630–645 (2018)
32. Wang, L., Ling, W.: Neural network-based abstract generation for opinions and arguments. In: Proceedings of the 2016 Conference of the North American Chapter of the Association for Computational Linguistics: Human Language Technologies, pp. 47–57 (2016)
33. Yoshida, Y., Suzuki, J., Hirao, T., Nagata, M.: Dependency-based discourse parser for single-document summarization. In: Proceedings of the 2014 Conference on Empirical Methods in Natural Language Processing, pp. 1834–1839 (2014)
34. Yu, N., Huang, M., Shi, Y., Zhu, X.: Product review summarization by exploiting phrase properties. In: Proceedings of the 26th International Conference on Computational Linguistics, pp. 1113–1124 (2016)

Extraction of Online Discussion Structures for Automated Facilitation Agent

Shota Suzuki(✉), Naoko Yamaguchi, Tomohiro Nishida, Ahmed Moustafa, Daichi Shibata, Kai Yoshino, Kentaro Hiraishi, and Takayuki Ito

Nagoya Institute of Technology, Gokiso-cho, Showa-ku, Nagoya-shi, Aichi-ken 466-8555, Japan
suzuki.shota@itolab.nitech.ac.jp,
http://www.itolab.nitech.ac.jp/itolab_hp_2017/

Abstract. This is an extension from a selected paper from JSAI2019. This paper proposes an approach that aims to extract the discussion structure from large-scale text-based online discussions. The ultimate goal is to develop an automated facilitation agent that is able to extract discussion structures from large-scale online discussions. Towards this end, we adopt the issue-based information system (IBIS), as a suitable format for structuring online discussions. In this context, we model the task of extracting an IBIS structure as it consists of node extraction and link extraction. Towards this end, a deep neural network based approach is employed in order to perform these two extraction subtasks. In order to evaluate the proposed approach, a set of experiments has been conducted on the data collected from the discussions in the online discussion support system called D-Agree. The experimental results show that the proposed approach is efficient for extracting online discussion structures.

Keywords: Agent · Deep learning · Facilitation · Information extraction · Natural language processing · Online discussion

1 Introduction

Internet-based democratic discussions have received great attention recently because they are the cornerstone of the next generation of open and public citizen forums [8,9]. Towards this end, Ito et al. [8] developed a consensus-building support system called COLLAGREE that has facilitator support functions. In this regard, COLLAGREE promotes online discussions that are adequately led by human facilitators who coordinate, lead, integrate, classify, and summarize these discussions to reach consensuses or agree on alternative solutions. In their experiments, the consensus-building support system successfully gathered many more opinions than comparable meetings, and the participants recognized the importance of facilitators.

Y. Ohsawa et al. (Eds.): JSAI 2019, AISC 1128, pp. 150–161, 2020.
https://doi.org/10.1007/978-3-030-39878-1_14

The issue-based information system (IBIS) [12] is an approach for structuring discussions. Thus, the IBIS is used by facilitators to lead discussions to reach better consensuses. The elements of the IBIS are "issues" (which need to be answered), "ideas" (which are possible answers), "pros" (which are arguments that support an idea), and "cons" (which are arguments against an idea). In this study, we use the term "nodes" to refer to the issues, ideas, pros, and cons; and "links" to refer to the relationships among these nodes. In addition, we designate the discussion structures that are created with the IBIS as IBIS structures.

In the fields of Internet-based democratic discussions and facilitation, our ultimate goal is to create an automated software agent that can function as intelligent facilitator. This automated facilitation agent has a number of advantages such as reacting quickly, facilitating the ongoing discussions in 24 h a day, and making fewer mistakes compared to human facilitators. Towards this end, we propose developing an automated facilitation agent that can facilitate online discussions with the IBIS in consensus-building support systems.

When the automated facilitation agent facilitates online discussions with the IBIS, the core function of the agent becomes extracting IBIS structures from these discussions. The reason is that facilitators need to understand what issues, ideas, pros, and cons are in the sentences of the online discussions and how many are there in order to facilitate these discussions. Although human facilitators can understand the above requirements by witnessing the discussions on consensus-building support systems, the automated facilitation agent cannot witness the discussions. Therefore, we address this extraction problem and propose a novel approach that enables the automated facilitation agent to recognize IBIS structures in online discussions.

To address the extraction problem, we initiate the two subtasks of node extraction and link extraction in IBIS structures. In this regard, node extraction means classifying sentences in online discussions into the elements of IBIS. At the same time, link extraction means predicting the relationships among the extracted nodes in the same way as for the classified sentences. Eventually, we obtain tree-like discussion structures that consist of the elements of IBIS and the relationships among these elements. In this paper, for both node extraction and link extraction, we propose a recurrent neural network (RNN) [4] based approach.

The results of our experiments, that are conducted on a Japanese online discussion dataset collected from a consensus-building support system called D-Agree, indicate that the proposed approach is efficient for extracting IBIS structures from online discussions. In this regard, these results are achieved by structuring online discussions with the IBIS and using RNNs. In addition, the accuracy values of our experimental results outperform those of the experimental results of related works in the fields of argumentation mining [15,16,20] and Japanese natural language processing [14].

As a result, we conclude that the proposed approach enables the automated facilitation agent to extract IBIS structures efficiently.

The contributions of this paper are as follows:

- We define two subtasks in extracting IBIS structures which are node extraction and link extraction. To the best of our knowledge, this paper is the first study to extract online discussion structures using the IBIS.
- We propose an approach that uses RNNs in order to carry out the two subtasks of node extraction and link extraction. Specifically, the proposed approach for link extraction adopts a regression technique even though many studies in the field of argumentation mining adopt a classification technique for link extraction or prediction.
- The results of our experiments indicate that the proposed approach is efficient for extracting IBIS structures from the online discussions. Therefore, the proposed approach enables the automated facilitation agent to extract IBIS structures efficiently.

2 Related Works

2.1 Argumentation Mining

Argumentation mining aims to identify the structure of different arguments in natural language texts, and therefore, is the research area that is most closely related to this study. For instance, many studies in the field of argumentation mining extract structures from essays [15,20], reviews [10], and legal texts [16] in the same way as we propose extracting structures from online discussions. Please note that among the studies in the field of argumentation mining, the subtasks of the component classification and the structure identification [21] are particularly related to our subtasks of node extraction and link extraction, respectively.

Component Classification. Firstly, Mochales-Palau and Moens [16] proposed an approach with a maximum entropy classifier and a support vector machine to classify argumentative clauses into premise or conclusion. They evaluated their approach with a dataset of documents that are extracted from legal texts of the European Court of Human Rights and achieved F1 scores of 0.6812 when classifying argumentative clauses into premises and 0.7407 when classifying argumentative clauses into conclusions. In addition, Kim [10] applied convolutional neural networks (CNNs) to some sentence classification tasks and achieved competitive results. Moreover, Wang et al. [23] proposed long short-term memory based RNNs and conducted a set of experiments on the Stanford Twitter Sentiment corpus. It is important to note that these studies classify argument components and sentences into the elements of arguments, essays, reviews, questions, and tweets. On the other hand, the proposed approach does classify sentences into the elements of IBIS.

Structure Identification. In the beginning, Mochales-Palau and Moens [16] proposed an approach based on context-free grammar for argumentation structure detection. In addition, Peldszus and Stede [18] divided the structure prediction into relation identification, central claim identification, role classification, and function classification. Towards this end, they proposed a model that predicts several aspects of the structure using an MST algorithm. In this context, they conducted a set of experiments on the arg-microtext corpus [17] and achieved an F1 score of 0.720. Although these studies considered argumentative relations as support or attack, and support or non-support, the proposed approach does extract the relationships among the elements of IBIS. Furthermore, while these studies address the structure identification tasks using a classification technique, the proposed approach does adopt a regression technique for link extraction.

2.2 Japanese Natural Language Processing

Eger et al. [3] argued that most of the approaches for argumentation mining are specifically designed for English and there are only few resources for other languages. This means that there are few studies for Japanese natural language processing. In this regard, a survey [14] of these few studies of Japanese natural language processing focused on discussion threads. In this study, inter-annotator agreement was evaluated with Fleiss' κ [5], and their dataset resulted in "moderate agreement" [13]. It is important to note that this study structurized online discussions with argument component types of claim, premise, and non-argumentative and their relationships of inner-post relations and inter-post interactions. On the other hand, the proposed approach does structurize online discussions with different elements of the IBIS and their relationships. In addition, the proposed approach is able to achieve higher accuracy values even though there are only few resources for Japanese.

3 Extracting IBIS Structures with Deep Learning

3.1 Node Extraction

The subtask of node extraction can be formulated as follows. Firstly, online discussions that are conducted in consensus-building support systems have a thread structure that consists of a set of threads such that $D = \{T_1, T_2, \ldots, T_N\}$. In addition, each thread consists of some submissions that are relevant to a certain topic in the form of Japanese text such that $T_n = \{S_1, S_2, \ldots, S_M\}$. After all, we classify the sentences in each submission $s \in S_m$ into four types, issues, ideas, pros, and cons, which are the elements of the IBIS. For example, Fig. 1 shows that we classify the sentence about "Let's discuss town development with IoT and AI." as issues, the sentence about "I propose signs that provide information." as ideas, and the sentence about "The signs are easy to understand." as pros. The reason we do not classify the submissions in a thread but rather classify the sentences in a submission is attributed to the fact that every submission often consists of more than two types of sentences. For example, let a submission consists of two sentences. The first sentence about "I propose signs that provide

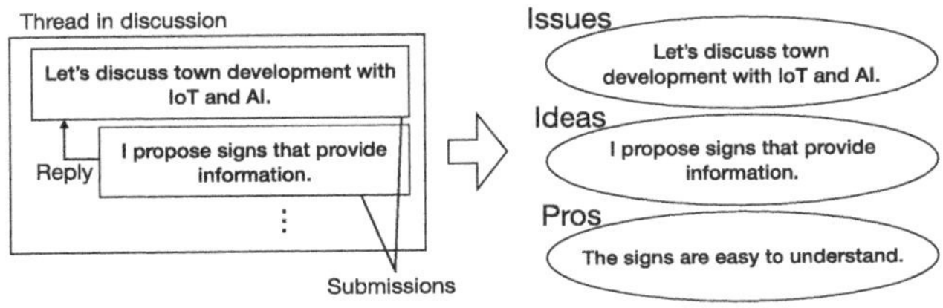

Fig. 1. Some examples of node extraction. Sentences of each submission are classified into four types, issues, ideas, pros, and cons.

information" that would be classified as ideas, while the second sentence about "The signs are easy to understand" that would be classified as pros.

Therefore, in order to promote node extraction, we propose the algorithm that works as follows. First, the proposed approach removes the noise from the sentences in the online discussions because the raw text in these sentences often includes noise. In this regard, we observe the frequent patterns and use regular expressions in order to remove the noise. Second, the proposed approach performs a morpheme analysis on these sentences to acquire their embedded representations. In specific, we ought to perform a morpheme analysis to acquire these embedded representations because Japanese texts are not written with a space between words. In this regard, in order to perform morpheme analysis, the proposed approach uses MeCab [11], which is an open source Japanese morpheme analysis engine that is developed through collaboration between Kyoto University and NTT. Next, in order to acquire the embedded representations in 400 dimensions of morphemes in each sentence, the proposed approach uses fastText [1] that is specially pretrained using Wikipedia corpus. In this regard, the proposed approach conducts zero padding in order to unify the size of the input. Then, the proposed approach classifies the sentences into their corresponding types of the IBIS. The implemented classification models are based on RNNs that focus on processing time series data. The inputs are the embeddings of words in each sentence that is obtained using fastText, while the outputs are their normalized probabilities. Finally, the proposed approach classifies each sentence into its relevant IBIS type that acquires the highest probability. This node extraction model is shown in Fig. 2.

3.2 Link Extraction

On the other hand, the subtask of link extraction is illustrated as follows. After extracting the nodes, we retrieve an issue node set, an idea node set, a pro node set, and a con node set from each thread. Then, we predict the relationships among the extracted nodes in each thread. In other words, we predict which head issue node each tail idea node links to, which head idea node each tail pro node links to, which head idea node each tail con node links to, and which head idea node each tail issue node links to. In this regard, please note that each tail node links to a head node that is extracted from the same or older submissions

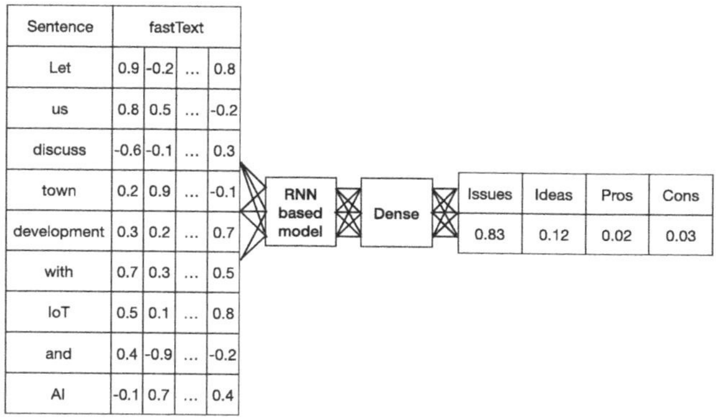

Fig. 2. Model for node extraction. Input of model is embedding of words in sentence obtained by using fastText. Output of model is normalized probability.

in the same thread because of the thread structure. For example, when an issue node that is set in Thread 1 consists of the three nodes of "Let's discuss town development with IoT and AI.", "What type of signs is efficient?", and "What do people in the city need?" and another issue node that is set in Thread 2 consists of one node about "How do we clean the city environment?", then we predict a tail idea node about "I propose signs that provide information." links to the head issue node about "Let's discuss town development with IoT and AI.".

Therefore, in order to promote link extraction, we propose the algorithm that works as follows. First, the proposed approach performs a morpheme analysis on the nodes in sets of tail nodes and head nodes and acquires their embedded representations in 400 dimensions of morphemes in these nodes in the same way as for the node extraction subtask. For the sake of link extraction, the proposed approach need not to remove noise from the candidate nodes because this noise has already been removed from these nodes in the node extraction subtask. Then, the proposed approach predicts which head node each tail node links to. Similarly, the deployed prediction model is based on RNNs like the classification model that is adopted for node extraction. In this regard, it is important to note that the proposed approach adopts a regression technique even though many studies [15,18,20] on argumentation mining adopt a classification technique for link extraction or prediction. Towards this end, the proposed approach calculates the cosine similarities between the output of the model and the candidate head nodes in order to select the head node that each tail node links to. The reason the proposed approach adopts a regression technique is attributed to the fact that the number of candidate head nodes varies during online discussions. Therefore, an approach with the classification technique cannot handle this varying number of candidate head nodes. Finally, the proposed approach considers that the input tail node points to the nearest head node to the output from the set of candidate head nodes. Please note that these candidate head nodes are extracted from the same or older submissions in the same thread as mentioned above. This link extraction model is shown in Fig. 3.

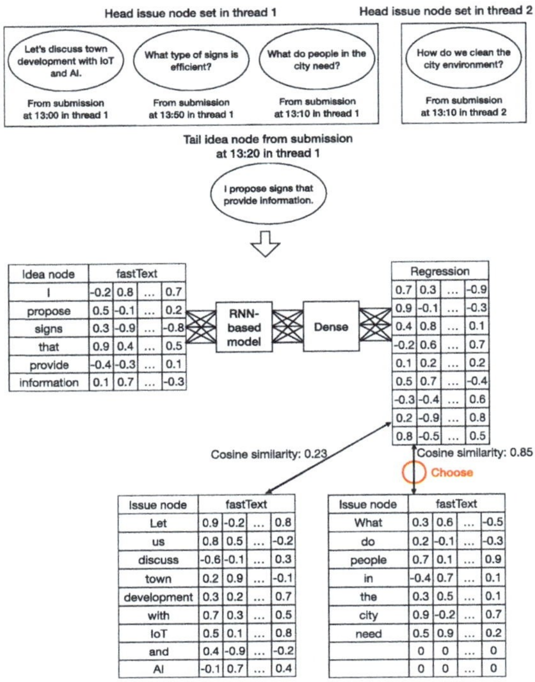

Fig. 3. Model for link extraction is shown. Input of model is embedded tail idea node obtained by using fastText. The proposed approach calculates cosine similarities between output and candidate head issue nodes. In this regard, the proposed approach excludes head issue node about "What type of signs is efficient?" because node is extracted from submission at 13:50, which is later than 13:20 and head issue node about "How do we clean the city environment?" because node is extracted from submission in thread 2, which is a different thread.

4 Experiments

4.1 Dataset

We gathered an experimental dataset from a number of Japanese online discussions (19 discussions) that were created in D-Agree [7,8], which is a consensus-building support system that is developed by the AI research center at the Nagoya Institute of Technology (NIT). We chose the themes of these discussions from the topics that are relevant to Nagoya City and city development and are easy for the participants to think about and discuss. For instance, some of the discussed themes are "How Nagoya City enhances interaction as a major city in the world?", "What is needed so that a city respects human rights and those who live in the city have the opportunity to be successful?", and "How should a city attract foreign people and companies?". Furthermore, 5 to 21 people are allowed to participate in these discussions. During the online discussions in

D-Agree, expert human facilitators, that are appointed by NIT as part of a large research project, facilitated these discussions in accordance with the IBIS.

After the end of discussions, the facilitators in D-Agree also annotate the experimental dataset. For node extraction, the annotators classify the sentences in the discussions into four types, issues, ideas, pros, and cons, which are the elements of the IBIS. On the other hand, for link extraction, the annotators draw out the relationships among the extracted nodes in the same way as for the classified sentences, and then signify which head issue node each tail idea node links to, which head idea node each tail pro node links to, which head idea node each tail con node links to, and which head idea node each tail issue node links to.

The agreement among the annotators in classifying the sentences in the online discussions is evaluated with Fleiss' κ [5]. In this regard, we choose one discussion, and three annotators classify the sentences in this discussion because classifying the sentences in all the discussions is too hard for the annotators. Eventually, κ is 0.667 among the annotators, and this number represents "substantial agreement" [13]. This result indicates our classification task with the IBIS is reasonable and the quality of the created dataset is high. On the other hand, the agreement value in extracting the relationships/links among nodes cannot be calculated with Fleiss' κ. The reason is attributed to the nature of link extraction; which is not a classification but a prediction of a head node that a tail node links to. Overall, the number of each type of nodes is given as follows: 150 issue nodes, 776 idea nodes, 405 pro nodes, and 402 con nodes. Therefore, the total number of nodes is 1,733. In this context, these nodes represent the classified sentences. On the other hand, this dataset contains: 488 links from ideas to issues, 392 links from pros to ideas, 366 links from cons to ideas, and 47 links from issues to ideas. Therefore, the total number of links is 1,450.

4.2 Experimental Settings

In order to perform node extraction and link extraction, we implement the algorithms of the proposed approach in Python. In our experiments, we adopt leave-one-out cross-validation [22] to evaluate the performance of the proposed approach. Towards this end, the test data is set to include one discussion, while the training data is set to include the other discussions. It is important to note that the automated facilitation agent needs to facilitate online discussions without prior knowledge of the discussion theme/topic. For example, if the automated facilitation agent needs prior knowledge about the theme of the discussions, we must hold discussions about the same theme with a human facilitator, then annotate the discussion dataset, and finally train the model with this annotated dataset. This is tedious work and means it is pointless creating the automated facilitation agent.

In order to evaluate the proposed approach, we chose six models: simple RNN, long short-term memory (LSTM) [6], bidirectional LSTM (Bi-LSTM) [19], gated recurrent unit (GRU) [2], bidirectional GRU (Bi-GRU), and CNN. These models, except for the CNN, focus on processing time series data. Therefore,

Table 1. Experimental results of node extraction

	Issues			Ideas			Pros			Cons		
	P	R	F	P	R	F	P	R	F	P	R	F
Simple RNN	0.176	0.141	0.112	0.485	0.680	0.539	0.214	0.125	0.131	0.156	0.110	0.104
LSTM	0.916	0.913	**0.905**	0.752	0.765	0.749	0.538	0.514	0.515	0.478	0.586	0.493
Bi-LSTM	0.924	0.894	0.901	0.741	0.783	0.751	0.551	0.513	0.516	0.499	0.573	**0.537**
GRU	0.898	0.902	0.895	0.758	0.782	**0.759**	0.576	0.534	**0.539**	0.511	0.624	0.534
Bi-GRU	0.868	0.872	0.864	0.744	0.773	0.748	0.520	0.471	0.484	0.482	0.549	0.520
CNN	0.860	0.774	0.796	0.681	0.722	0.690	0.386	0.370	0.370	0.439	0.454	0.451

Table 2. Experimental results of link extraction

	Ideas → Issues	Pros → Ideas	Cons → Ideas	Issues → Ideas
	P	P	P	P
Simple RNN	0.805	0.865	**0.913**	**0.962**
LSTM	0.795	**0.868**	0.900	0.955
Bi-LSTM	0.811	0.858	0.892	0.955
GRU	0.798	0.858	0.909	0.942
Bi-GRU	**0.825**	0.855	0.900	0.942
CNN	0.817	0.853	0.895	0.942

we choose the CNN as a benchmark to evaluate the performance of the other five RNN based models. In this regard, it is worth noting that we implement all these models in Keras, which is a high-level neural network API.

It is also important to note that, we use the same hyperparameters for training all the models. In specific, we train the models for 30 epochs with a batch size of 32 in all the experiments of node extraction. Similarly, we train the models for 50 epochs with a batch size of 64 in all the experiments of link extraction.

5 Results and Discussion

5.1 Results

We conducted a set of experiments in order to evaluate the node extraction performance of the proposed approach. The results of these experiments are shown in Table 1. These results represent the values of precision, recall, and F1 score for the sentence classification using the models of simple RNN, LSTM, Bi-LSTM, GRU, Bi-GRU, and CNN. Please note that each value is the average of the leave-one-out cross-validation. In specific, the results in Table 1 show F1 scores of 0.905 when classifying sentences into issues using LSTM, 0.759 when classifying sentences into ideas using GRU, 0.539 when classifying sentences into pros using GRU, and 0.537 when classifying sentences into cons using Bi-LSTM.

In this regard, please note that LSTM, GRU, and Bi-LSTM are not CNN but RNN based models.

In addition, we conducted a second set of experiments in order to evaluate the link extraction performance of the proposed approach. The results of these experiments are shown in Table 2. These results represent the accuracy of predicting which head node each tail node links to using the models of simple RNN, LSTM, Bi-LSTM, GRU, Bi-GRU, and CNN. Please note that each value is the average of the leave-one-out cross-validation as with the results of node extraction. In this regard, it is worth noting that the values of recall and F1 score could not be calculated because the proposed approach aims to predict a head node that a tail node links to, not to classify the head node. This is the same way as for calculating an agreement value in Subsect. 4.1. The results in Table 2 show precision values of 0.825 when predicting which head issue node each tail idea node links to using Bi-GRU, 0.868 when predicting which head idea node each tail pro node links to using LSTM, 0.913 when predicting which head idea node each tail con node links to using simple RNN, and 0.962 when predicting which head idea node each tail issue node links to using simple RNN. In this regard, please note that Bi-GRU, LSTM, and simple RNN are not CNN but RNN based models.

5.2 Discussion

To summarize, the experimental results demonstrate the efficiency of the proposed approach in extracting IBIS structures from online discussions. In specific, the results of node extraction with LSTM and GRU show high F1 scores when classifying sentences into issues and ideas, respectively. These results outperform their counterparts of the related works [16,20] that are mentioned in Subsect. 2.2. In specific, the F1 scores of 0.905 when classifying sentences into issues and 0.759 when classifying sentences into ideas are higher than the F1 score of 0.715 when classifying argument components into premise in a study that was proposed by the authors in [14] as mentioned in Subsect. 2.2. Contrarily, the results of node extraction show lower F1 scores when classifying sentences into pros and cons. In specific, the F1 scores of 0.539 when classifying sentences into pros and 0.537 when classifying sentences into cons are lower than the F1 score of 0.581 when classifying argument components into claims in a study that was proposed by the authors in [14] as mentioned in Subsect. 2.2. These results are attributed to the existence of sentences that are classified with not only their texts but also their contexts. For example, a number of sentences are sometimes classified as pro nodes and other times are classified as con nodes even though their texts are the same. Therefore, it is concluded that the proposed approach, which does not use contexts as a feature, is not able to classify sentences into pros and cons. An approach that regards contexts as features in order to classify the sentences [15] is expected to improve the results.

On the other hand, the results of link extraction show high precision values in predicting all the types of links. These results outperform their counterparts of the related works [15,20] that were mentioned in Subsect. 2.2. In particular, the precision values in predicting all the types of links are higher than the precision value of 0.458 when predicting inner-post relation in a study that was proposed by the authors in [14] as mentioned in Subsect. 2.2. In addition, the few differences of the precision values among the six models indicate that adopting a regression technique and limiting the number of candidate head nodes are efficient for link extraction. In this regard, it is worth mentioning that the precision values of predicting a head issue node that a tail idea node links to are slightly less than the other precision values. This result is attributed to the fact that similar head issue nodes are extracted from a single discussion theme/topic where the range of issues is narrow. Therefore, since we cannot further narrow down the number of candidates head nodes, we propose to regard context as a feature of the prediction model.

6 Conclusion and Future Work

This paper proposes an approach that aims to extract IBIS structures from the online discussions in consensus-building support systems. In this regard, the proposed approach addresses the two challenges of node extraction and link extraction in order to extract IBIS structures. To evaluate the performance of the proposed approach, a set of experiments are conducted on the discussions dataset created in D-Agree that is a consensus-building support system. The experimental results demonstrate that the proposed approach is efficient for extracting IBIS structures from online discussions. Mainly, these results are achieved by structuring these discussions with the IBIS and using RNNs. Therefore, the proposed approach enables the automated facilitation agent to extract IBIS structures, efficiently. The future work involves improving the accuracy of extracting IBIS structures and enabling the usage of natural languages other than Japanese. Another direction includes studying how the automated facilitation agent intervenes in online discussions.

Acknowledgments. This work was supported by JST CREST Grant Number JPMJCR15E1, Japan.

References

1. Bojanowski, P., Grave, E., Joulin, A., Mikolov, T.: Enriching word vectors with subword information. arXiv preprint arXiv:1607.04606 (2016)
2. Cho, K., Van Merriënboer, B., Gulcehre, C., Bahdanau, D., Bougares, F., Schwenk, H., Bengio, Y.: Learning phrase representations using RNN encoder-decoder for statistical machine translation. arXiv preprint arXiv:1406.1078 (2014)
3. Eger, S., Daxenberger, J., Stab, C., Gurevych, I.: Cross-lingual argumentation mining: machine translation (and a bit of projection) is all you need! arXiv preprint arXiv:1807.08998 (2018)
4. Elman, J.L.: Finding structure in time. Cogn. Sci. **14**(2), 179–211 (1990)
5. Fleiss, J.L.: Measuring nominal scale agreement among many raters. Psychol. Bull. **76**(5), 378 (1971)

6. Hochreiter, S., Schmidhuber, J.: Long short-term memory. Neural Comput. **9**(8), 1735–1780 (1997)
7. Ito, T.: Towards agent-based large-scale decision support system: the effect of facilitator. In: Proceedings of the 51st Hawaii International Conference on System Sciences (2018)
8. Ito, T., Imi, Y., Ito, T., Hideshima, E.: COLLAGREE: a faciliator-mediated large-scale consensus support system. In: Collective Intelligence 2014 (2014)
9. Ito, T., Imi, Y., Sato, M., Ito, T., Hideshima, E.: Incentive mechanism for managing large-scale internet-based discussions on COLLAGREE. In: Collective Intelligence 2015 (2015)
10. Kim, Y.: Convolutional neural networks for sentence classification. arXiv preprint arXiv:1408.5882 (2014)
11. Kudo, T., Yamamoto, K., Matsumoto, Y.: Applying conditional random fields to Japanese morphological analysis. In: Proceedings of the 2004 Conference on Empirical Methods in Natural Language Processing (2004)
12. Kunz, W., Rittel, H.W.: Issues as elements of information systems, vol. 131. Citeseer (1970)
13. Landis, J.R., Koch, G.G.: The measurement of observer agreement for categorical data. Biometrics **33**, 159–174 (1977)
14. Morio, G., Fujita, K.: End-to-end argument mining for discussion threads based on parallel constrained pointer architecture. In: Proceedings of the 5th Workshop on Argument Mining, pp. 11–21 (2018)
15. Nguyen, H., Litman, D.: Context-aware argumentative relation mining. In: Proceedings of the 54th Annual Meeting of the Association for Computational Linguistics (Volume 1: Long Papers), vol. 1, pp. 1127–1137 (2016)
16. Palau, R.M., Moens, M.F.: Argumentation mining: the detection, classification and structure of arguments in text. In: Proceedings of the 12th International Conference on Artificial Intelligence and Law, pp. 98–107. ACM (2009)
17. Peldszus, A., Stede, M.: An annotated corpus of argumentative microtexts. In: Proceedings of the First Conference on Argumentation, Lisbon, Portugal (June 2015, to appear)
18. Peldszus, A., Stede, M.: Joint prediction in MST-style discourse parsing for argumentation mining. In: Proceedings of the 2015 Conference on Empirical Methods in Natural Language Processing, pp. 938–948 (2015)
19. Schuster, M., Paliwal, K.K.: Bidirectional recurrent neural networks. IEEE Trans. Sig. Process. **45**(11), 2673–2681 (1997)
20. Stab, C., Gurevych, I.: Identifying argumentative discourse structures in persuasive essays. In: Proceedings of the 2014 Conference on Empirical Methods in Natural Language Processing (EMNLP), pp. 46–56 (2014)
21. Stab, C., Gurevych, I.: Parsing argumentation structures in persuasive essays. Comput. Linguist. **43**(3), 619–659 (2017)
22. Stone, M.: Cross-validatory choice and assessment of statistical predictions. J. Roy. Stat. Soc. Ser. B (Methodol.) **36**, 111–147 (1974)
23. Wang, X., Liu, Y., Chengjie, S., Wang, B., Wang, X.: Predicting polarities of tweets by composing word embeddings with long short-term memory. In: Proceedings of the 53rd Annual Meeting of the Association for Computational Linguistics and the 7th International Joint Conference on Natural Language Processing (Volume 1: Long Papers), vol. 1, pp. 1343–1353 (2015)

Variables Extraction in Natural (English) Language Through Possessive Relationships

Danilo Eidy Miura(✉), Yukio Ohsawa, and Teruaki Hayashi

The University of Tokyo, Hongo 7-3-1, Bunkyo, Tokyo 113-8656, Japan
daneidyucb@gmail.com

Abstract. This is an extension from a selected paper from JSAI2019. The already highlighted importance of the 'flow' of data in the Market of Data brings needs of development of ways to better explore the utilization of data. Aware of the existence of rich knowledge stored and shared in text format, this paper aims to propose a method of representation of variable names that can be identified in natural language written knowledge. With the use of possessive relationships between words in Noun Phrases, we supported the representation of variable name relating a variable to a thing or event. A simple experiment was performed to demonstrate the efficacy of the proposed representation supported by Data Jacket Store, where we can find well-form variable names under the name of Variable Labels.

Keywords: Market of Data · Possessive Noun Phrases · Data Jacket

1 Introduction

The Chance Discovery in the Market of Data is a field of research that aims to design the flow of data in the society through creative methods and find hidden patterns. From the generation, through supply, processing, distribution to utilization of data, discoveries can enhance the demands and pull the consumption of data in the society. Therefore, the perception of the value of data through its utilization is the essential force to innovate in market. Although data is offered and advertised in online repositories, such as Data Jacket Store, potential users of data may not be aware of the potential utilization of the offered data. Knowledge of data analysis and processing is required to explore the applications of data.

In order to support data users or suppliers to assess the value of their data, the exploration of potential utilization is the basis for valuing what is not in use yet. To support users in the exploration of potential use of data, we aim to recognize potential use of data from repositories of knowledge recorded in texts, such as research papers and data analytics reports. In this paper we explain the formal representation of data as a well formed variable name (WFVN) that can be used to discover potential new variable labels to name data.

Y. Ohsawa et al. (Eds.): JSAI 2019, AISC 1128, pp. 162–169, 2020.
https://doi.org/10.1007/978-3-030-39878-1_15

In the field of Knowledge Engineering, the recognition of variables is an essential step to design the knowledge-based system. Implemented systems will contain well-formed variables and knowledge representation. But in natural language, the variables may be expressed in free style, enabling a direct codification to computer-readable language. In order to get the advantage of accumulated knowledge written in natural language, the variable identification and its formal representation may allow knowledge-based systems users to explore possibilities of data utilization.

2 Related Works

In previous works in knowledge representation [1,2], we learned that the representation of the knowledge depends on the intended task to perform, giving the limited capacity to codify the complete reality of the represented knowledge. Therefore, the representation of the knowledge should be limited and defined according to convenience to the given task. In the definition of the representation, the level of formalism of the representation may enable the use of the represented knowledge in different ways [3]. In this study, the functional approach to representation was adopted to enable further analysis of the represented knowledge, as defined in later section.

In this investigation, Predicate Logic will be used to formalize a representation of the variable data, Techniques of Named Entity Recognition will support mining linguistic features from text and a Data Catalog, Data Jacket Store, will be used to support the experiment to demonstrate the Variable Extraction from texts.

2.1 Variable Data and Predicate Logic

The aim of this study is to identify potential variable data that satisfy conditions for the integration of natural language and computer programing. In computer programing, a variable is a storage location paired with an associated symbolic name, which contains some known or unknown quantity of information referred to as a value. In order to satisfy this definition, the PNP should provide enough information to identify the association, the symbolic name and value. In this study, the value will not be included in the representation we will propose, for being unnecessary information in Data Jacket. In addition, in science and research, an attribute is a characteristic of an object. This definition helps us to define what feature is necessary to identify in PNP.

The predicate logic is a collection of formal systems used in mathematics, philosophy, linguistics, and computer science. And it is used to represent information and validations. Due to its formality, the logic of predicates [4] can be use to represent variable data in a way that can be computed and understood as natural language [5]. The representation of variable data here is proposed in the form of a predicate that states the relationship between objects and their attributes. In this system, we use symbols to denote relations or functions between valence (or arguments) of the statement.

Definition 1. *Here we propose the concept of objective possession and abstract possession. Objective possession is the possession of concrete entities, such as dog, desk, car, wallet and others. Abstract possession is the possession of abstractions which can be intrinsic or attributed to the possessor object, such as name, counts, frequency, measurement, and others. Using symbols to represent the variable data, we establish rules to define a possessive relationship in conditions to be a variable:*

$$hasvariable(x, y) \leftarrow has(x, y), attribute(y) \tag{1}$$

In addition, let us emphasize a variable characteristic should not be a name of an object. Concrete possession will describe non variable facts, or propositions, which cannot represent multiple possibilities of data but a single True or False statement.

2.2 Entity Recognition

In Natural Language Processing (NLP), Named Entity recognition and Classification (NERC) is a kind of task that identifies named entities according to predefined classes [6] (persons, locations, organizations, ...). In this study, the target entity is not a proper name (Named Entity) but an entity as flaccid designator that may not be a singular entity in all possible worlds, but distinct enough to designate an object or an event. For example, *water* is a designator of a specific substance that will not be confused with other liquids.

The use of textual features support the identification of entities. Similarly, markings of Possessive Noun Phrase [7] are used as feature to identify potential variable names in natural language. Example of variable name 'temperature of water', representing a possessive relation between an entity and its property.

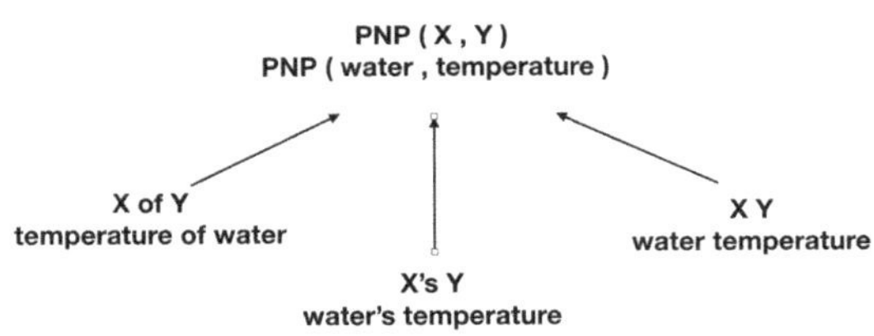

Fig. 1. PNP Extraction with linguistic feature

Figure 1 shows three different forms of PNP to be represented as predicate.

In Linguistics, Possessive Noun Phrases (PNP) are the expressions for possessive relationships possessor-possessed without verbal expression, such as "water *has* temperature". And different of possessive pronouns, possessor and possessed don't change its forms (my, your, his, her, its, our, their, ours, hers, theirs), keeping the noun name as original. With the use of textual features, such as markers and syntax , relationships between elements of the sentence can be identified and named. Possession relation may be marked with forms *'s* or preposition *of*.

2.3 Data Jacket

Data Jacket is a method for labeling data and provide relevant information for the communication between providers and users [8–12]. The method includes fields to describe the data in free style text, a list of variable names of the data and others. Data Jackets are currently stored in the Data Jacket Store [13, 14], a catalog of data using the method Data Jacket. This catalog contains more than 1,000 Data Jackets written in English language structured with RDF vocabulary and stores contents in natural language text.

Title of the dataset

Numbeo - Cost of Living of Countries

Outline of data

Numbeo provides current and timely information on world living conditions including cost of living. This data has indicators and data related to cost of living in many countries, including necessities and other daily costs.

Variable labels of dataset (the name of the variables)

Index of cost | price of food | price of service | basket of goods | basket of services | ranking of good by price | cost of living | comparison of cost of living of countries | historical data in a city by year

Fig. 2. Adapted example of Data Jacket

Figure 2 shows an example of Data Jacket.

As an informative label that describes a dataset, Data Jacket provides knowledge to represent the existence of data [16]. The description provides extended information about the data that provider wants to highlight, including context of the data, provider, common use, proposals and others. This field may provide PNPs that is not present in the field of variable names, which is the current content of the data. Instead, PNPs may suggest potential variables [15] related to the described data in Data Jacket.

3 Experiment

The purpose of this study is to propose a method to recognize potential variable data in natural English language, using the Possessive Noun Phrase. According to the definitions of PNP and variable data, a PNP provides partial necessary information to represent a variable data for computer programming. A simple conversion of PNP to variable name raises a problem to satisfy the conditions. The representation must represent object and attribute, but PNPs may include other compositions, such as "man's dog", representing object of object. A possession may also be an object which does not represent a variable data, but an object of the world possessed by another object.

To solve this problem, the following experiment describes a method to train the detector to distinguish the words that may represent a variable or an object in the possessive relation.

3.1 Procedures

Preparation of a List of Variable Names. From Data Jacket Store, 1,032 Data Jackets were collected and the total of 5,836 variable names, from which 391 presented possessive marking, extracted and prepared for the experiment.

The attributes and objects from 391 variable names with possessive marking were separated in distinguished lists. In total of 298 different terms were distributed accordingly to their appearance as an attribute or object.

The representativeness of each term to be an attribute was defined as the Naive Bayes probability of the term to be in the variable list in a range of 0 to 1, as follows:

Definition 2. *Let t be a term in the text and v be the class Variable. Representativeness of Attribute in a certain term is defined by the probability of the term be defined as attribute over possibility of being attribute or object*

$$P(v|t) = P(t|v) * P(v)/P(t) \tag{2}$$

The table below shows the distribution of terms according to their representativeness (Table 1).

Table 1. Distribution of terms according to representativeness

Representativeness of attribute	Frequency
1	99
between 0.5 and 1	12
between 0 and 0.5	7
Total	298

Extracting Possessive Noun Phrases. From the Data Jackets, the descriptions were used as source of potential variables, written in free style natural language. In total of 1379 different PNPs marked by *of* , were extracted to be tested according to the representativeness of the possession to be an attribute.

The table below shows the distribution of PNPs according to the representativeness of the possession to bee attribute (Table 2).

Table 2. Distribution of PNPs according to representativeness

Representativeness of attribute	Frequency
1	264
between 0.5 and 1	182
between 0 and 0.5	26
0	77
Unknown	830
Total	1379

The test extracted 264 possessions with the highest representativeness 1 from PNPs that do not exist in the list of variable labels in the Data Jacket Store. Also, 830 PNPs were not recognized by terms extracted from the Variable Names during the previous procedure.

The table below shows a few examples of potential variable names with the highest representativeness and example of existing variable names from Data Jacket Store (Table 3).

Table 3. Examples of potential variable labels and existing variables

Potential variables	Examples of existing variables
History of product	History of buyer, history of pregnancy, history of service
Degree of depression	Degree of discomfort, degree of damage, degree of employee
Area of implementation	Area of residence, area of building, area of target
Place of occurrence	Place of employment, place of contact, place of study, place of residence, place of injury
Type of designation	Type of measurement, type of instrument, type of residence, type of student, type of business
Type of data	Type of business, type of ceiling, type of gait, type of clothing, type of instrument
Amount of data	Amount of exercise, amount of drift, amount of fluid, amount of wages, amount of money
Type of cycling	Type of business, type of station, type of gait, type of student, type of clothing
Amount of cultivate	Amount of fluid, amount of guarantee, amount of wages, amount of exercise, amount of drift

See how the type of data for the term for the attribute can be analogically inferred from examples of variable names from Data Jackets. For example, the *history* of buyer is formatted as an activity log of a customer when buying goods, similarly, *history* of product can be formatted as log of sales. The same logic can be applied to other terms of attribute: type (predefined named classes), degree (predefined scales or measurements), area (2D measurements), amount (numerical quantification), place (named location), and so on.

4 Discussion and Future Work

The extraction of Potential Variables by matching the highest representativeness of attribute to possessions in PNPs show an analogical interpretation of the PNP as existing variables. Succeeding, the type of the existing data can be transferred to the extracted PNP, enabling computation of the variable as data. Data Jackets provides a coverage to certain qualitative attributes, such as *number* , *amount* , and qualitative attributes, such as *type* , *history* . The expectation of increasing variety Variable Names in Data Jackets will increase the coverage of potential Variable Names in free texts.

The presented method has limitations on the coverage of potential Variable Names in free texts that should solved in future work:

(1) Non-marked PNP. This method included only PNPs marked by *of* , which limits the coverage to limited number of PNPs, but keep focus on the possessive relations. Non-marked PNPs can also be found with the pattern Noun Noun, such as *purchase history* or *water temperature* , but the pattern Noun Noun includes many other relations rather than possession, which requires more cautious extraction.

(2) Lack of definition of the type of attribute. The attributes in Variable Names are considered as is. Some of the terms represent ambiguous type of data, such as *amount* as volume or counter, and *location* as coordinates or name of a place. These attributes require some attention to properly define classes of attributes and respective types.

The experiment revealed a limitation on the predictability of the attribute. It was not possible to make a definition of variable name for the terms which has a representativeness between 0 and 1. Those terms have the possibility to be take role of attribute or object in the predication of PNPs. This may happen due to the level of abstraction of attributes objects may have. Such as the example *number of frequency of use of product* can show three levels of abstraction: the first *use* is the event of application of the product. The occurrence of the event can be counted being qualified by *frequency* . And the counters of the frequency of events can be explicitly expressed by *number* . Terms in PNPs may present different levels of abstractions and the intermediates may take role of both possession or possessor.

Acknowledgements. This study was partially based on results obtained from a project commissioned by the New Energy and Industrial Technology Development Organization (NEDO), and JSPS KAKENHI Grant Numbers JP16H01836. Also, we would like to thank Kyodo Printing Co., Ltd., Artificial Intelligence Research Promotion Foundation, and Quantum Leap Flagship Program of MEXT.

References

1. Studer, R., Benjamins, V.R., Fensel, D.: Knowledge engineering: principles and methods. Data Knowl. Eng. **25**(1–2), 161–197 (1998)
2. Davis, R., Shrobe, H., Szolovits, P.: What is a knowledge representation? AI Mag. **14**(1), 17–17 (1993)
3. Guarino, N.: Formal ontology, conceptual analysis and knowledge representation. Int. J. Hum.-Comput. Stud. **43**(5–6), 625–640 (1995)
4. Nosek, J.T., Roth, I.: A comparison of formal knowledge representation schemes as communication tools: predicate logic vs semantic network. Int. J. Man-Mach. Stud. **33**(2), 227–239 (1990)
5. Baral, C., Gelfond, M.: Logic programming and knowledge representation. J. Log. Program. **19**, 73–148 (1994)
6. Nadeau, D., Sekine, S.: A survey of named entity recognition and classification. Lingvisticae Investigationes **30**(1), 3–26 (2007)
7. Abney, S.P.: The English noun phrase in its sentential aspect (Doctoral dissertation, Massachusetts Institute of Technology) (1987)
8. Ohsawa, Y., Kido, H., Hayashi, T., Liu, C.: Data jackets for synthesizing values in the market of data. Procedia Comput. Sci. **22**, 709–716 (2013)
9. Ohsawa, Y., Liu, C., Hayashi, T., Kido, H.: Data jackets for externalizing use value of hidden datasets. Procedia Comput. Sci. **35**, 946–953 (2014). Chicago
10. Hayashi, T., Ohsawa, Y.: Comparison between utility expectation of public and private data in the market of data. Procedia Comput. Sci. **96**, 1267–1274 (2016)
11. Abe, A.: Curating and mining (big) data. In: 2013 IEEE 13th International Conference on Data Mining Workshops, pp. 664–671. IEEE, December 2013
12. Liu, C., Ohsawa, Y., Suda, Y.: Valuation of data through use-scenarios in innovators' marketplace on data jackets. In: 2013 IEEE 13th International Conference on Data Mining Workshops, pp. 694–701. IEEE, December 2013
13. Hayashi, T., Ohsawa, Y.: Knowledge structuring and reuse system design using RDF for creating a market of data. In: 2015 2nd International Conference on Signal Processing and Integrated Networks (SPIN), pp. 607–612. IEEE, February 2015
14. Hayashi, T., Ohsawa, Y.: Data jacket store: structuring knowledge of data utilization and retrieval system (2016)
15. Hayashi, T., Ohsawa, Y.: Matrix-based method for inferring variable labels using outlines of data in data jackets. In: Pacific-Asia Conference on Knowledge Discovery and Data Mining, pp. 696–707. Springer, Cham, May 2017
16. Hayashi, T., Ohsawa, Y.: VARIABLE QUEST: network visualization of variable labels unifying co-occurrence graphs. In: 2017 IEEE International Conference on Data Mining Workshops (ICDMW), pp. 577–583. IEEE, November 2017

GTransE: Generalizing Translation-Based Model on Uncertain Knowledge Graph Embedding

Natthawut Kertkeidkachorn[1(✉)], Xin Liu[1], and Ryutaro Ichise[1,2]

[1] National Institute of Advanced Industrial Science and Technology, Tokyo, Japan
{n.kertkeidkachorn,xin.liu}@aist.go.jp
[2] National Institute of Informatics, Tokyo, Japan
ichise@nii.ac.jp

Abstract. This is an extension from a selected paper from JSAI2019. Knowledge graphs are useful for many AI applications. Many recent studies have been focused on learning numerical representations of a knowledge graph in a low-dimensional vector space. Learning representations benefits the deep learning framework for encoding real-world knowledge. However, most of the studies do not consider uncertain knowledge graphs. Uncertain knowledge graphs, e.g., NELL, are valuable because they can express the likelihood of triples. In this study, we proposed a novel loss function for translation-based models, GTransE, to deal with uncertainty on knowledge graphs. Experimental results show that GTransE can robustly learn representations on uncertain knowledge graphs.

Keywords: Knowledge graph · Uncertainty · Knowledge representation

1 Introduction

A knowledge graph (KG) stores a set of triples. A triple (h, r, t), where h and t are entities and r is a relation directed from h to t, expresses a fact. For example, $(Tokyo,\ capital\_of,\ Japan)$ describes that Tokyo is the capital city of Japan. Currently, there are many available KGs, e.g. DBpedia, YAGO, and Freebase.

To deal with vast knowledge, automatic knowledge graph construction approaches [2,9] are introduced. Automated construction process often results in noisy KG containing inaccurate facts, whose validity can only be expressed by confidence. Consequently, we refer to such KG as an uncertain KG. An uncertain KG stores the facts with their confidence.

Learning representation of KG, known as KG embedding, aims to capture latent representations of entities and the relations. One of the popular models is the translation-based model. The translation-based models learn representations by computing a dissimilarity score of triples [1]. Nonetheless, the translation-based models are optimized by the margin loss function, where uncertainty is

Y. Ohsawa et al. (Eds.): JSAI 2019, AISC 1128, pp. 170–178, 2020.
https://doi.org/10.1007/978-3-030-39878-1_16

ignored. In uncertain KGs, reliability of a fact is represented by confidence. Ignoring such confidence may result in low-quality representations.

In this paper, we therefore introduce a novel approach for learning the representation of uncertain KGs by generalizing the translation-based model for knowledge graph embedding.

2 Problem Definition

In this section, we define the learning representation on uncertain KGs.

Definition 1. *An Uncertain KG is a directed graph denoted by* $G = (E, R, Q)$, *where* E, R, *and* Q *are the entity set, relation set, and fact set, respectively. A fact is represented by a quadruple* $q = (h, r, t, s)$, *where* $h, t \in E$, $r \in R$, *and* $s \in \mathbb{R}_{[0,1]}$. *It indicates that entities* h *and* t *are connected by a relation* r *with confident score* s. *A higher* s *means it is more likely* h *and* t *are connected by* r.

Definition 2. *Given an uncertain KG, the problem of learning representation is to learn numerical representations of an entity* $\mathbf{e} \in \mathbb{R}^K$ *for each* $e \in E$ *and a relation* $\mathbf{r} \in \mathbb{R}^K$ *for each* $r \in R$ *such that for each* $(h, r, t, s) \in Q$; $L_A(h, r, t) \propto s$, *where* L_A *is the loss function with respect to a learning algorithm* A., *i.e. the facts can be preserved in* $\mathbb{R}^K$ *while considering the score* s.

3 Related Work

KG embedding is classified into three models: tensor-factorization-based, neural-network-based and translation-based [5]. In this study, we mainly focused on a translation-based model.

The translation-based models learn embedding representations by considering the relation r from the head entity h to the tail entity t as distance or dissimilarity. They commonly use a margin-based loss function. Inspired by the skip-gram model [13], where surrounded words are used to learn the representation of the word, TransE [1] translates the triple (h, r, t) as $\mathbf{h} + \mathbf{r} = \mathbf{t}$. With this translation, it can capture the first-order rules. TransH [14] extends TransE by projecting entities on a hyperplane to a specific relation space in order to select embedding components for each relation. TransR [12] further generalizes TransH by using a linear transformation to map between relations and corresponding entities. TransD [7] improves on TransR by employing two vectors for representing an entity and a relation. TransG [15] addresses multiple relation semantics by introducing multiple vectors. TransA [8] introduce a proper way to find a suitable margin. pTransE [11] considers relation paths by taking the sum of relations on the path when computing the score. CKRL [16] improves TransE by computing triple scores on KG as the penalty when learning representations. TorusE [5] solves the regularization problem of TransE by embedding entities and relations into a torus space.

In an uncertain KG, the level of validity of a fact is expressed by a confidence s. In practice, we can omit the confidence of the facts and directly learn the embedding as the traditional methods. Nevertheless, without confidence as an indicator, noisy facts could degrade the quality of the embedding representations. So far, there is a study on uncertain knowledge graph embedding [3]. Nevertheless, the translation-based model for uncertain KGs has not been studied yet.

4 Generalizing Translation-Based Model

The translation-based model learns embeddings by computing a score representing the distance of a triple (h, r, t). In TransE, the main assumption is that $\mathbf{h} + \mathbf{r} = \mathbf{t}$, if $(h, r, t) \in T$, where T is a triple set in KG. To measure the fitness of an embedding, a scoring function $f(h, r, t)$ is computed. The main difference between the different translation-based model approaches is in how to compute the scoring function. In TransE, the scoring function f(h,r,t) is measured using the L-1 or L-2 norm of $\mathbf{h} + \mathbf{r} - \mathbf{t}$.

Because of its score function, TransE could learn a trivial solution, where all entity and relation embeddings are equal. To better learn the embeddings, negative examples are needed. KG usually stores only positive examples. To generate negative examples, TransE uses a negative sampling method that corrupts the head or tail entity of a positive triple by a random entity. The negative sampling method defines a set of negative triples as $T'_{(h,r,t)} = \{(h', r, t)|(h, r, t) \in T, h' \in E, h' \neq h\} \cup \{(h, r, t')|(h, r, t) \in T, t' \in E, t' \neq t\}$.

Like other translation-based models, TransE uses a margin-based loss function to learn the embedding representations. The loss function is as follows:

$$L = \sum_{(h,r,t)\in T} \sum_{(h',r,t')\in T'} [f(h, r, t) - f(h', r, t') + M]_+, \tag{1}$$

where $[x]_+$ is the positive part of x, $f(\cdot)$ is a score function, M is a margin, and (h', r, t') is a negative sample in T'. In TransE, the confidence s is not introduced to the method yet.

The generalized translation-based model (GTransE) improves the margin-based loss function in order to support an confidence on the quadruple q. As shown in Eq. 1, the margin-based loss function maximizes the discriminative margin between the positive examples and the negative examples such that $f(h', r, t')$ becomes larger with the margin M while $f(h, r, t)$ becomes smaller (ideally, 0). Intuitively, this means maximizing the distance between the positive entity and the negative entity of the head or tail entities with respect to the margin M. Suppose that the L-1 norm, $f(h, r, t) = |\mathbf{h} + \boldsymbol{r} - \mathbf{t}|$, is the score function. Then, the condition, $|\mathbf{e} - \mathbf{e}'| \approx M$ for each $(e, e') \in \{(h, h'), (t, t')\}$ and $e \in E$, holds on the margin-based loss function, as is verified in [6].

However, an uncertain KG contains confidence of quadruples. Due to the confidence, the influence of each quadruple is not uniform when learning the embeddings. Consequently, the global margin might not be able to capture the quadruple's influence. To capture the confidence of the quadruples, the margin M should be varied. On the usual KG, a margin M is set as a global constant

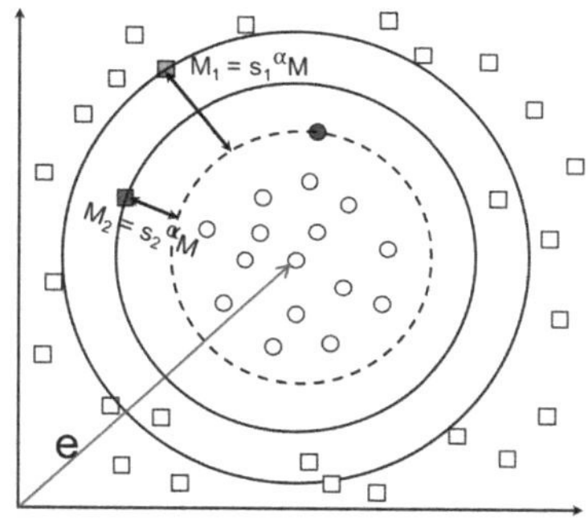

Fig. 1. The illustration of the entity embedding space with the different margin due to the uncertainty of quadruple. Circles and rectangles are positive and negative entities.

because all triples have the same confidence ($s = 1.0$). In an uncertain KG, each quadruple q denoted by (h, r, t, s) has its own confidence s. The condition $|\mathbf{e} - \mathbf{e}'| \approx M$ should be varied in accordance with the uncertainty of the quadruple s. Intuitively, the higher the uncertainty of the quadruple, the less margin should be used to keep the relation because (e, e') is likely to be noise. On each quadruple q, M should be then deducted when the uncertainty gets higher. This intuition behind is that if (h, r, t, s_i), (h, r, t, s_j) and $s_i \leq s_j$, then $s_i \cdot |\mathbf{h} - \mathbf{h}'| \leq s_j \cdot |\mathbf{h} - \mathbf{h}'|$ and $s_i \cdot |\mathbf{t} - \mathbf{t}'| \leq s_j \cdot |\mathbf{t} - \mathbf{t}'|$. The interpretation of confidence margins is motivated by the large and optimal margin [6,8]. In Fig. 1, we should maximize the larger margin of the high confidence quadruple (the green rectangle) rather than the low confidence quadruple (the red rectangle) because it is more likely that $|\mathbf{e} - \mathbf{e}'| \approx M$ holds on the green one. Therefore, the confidence margin-based loss function is derived as follows.

$$L = \sum_{(h,r,t,s) \in Q} \sum_{(h',r,t',s) \in Q'} [f(h, r, t) - f(h', r, t') + s^{\alpha} M]_{+}, \tag{2}$$

where $[x]_{+}$ is the positive part of x, $f(\cdot)$ is the score function for (h, r, t) of the quadruple q, M is the margin, $(h', r, t', 1.0)$ is a negative sample in Q' generated in the same way as T', s is the confidence of the quadruple, and α is a hyperparameter for adjusting the influence of confidence on the quadruple $\alpha \in \{x \in \mathbb{R} : x \geq 0\}$. As α becomes larger, the influence of confidence is amplified. When α equals 0, the influence of confidence is neutralized, and becomes the same as in the traditional translation-based model. Therefore, we can consider our loss function to be a generalization of the margin-based loss function.

5 Experiments and Results

To evaluate our approach, we conduct knowledge graph completion on uncertain KGs. Given an incomplete uncertain KG G, the task is to fill in G by predicting the set of missing quadruples $Q' = \{(h, r, t, \cdot) \mid h, t \in E, r \in R, (h, r, t, \cdot) \notin Q\}$. Note that it is difficult to measure confidence of quadruples. We thus decided not to try to predict the confidence of quadruples in this experiment.

5.1 Dataset

We constructed two new datasets: a synthetic dataset and a real-world dataset.

In the synthetic dataset, we derived an uncertain KG from FB15K-237. We selected FB15K-237, because it has no redundancy and has corrected the flaws of FB15K as reported in [4] and is more practical than WN18RR. We first sample a ρ proportion size triples and form a set T_c, where $T_c \subset K$ and $|T_c| = \rho|K|$. For each triple t; $t \in T_c$, we randomly generate a confidence score $s \sim uniform(0,1)$ and form the quadruple (h, r, t, s); for the other triples t; $t \in T - T_c$, we associate them with a confidence score $s = 1.0$; as $(h, r, t, 1.0)$. Then, we try to corrupt each fact with the probability p. Specifically, (1) For each quadruple, we randomly generate probability $p \sim uniform(0,1)$. (2) If $s < p$, we corrupt the triple by the negative sampling method, i.e., randomly replacing h, r, or t such that $h', t' \in E; h \neq h'; t \neq t'; r' \in R$ and $r \neq r'$ in order to create a quadruple (h', r', t', s); (3) If $s \geq p$, we leave the quadruple unchanged. Intuitively, the quadruple that has low confidence s are more likely to be corrupted, and the new created "negative" quadruple (h', r', t', s) has low confidence as our expectation. Statistically, $\int_0^1 p\,dp = 50\%$ of the quadruples in T_c will be corrupted. Note that, ρ plays a significant role in the uncertain KG because it indicates how uncertain KG is. As ρ increased, the likelihood, of which the uncertain KG contains noisy facts, increases. In uncertain KGs, we can regard ρ as the degree of uncertainty. We generate the KGs with the different ρ at 0.2, 0.4, 0.6, 0.8 and 1.0 to study uncertain KGs in various conditions.

In the real world dataset, we constructed the dataset from a real KG, NELL [2]. NELL provides a confidence score for each triple. To build our dataset, we first collected quadruples form NELL at the 995th iteration. Then, we followed the cleaning process described in [17]. However, we did not add the inverse relation to the dataset as was done in that study. Adding the inverse relation would have made the dataset redundant. Therefore, our dataset is more difficult than the dataset in the study [17].

The statistical details of the two datasets are as follows. FB15K-237 contains 14,541 entities, 237 relations, 272,115, 17,536, 20,467 training, validation, and testing quadruples, respectively, while NELL comprises 75491 entities, 200 relations, 134,213, 10,000, 10,000 training, validation, and testing quadruples.

5.2 Experimental Setup

The evaluation protocol was similar to that of TransE [1]. In this task, a head or a tail entity of test triples is replaced by each entity in the knowledge graph and the score of each replaced triple is computed. After the calculation, the scores of the triples are used to rank the entities in what is referred to as a *raw ranking*.

Table 1. Results of KG completion on generated uncertain KGs from FB15K-237 with $\rho = 0.2, 0.4, 0.6, .08$ and 1.0

Method	FB ($\rho = 0.2$)			FB ($\rho = 0.4$)			FB ($\rho = 0.6$)			FB ($\rho = 0.8$)			FB ($\rho = 1.0$)		
	% Hit@		MR	%Hit@		MR	%Hit@		MR	%Hit@		MR	%Hit@		MR
	1	10		1	10		1	10		1	10		1	10	
$\text{TransE}_{s\geq 0}$	12.6	37.5	0.21	9.7	34.9	0.18	7.4	31.6	0.15	5.1	28.4	0.13	3.5	24.5	0.10
$\text{TransE}_{s\geq 0.2}$	13.0	37.6	0.21	11.2	35.5	0.19	9.3	33.4	0.17	6.9	30.4	0.15	4.9	27.3	0.12
$\text{TransE}_{s\geq 0.4}$	13.8	38.0	0.22	12.0	35.7	0.20	10.7	34.1	0.18	8.7	31.8	0.16	7.0	29.0	0.14
$\text{TransE}_{s\geq 0.6}$	13.8	37.7	0.22	12.8	35.7	0.20	11.6	33.8	0.19	10.1	31.7	0.17	9.0	26.9	0.15
$\text{TransE}_{s\geq 0.8}$	14.0	38.0	0.22	12.9	35.3	0.20	11.9	33.1	0.19	10.6	30.7	0.17	8.8	25.5	0.14
$\text{TransE}_{s=1.0}$	14.0	37.9	0.22	12.7	34.9	0.20	11.6	32.2	0.18	9.3	27.5	0.15	1.0	10.1	0.04
$\text{GTransE}_{\alpha=1}$	14.3	38.9	0.22	13.1	37.4	0.21	12.0	36.1	0.20	9.8	34.3	0.18	9.2	30.3	0.16
$\text{GTransE}_{\alpha=2}$	**14.5**	39.1	**0.23**	**13.6**	37.7	0.21	12.6	**36.4**	0.20	10.8	34.7	0.19	10.5	**31.4**	0.17
$\text{GTransE}_{\alpha=3}$	14.2	41.1	**0.23**	13.5	37.8	**0.22**	**13.3**	35.8	**0.21**	**12.9**	**35.0**	**0.20**	**10.8**	31.3	**0.18**
$\text{GTransE}_{\alpha=4}$	14.2	**41.2**	**0.23**	13.1	**38.0**	0.21	13.2	35.8	**0.21**	12.8	34.9	**0.20**	**10.8**	30.8	**0.18**

Table 2. Results of KG completion on NELL dataset

Method	% Hit@		MR
	1	10	
$\text{TransE}_{s\geq 0.90}$	10.44	30.15	0.18
$\text{TransE}_{s\geq 0.95}$	9.88	29.05	0.17
$\text{TransE}_{s=1.00}$	4.39	15.20	0.08
$\text{GTransE}_{\alpha=1}$	11.11	30.47	0.18
$\text{GTransE}_{\alpha=2}$	12.08	31.22	**0.19**
$\text{GTransE}_{\alpha=3}$	12.20	31.49	**0.19**
$\text{GTransE}_{\alpha=4}$	**12.21**	**31.81**	**0.19**

However, the *raw ranking* is not a fair measure of performance, as reported in [1]. We therefore evaluated the performance by using a filtered ranking [1]. A filtered rank re-ranks *raw rank* by eliminating entities whose corresponding triples are contained in the training, validation, and test datasets. After obtaining the filtered ranking, we employed three evaluation metrics: Hit@1, Hit@10 and mean reciprocal rank (MR). Hit@1 and Hit@10 are the percentages of test triples whose entities are ranked in the top 1 and 10, respectively. MR averages the inverses of the ranks of the correct entities.

The baseline in the experiment was chosen to be TransE, because the objective of the experiment was to evaluate the performance of the confidence margin-based loss function. Although GTransE can use any score functions for the translation-based models, we limited the score function to L-1 norm of TransE, in order to focus on the performance of the loss function. However, using TransE in the traditional way is not fair, because GTransE utilizes the confidence of the quadruples to learn the representation. We thus decided to implement an intuitive way for TransE utilizing confidence s. The idea is that quadruples whose s is less than the threshold are eliminated from the training data. The experiment set six thresholds for TransE: $s \geq 0.0$, $s \geq 0.2$, $s \geq 0.4$, $s \geq 0.6$, $s \geq 0.8$, and $s = 1.0$. Note that we conducted the experiment by setting the threshold for the NELL dataset at $s \geq 0.90$, $s \geq 0.95$ and $s = 1.0$ because the range of confidence scores in NELL lies between 0.9 and 1.0.

The implementations of TransE and GTransE both used the grid search algorithm to find appropriate parameters. The dimension was selected from {20, 50, 100, 200}. The search range for the margin M was set at {1, 5, 10, 50, 100}. The learning rate was selected from {0.1, 0.001, 0.0001}. The best parameters were selected by the best MR on the validation set. We used the **Bern** method [14] to generate the negative samples. Normalization was used as the regularization. TransE and GTransE were optimized with the Adam algorithm [10]. All methods were trained in 2000 epochs. For GTransE, we set the α parameter to 1, 2, 3, and 4 to study the effect of varying α.

5.3 Results

The experimental results are presented in Tables 1 and 2. They show that GTransE outperforms TransE on all datasets.

As α increases, the margin of quadruples with low uncertainty (high confidence s) becomes larger for learning the representations. GTransE then maintains performance, even though ρ, representing the degree of uncertainty, increases when compared with all TransE, as shown in Table 1. This confirms the effect of the large margin [6]. Regarding the results of TransE in this table, we found that when ρ is low, filtering quadruples that have low confidence tends to yield better results. This is true because it makes more sense to remove them from the training than to keep them. However, a ρ increases, the portion of low-confidence quadruples drastically increases; consequently, too many quadruples are removed. As a result, TransE does not have enough relationships to learn the representation. This is why $\text{TransE}_{s=1.0}$ gives the worst results when $\rho = 1.0$.

The uncertainty is lower in the NELL dataset, than in the synthetic dataset. Accordingly, $\text{GTransE}_{\alpha=1}$ yields similar results as $\text{TransE}_{s=0}$ in Table 2. However, an improvement over TransE appears as α increases. The reason is that the confidence of quadruples is amplified. Such amplification makes the margin become relatively larger for high confidence facts. GTransE then can better adjust the margin for such facts. The results clearly show the effect of α.

In summary, removing low-confidence quadruples enables TransE to learn a comparatively good representation; however, removing too many quadruples becomes a double-edged sword. GTransE simply uses all training data and directly uses the confidence of quadruples to determine the margin when learning embedding representations. GTransE thus does not suffer when the amount of training data is small and it still captures the effect of uncertainty.

6 Conclusion

We introduced learning representation on uncertain knowledge graphs. To deal with the confidence score, we proposed a new margin loss function for translation-based models. The experiment shows that our approach is promising to capture confidence in knowledge graphs. In the future, we plan to investigate other score functions in the other translation-based models.

Acknowledgment. This work was partially supported by the New Energy and Industrial Technology Development Organization (NEDO).

References

1. Bordes, A., Usunier, N., Garcia-Duran, A., Weston, J., Yakhnenko, O.: Translating embeddings for modeling multi-relational data. In: Proceedings of Advances in Neural Information Processing Systems (NIPS), pp. 2787–2795 (2013)
2. Carlson, A., Betteridge, J., Kisiel, B., Settles, B., Hruschka Jr, E.R., Mitchell, T.M.: Toward an architecture for never-ending language learning. In: Proceedings of the AAAI Conference, pp. 1306–1313 (2010)
3. Chen, X., Chen, M., Shi, W., Sun, Y., Zaniolo, C.: Embedding uncertain knowledge graphs. In: Proceedings of the AAAI Conference, vol. 33, pp. 3363–3370 (2019)
4. Dettmers, T., Minervini, P., Stenetorp, P., Riedel, S.: Convolutional 2D knowledge graph embeddings. In: Proceedings of the AAAI Conference, pp. 1811–1818 (2018)
5. Ebisu, T., Ichise, R.: TorusE: knowledge graph embedding on a lie group. In: Proceedings of the AAAI Conference, pp. 1819–1826 (2018)
6. Fan, M., Zhou, Q., Zheng, T.F., Grishman, R.: Large margin nearest neighbor embedding for knowledge representation. In: 2015 IEEE/WIC/ACM International Conference on Web Intelligence and Intelligent Agent Technology (WI-IAT), vol. 1, pp. 53–59, December 2015
7. Ji, G., He, S., Xu, L., Liu, K., Zhao, J.: Knowledge graph embedding via dynamic mapping matrix. In: Proceedings of Annual Meeting of the Association for Computational Linguistics, pp. 687–696 (2015)
8. Jia, Y., Wang, Y., Lin, H., Jin, X., Cheng, X.: Locally adaptive translation for knowledge graph embedding. In: Proceedings of the AAAI Conference, pp. 992–998 (2016)
9. Kertkeidkachorn, N., Ichise, R.: An automatic knowledge graph creation framework from natural language text. IEICE Trans. Inf. Syst. **101**(1), 90–98 (2018)
10. Kingma, D., Adam, J.B.: A method for stochastic optimization. In: Proceedings of International Conference on Learning Representations (2015)
11. Lin, Y., Liu, Z., Luan, H., Sun, M., Rao Liu, S.: Modeling relation paths for representation learning of knowledge bases. In: Proceedings of Conference on Empirical Methods in Natural Language Processing, pp.705–714 (2015)
12. Lin, Y., Liu, Z., Sun, M., Liu, Y., Zhu, X.: Learning entity and relation embeddings for knowledge graph completion. In: Proceedings of the AAAI Conference, pp. 2181–2187 (2015)
13. Mikolov, T., Sutskever, I., Chen, K., Corrado, G.S., Dean, J.: Distributed representations of words and phrases and their compositionality. In: Proceedings of Advances in Neural Information Processing Systems (NIPS), pp. 3111–3119 (2013)

14. Wang, Z., Zhang, J., Feng, J., Chen, Z.: Knowledge graph embedding by translating on hyperplanes. In: Proceedings of the AAAI Conference, pp. 1112–1119 (2014)
15. Xiao, H., Huang, M., Zhu, X.: TransG: a generative model for knowledge graph embedding. In: Proceedings of Annual Meeting of the Association for Computational Linguistics, pp. 2316–2325 (2016)
16. Xie, R., Liu, Z., Lin, F., Lin, L.: Does William Shakespeare really write hamlet? Knowledge representation learning with confidence. In: Proceedings of the AAAI Conference (2018)
17. Xiong, W., Hoang, T., Wang, W.Y.: DeepPath: a reinforcement learning method for knowledge graph reasoning. In: Proceedings of Conference on Empirical Methods in Natural Language Processing, pp. 564–573 (2017)

Latent-Space Data Augmentation for Visually-Grounded Language Understanding

Aly Magassouba(✉), Komei Sugiura, and Hisashi Kawai

National Institute of Information and Communications Technology, 3 Chome-5 Hikaridai, Soraku District, Seika, Kyoto Prefecture 619-0237, Japan
aly.magassouba@nict.go.jp

Abstract. This is an extension from a selected paper from JSAI2019. In this paper, we study data augmentation for visually-grounded language understanding in the context of picking task. A typical picking task consists of predicting a target object specified by an ambiguous instruction ,e.g., "Pick up the yellow toy near the bottle". We specifically show that existing methods for understanding such an instruction can be improved by data augmentation. More explicitly, MCTM [1] and MTCM-GAN [2] show better results with data augmentation when specifically considering latent space features instead of raw features. Additionally our results show that latent-space data augmentation can improve better a network accuracy than regularization methods.

Keywords: Multimodal language understanding · Domestic service robots

1 Introduction

Endowing robots with natural communication ability is particularly valuable for domestic service robots (DSRs) [3] that are expected to interact with non-expert users. These robots are expected to provide support to elderly and disable people with functions such as object retrieval or placing solely.

Given this background, we address the context of fetching task, which is one of the most crucial manipulation tasks, from ambiguous instructions. This task consists of picking up a target object instructed by a user. However, understanding and grounding the fetching instruction is particularly complex because it does not follow any predefined rule: the information may be truncated, hidden, or expressed in a multitude of ways. The unpredictability and richness of language make this task difficult to solve for DSRs that are required to infer the user's intention.

Data-driven methods [4,5] aim to solve such tasks by combining visual and linguistic knowledge in multimodal language understanding models. Inspired by these approaches, we proposed a multimodal target source classifier (MTCM) in [6] that was extended with generative adversarial nets (MTCM-GAN) in [2].

Y. Ohsawa et al. (Eds.): JSAI 2019, AISC 1128, pp. 179–187, 2020.
https://doi.org/10.1007/978-3-030-39878-1_17

The current study focuses on data augmentation for the former two approaches. Sophisticated models that address multimodal language understanding (MLU) requires large-scale data. Unfortunately data collection and labelling in robotics is particularly tedious and time-consuming. Data augmentation provides a solution to overcome this limitation. This study introduces and evaluates several methods of data augmentation for MLU.

2 Related Work

Many robotic studies have addressed visually-grounded communication with robots (e.g., [1,5,7]). Like the authors of many studies in the field of robotics, we are interested in fetching tasks in daily-life environments. Recent studies have handled multimodal language understanding using multimodal similarity-based integration [4,8–10]. The approach proposed in [8] uses an LSTM to learn the probability of a referring expression, while a unified framework for referring expression generation and comprehension was proposed in [4]. A method handling unconstrained spoken language instructions with dialogues was proposed in [9] and [10]. In [2], we extended multimodal similarity-based integration with data augmentation by using a GAN classifier.

GAN-based approaches have been widely used to address classification problems [11–13]. The latter cited studies proposed to improve the classification task by exploiting the data augmentation property of a GAN. In our previous works [5,12], we introduced simultaneous data augmentation and classification to multimodal language understanding for manipulation tasks.

Alternatively, to overcome the lack of data, recent studies [14] have used simulated environments. Synthetic data are becoming more photo-realistic. Furthermore, robotic tasks such as grasping [15] or navigation [16] can be repeated in a wide range of environments and conditions at a low cost.

3 Problem Statement

Let us recall the background of multimodal language understanding for DSRs. We aim to understand instructions such as "Give me the pink doll near the tea pot on the desk". Considering environments in daily life, several grounding challenges arise regarding understanding instructions. In particular, users tend to use referring expressions to describe an object. For instance, the target object in the previous example "pink doll" is characterized by its color. and neighbor landmarks "tea pot". Similarly the source of the target object may be mentioned or not depending on the context. For instance, "Give me the pink doll" is a likely but ambiguous instruction. One or several of these grounding challenges may appear in a single instruction. To solve this problem and the related grounding challenges, we consider the following inputs and outputs for our system:

- **Inputs:** Linguistic instructions and an image as input
- **Output:** Most likely target object.

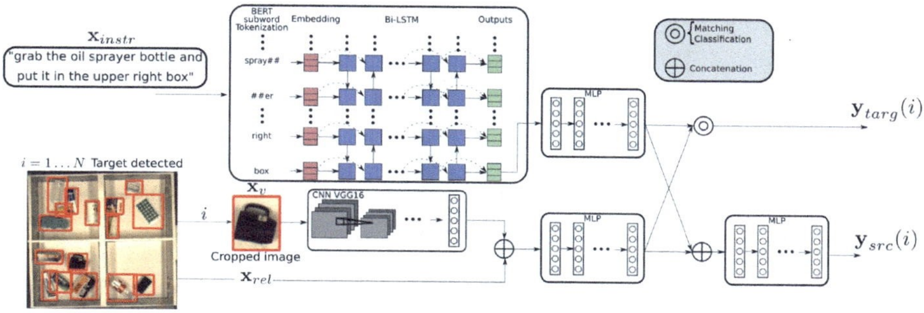

Fig. 1. Proposed method framework: the MTCM is based on a CNN-LSTM architecture to process linguistic and visual inputs. The model predicts the likelihood of a target object using binary classification, in addition to the source of the target object. For comparison, we also implement a matching function to directly predict the object, as performed in [9]

4 Multimodal Target-Source Classifier Model with Data Augmentation

Inspired by the latest advancements in image grounding [4], we proposed a multimodal target-source classifier model (MTCM) in [1] (see Fig. 1). This model combines a convolutional neural network (CNN) in addition to a long short-term memory (LSTM) network that process the visual and linguistic inputs, respectively. The set of inputs of the model is $\{\mathbf{x}_{instr}, \mathbf{x}_v, \mathbf{x}_{rel}\}$, where $\mathbf{x}_v$ denotes the visual inputs, $\mathbf{x}_{instr}$ denotes the linguistic inputs, and $\mathbf{x}_{rel}$ denotes the relational feature inputs. Input $\mathbf{x}_{rel}$ denotes the relational feature between the target object and the environment, that is, the position in the scene, position within the source, and position with respect to neighboring objects.

Visual inputs $\mathbf{x}_v$ correspond more explicitly to the cropped image of target object $\mathbf{y}$. A CNN is used to process image $\mathbf{x}_v$. In our approach, we consider the 16-layer network VGG16 [17] to encode each image. The output of the fully connected layer (FC7) is used to extract visual features.

By contrast, the linguistic features are embedded and then encoded by a multi-layer bidirectional LSTM (Bi-LSTM) network. Instead of directly training an embedding model from scratch, we use a pre-trained sub-word embedding model, BERT [18], to initialize the embedding vectors. The concatenation of the last hidden layers of the forward layer and backward layer of the Bi-LSTM is extracted to encode the linguistic inputs. After encoding the visual, relational, and linguistic inputs, a common latent representation is required to compare the extracted features from the CNN and LSTM. Two multi-layer perceptrons (MLPs) are used for this purpose. In parallel, an MLP is used to predict the source of the target object based on the output of the linguistic and visual MLPs.

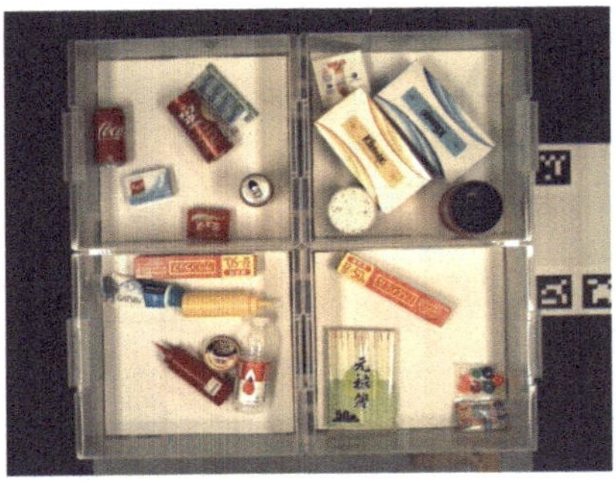

(a) Original

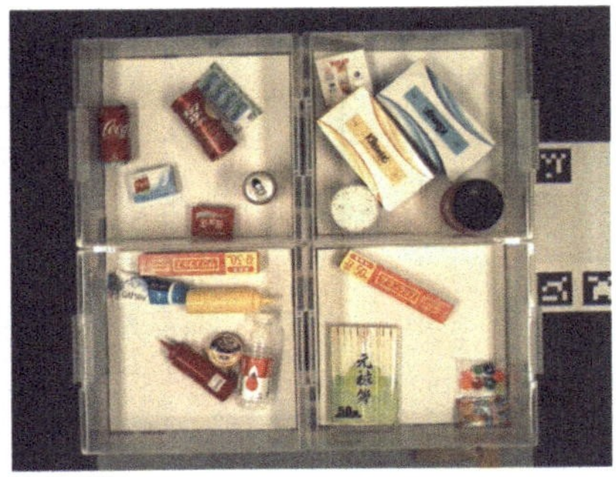

(b) Gaussian noise

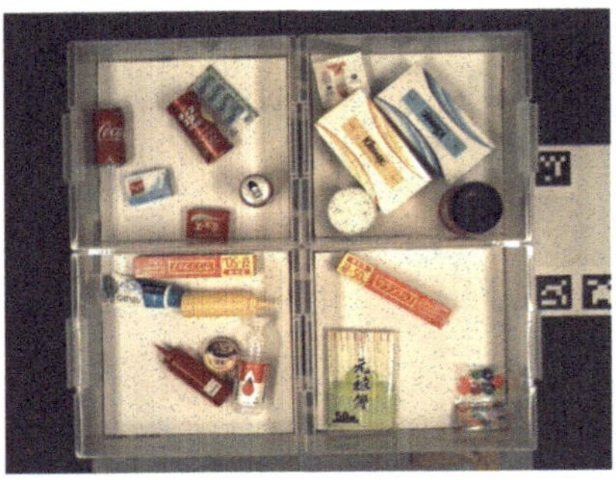

(c) Salt-pepper noise

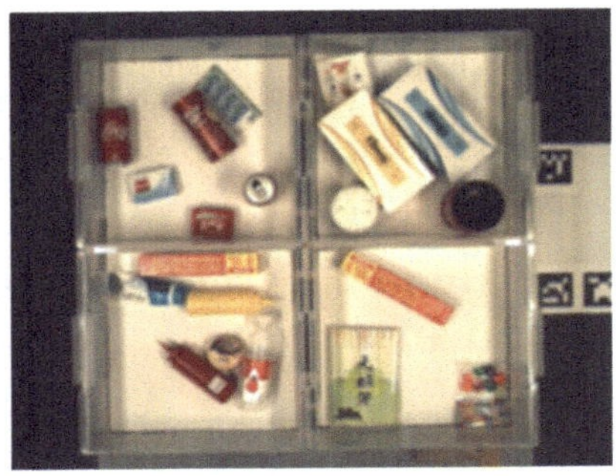

(d) Blurring

Fig. 2. Samples of augmented visual inputs in the raw space

Finally, the output of the model is given by $\mathbf{y}_R = \{\mathbf{y}_{targ}, \mathbf{y}_{src}\}$, where y_{targ} is the likelihood of the target object and y_{src} is the predicted class of the target source

In the case of the binary classification, the prediction task is solved by minimizing a cross-entropy function so that J is

$$J(\mathbf{y}) = -\sum_{n}\sum_{j} y^*_{nj} \log p(y_{nj}), \tag{1}$$

where y^*_{nj} denotes the label of the j-th dimension of the n-th sample. The loss function J_M of the network is then given by:

$$J_M = \lambda_1 J_{targ} + \lambda_2 J_{src} \tag{2}$$

where $J_{src} = J(\mathbf{y}_{src})$ and target $J_{targ} = J(\mathbf{y}_{targ})$ from (1), while λ_1 and λ_2 are some weighting parameters. On the other hand, a Hinge loss function is used for J_{targ} when the tasks consists in matching the most likely object with the initial instruction. This loss consists in increasing the similarity between correct pairs of linguistic and visual/relational features and the dissimilarity between incorrect pairs. With s_i as an instruction and y_i a target object, the cost function J_{targ} becomes:

$$J_{targ} = \sum_n \max(0, M + f(s_n, y_m) - f(s_n, y_n)) + \max((0, M + f(s_k, y_n) - f(s_n, y_n)), \quad (3)$$

where M is the margin, and $f(.)$ is the similarity function (*e.g.* cosine similarity). The incorrect target object (y_m) and sentences (s_k) are randomly sampled from the same image as the real target object.

Two types of data augmentation are considered in this study: data augmentation on the raw and latent spaces. In the first configuration the input images $\mathbf{x}_v$ are augmented with well-know image processing methods as illustrated on Fig. 2. We considered mainly additive Gaussian (GN), Salt-Pepper (SP) and Blurring noises. The effects are observed with similarity-based as well as binary prediction as described above.

In the second configuration, the latent features are augmented instead. Such a method was first introduced in MTCM-GAN [2] where a GAN-based simultaneous classification and data augmentation is performed. In this approach, a generator network is used to mimic the distribution of the actual latent space features. A classification is then performed on both the actual and generated latent features. Considering the MTCM given in Fig. 1, the latent space features are extracted from the MLPs that encode the visual and linguistic features. In this study we extend the latent space data augmentation with Gaussian and salt-pepper noises.

Table 1. Mean validation top-1 and binary accuracy on the PFNPIC data set considering different MLU methods and data augmentation approaches. These results are based on five trials.

Augmentation Space	Method	Accuracy	
		Top-1	Binary
–	Baseline (Hatori et al. [9])	88.0	–
–	MTCM [1]	88.8	94.5
Raw	MTCM + Gaussian Noise	88.0	93.2
	MTCM + Salt-Pepper	88.2	94.0
	MTCM + Blurring	88.0	93.5
Latent	MTCM-GAN [2]	–	95.7
	MTCM-WGAN	–	95.9
	MTCM-GAN + Gaussian Noise	–	96.1
	MTCM-GAN + Salt-Pepper	–	96.1
	MTCM-GAN + Spectral Normalization	–	95.9

5 Experiments

To study the effect of data augmentation in a real-word scenario, we evaluated several MTCM and MTCM-GAN on the PFN-PIC dataset [9]. We used the same dataset as that in their original paper, with 89,861 sentences and 25,517 bounding boxes in the training set, and 898 sentences and 352 bounding boxes in the validation set. In each image, target objects were placed randomly in four boxes (see Fig. 2). These boxes were the target sources.

For the linguistic processing, each sub-word was first encoded as a 1,024-sized vector using BERT. We used the largest version of pre-trained BERT (24 layers) considering uncased words. The embedded vectors were input into a three-layer Bi-LSTM, with 1,024-sized cells. The last hidden state of the Bi-LSTM was eventually extracted. In parallel, the images were processed in a CNN. We used a VGG16 pre-trained model and extracted the output of the seventh fully connected (FC7) layer. Both linguistic and visual/relational features were transformed by three-layer MLPs with dimension $d = 1024$. We applied batch normalization and the ReLU activation function for each layer of the MLP. The third MLP that predicted the source also had three layers and dimension $d = 2048$. Similar to the previous MLPs, the ReLU activation function was used, except for the last layer, which used a softmax function for the prediction. Finally, the network was trained using the Adam optimizer with an initial learning rate of $2e^{-4}$. The weighting parameters of the loss function were set to $\lambda_1 = 1$ and $\lambda_2 = 0.7$

The results of our experiments are reported in Table 1. In the first column, the models are evaluated on a top-1 accuracy basis while the second column refers to the binary accuracy. Given $y \in \{0, 1\}$, the binary accuracy is defined as:

$$\sum \frac{N_{\hat{y}=y}}{|X_{pairs}|}, \tag{4}$$

where $|X_{pairs}|$ characterizes the potential pairs of instructions/target object with

$$\begin{aligned} X_{pairs} = &\{x(i,j,k) | i = 1, ..., |S_{img}|, \\ &j = 1, ..., |S_{target}|, k = 1, ..., |S'|\} \end{aligned} \tag{5}$$

In (5), a pair $x(i, j, k)$ depends on the observable scene or image i among the set of images S_{img}, a target object j among the set S_{target} of target candidates in the image, and an instruction k among the set of instructions $S^{'}$ related to the objects. It should be emphasized that the set $S^{'}$ is defined as the concatenation of correct S_+ and incorrect sentences S_- with respect to a given target object so that $S' = S_+ \cup S_-$.

In the first rows, we noticed that data augmentation on the raw space did not improve both the top-1 accuracy and binary accuracy. Although the results are above the baseline method [9], they remained below the MTCM accuracy without additional noise. In particular, the Gaussian noise drastically decreased the accuracy by 0.8% for top-1 and by 1.3% for binary accuracy.

In the configuration of latent space data augmentation, MTCM was extended with MTCM-GAN. As reported in [2] latent space data augmentation increased the accuracy by 1.2%. Similarly to the previous configuration we analyze the Gaussian, Salt-pepper noise. In addition, regularization methods that are Wasserstein GAN [19] and spectral normalization were also studied. The results showed that data augmentation with Gaussian noise or salt-pepper noise yielded the best results in addition to MTCM-GAN. Regularization methods such as WGAN and spectral normalization [20] also improved the binary accuracy but remained latent-space data augmentation with Gaussian or salt-pepper noise.

These results emphasize that latent-space data augmentation is more efficient than raw space data augmentation. In particular, while Gaussian and salt-pepper noises decrease the accuracy in the raw space, they have a beneficial effect in the latent space.

Additionally, the qualitative results of MCTM-GAN are shown in Fig. 3, which illustrates typical binary predictions. In the two samples, the likelihood of each object in the image was reported given the initial instruction: overall accurate results were obtained.

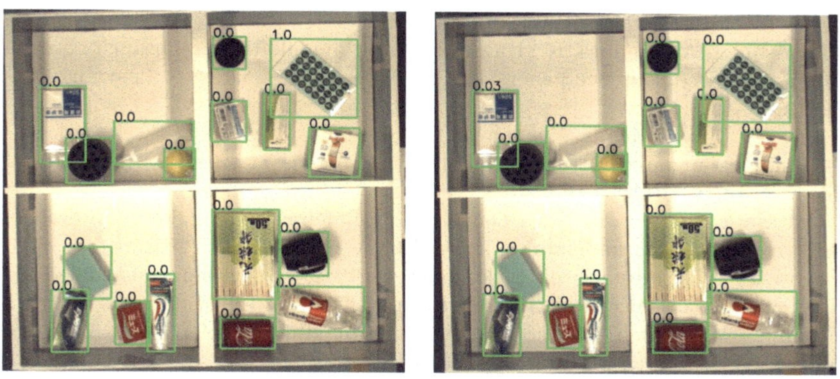

(a) Pick the greenish numbered sticker and put it into lower left box

(b) Pick up the white tube of toothpaste and drop it into the top right box

Fig. 3. Binary prediction samples of MTCM-GAN

6 Conclusion

We evaluated several methods of data augmentation for visual grounding models, which can predict the likelihood of target objects and their respective source given ambiguous instructions for a picking task. Our results showed that it is more beneficial to augment data in the latent space than in the raw space. Indeed, by augmenting data in the raw space we noticed a decrease in performance by 1.3% while an increase in performance by 1.6% was observed in the latent space.

Acknowledgement. This work was partially supported by JST CREST, SCOPE and NEDO.

References

1. Magassouba, A., Sugiura, K., Kawai, H.: A multimodal target-source classifier model for object fetching from natural language instructions. In: Proceedings of the National Congress of the Japanese Society for Artificial Intelligence, pp. 2D3E403–2D3E403 (2019). (in Japanese)
2. Magassouba, A., Sugiura, K., Trinh Quoc, A., Kawai, H.: Understanding natural language instructions for fetching daily objects using GAN-based multimodal target-source classification. IEEE RA-L **4**(4), 3884–3891 (2019)
3. Iocchi, L., Holz, D., Ruiz-del Solar, J., Sugiura, K., Van Der Zant, T.: RoboCup@Home: analysis and results of evolving competitions for domestic and service robots. Artif. Intell. **229**, 258–281 (2015)
4. Yu, L., Tan, H., Bansal, M., Berg, T.L.: A joint speaker listener-reinforcer model for referring expressions. In: CVPR, vol. 2 (2017)
5. Magassouba, A., Sugiura, K., Kawai, H.: A multimodal classifier generative adversarial network for carry and place tasks from ambiguous language instructions. IEEE RA-L **3**(4), 3113–3120 (2018)
6. Magassouba, A., Sugiura, K., Kawai, H.: Multimodal attention branch network for perspective-free sentence generation. In: Conference on Robot Learning (CoRL) (2019)
7. Cohen, V., Burchfiel, B., Nguyen, T., Gopalan, N., Tellex, S., Konidaris, G.: Grounding language attributes to objects using bayesian eigenobjects. In: Proceedings IEEE IROS 2019 (2019)
8. Nagaraja,V.K., Morariu, V.I., Davis, L.S.: Modeling context between objects for referring expression understanding. In: ECCV, pp. 792–807 (2016)
9. Hatori, J., et al.: Interactively picking real-world objects with unconstrained spoken language instructions. In: IEEE ICRA, pp. 3774–3781 (2018)
10. Shridhar, M., Hsu, D.: Interactive visual grounding of referring expressions for human-robot interaction. In: RSS (2018)
11. Springenberg, J.T.: Unsupervised and semi-supervised learning with categorical generative adversarial networks. In: Proceedings ICLR 2015 (2015)
12. Sugiura, K., Kawai, H.: Grounded language understanding for manipulation instructions using GAN-based classification. In: IEEE ASRU (2017)
13. Odena, A., Olah, C., Shlens, J.: Conditional Image Synthesis with Auxiliary Classifier GANs. In: ICML, pp. 2642–2651 (2017)
14. Anderson, P., Wu, Q., Teney, D., Bruce, J., Johnson, M., Sünderhauf, N., Reid, I., Gould, S., van den Hengel, A.: Vision-and-language navigation: Interpreting visually-grounded navigation instructions in real environments. In: ECCV, pp. 3674–3683 (2018)
15. Bousmalis, K., et al.: Using simulation and domain adaptation to improve efficiency of deep robotic grasping. In: Proceedings of IEEE ICRA, pp. 4243–4250 (2018)
16. Tan, J., Zhang, T., Coumans, E., Iscen, A., Bai, Y., Hafner, D., Bohez, S., Vanhoucke, V.: Sim-to-real: learning agile locomotion for quadruped robots. In: RSS (2018)
17. Simonyan, K., Zisserman, A.: Very deep convolutional networks for large-scale image recognition. In: Proceedings ICLR 2015 (2014)

18. Devlin, J., Chang, M.-W., Lee, K., Toutanova, K.: BERT: Pre-training of deep bidirectional transformers for language understanding. In: ACL, pp. 4171–4186 (2019)
19. Gulrajani, I., Ahmed, F., Arjovsky, M., Dumoulin, V., Courville, C.: Improved training of wasserstein GANs. In: NIPS, pp. 5769–5779 (2017)
20. Miyato, T., Kataoka, T., Koyama, M., Yoshida, Y.: Spectral normalization for generative adversarial networks. In: ICLR (2018)

Machine Learning and Data Mining

A Community Sensing Approach for User Identity Linkage

Zexuan Wang(✉), Teruaki Hayashi, and Yukio Ohsawa

The University of Tokyo, 7-3-1 Hongo, Bunkyo-ku, Tokyo, Japan
wangzexuan@g.ecc.u-tokyo.ac.jp

Abstract. This is an extension from a selected paper from JSAI2019. User Identity Linkage (UIL) aims to detect the same individual or entity across different Online Social Networks, which is a crucial step for information diffusion among isolated networks and information transfer between different domains. While many pair-wise user linking methods have been proposed on this important topic, the community information naturally exists in the network is often discarded during process. In this paper, we proposed a novel embedding-based approach that considers and utilizes both individual similarity and community similarity by jointly optimizing them in a single loss function. Experiments conducted on real datasets obtained from Foursquare and Twitter illustrate that the proposed method outperforms other commonly used baselines in UIL, which only consider the individual similarity between users or entities.

Keywords: User identity linkage · Clustering · Multi-task learning

1 Introduction

In recent years, Online Social Networks (OSNs) such as Twitter, Facebook and Foursquare tend to become the central platform of people's social life. Tons of contextual (e.g. tweets, photos) and network structure related (e.g. users' profiles, relations) data is created every day on these OSNs, which is an important resource for many valuable applications such as user behavior prediction, community analysis, and cross-domain recommendation.

While the invention of World Wide Web (WWW) aims to link information together and make them flow freely to most part of the world, today's networks are often prevented from interacting with each other and therefore formed many isolated islands of information. Based on this circumstance, all such applications mentioned above require a crucial step called User Identity Linkage (UIL) [13], which aims to utilize the open information revealed in data to find the opportunity to transfer information as well as knowledge across these different OSNs to identify and link the same person or entity, then finally break this isolated situation. These linkages are also called Anchor Links as they help align different networks under the common scenario that users usually don't explicitly claim the

Y. Ohsawa et al. (Eds.): JSAI 2019, AISC 1128, pp. 191–202, 2020.
https://doi.org/10.1007/978-3-030-39878-1_18

ownership of their multiple accounts in different networks, and due to privacy protection rules, personal information is always restricted inside each isolated OSNs. The requirements on both discovering new knowledge and respecting privacy policies have driven us to take more advantage of the hidden information existing in the original network and its community structures.

The rest of this paper is organized as following: Sect. 2 reviews the related work of this study, followed by Sect. 3 where we discuss formalized problem statements and fundamental definitions. In Sect. 4, we present the overview and technically detail of our approach on UIL problem. We outline our experimental setting, evaluation metrics and their results in Sect. 5. Section 6 concludes the paper and outlines some possible future directions.

2 Related Work

Abundant literature has been focusing on the UIL problem, and the majority of them may fall into following two categories:

2.1 Structure-Based Approaches

These approaches focus directly on the structural features of a social network, for example, user names, following relationship and common neighbors between different users [2,4], while the problem of those approaches lies on the difficulty to find an optimal distance function between nodes to evaluate their similarity based on structural features as networks are not presented in the Euclidean space [14].

2.2 Embedding-Based Approaches

Network embedding [1,7,11] is a new way of network representation that is able to encode the network in a continuous low-dimensional vector space while effectively preserving the network structure, for example, [17] proposed a dual-learning embedding paradigm to learn latent semantics of both user activities and network structure to improve the linking result.

2.3 Multi-task Learning

In recent years, many machine learning techniques require a large amount of data to achieve a model with high accuracy, however, in some real-world applications such as medical image analysis or high precision physical experiments, enough data samples in good quality are often difficult to collect [16]. Meanwhile, to solve problems in a complex system, it requires cross-discipline cooperation instead of limited knowledge from a single domain. Multi-Task Learning (MTL) is inspired by human learning activities when people often tend to apply the knowledge obtained from previous tasks to help with working on a new but related task. MTL has been used successfully in many applications including computer vision [10] and drug discovery [9]. For our study, this idea is adopted by sharing information (representation, similarity of user nodes in OSNs) between multiple related tasks, knowledge from a different domain could be used to enhance the non-optimal model of a single task due to the lack of data samples or external knowledge.

While existing methods on UIL haven't paid enough attention to the social communities naturally formed by people in the real world. For example, users who have limited profile information could be evaluated easier when they are located in interest groups together with their close neighbors because they may share many same labels. To better resolve the UIL problem, we proposed a novel approach called Community Sensing User Identity Linkage (CSUIL), which takes advantage of both structural and embedded features of a network by designing a jointly learning model. It aids user mapping process by not only driving given user node closer to its corresponding pair-wise target but also increasing the similarity of the same community they belongs to, which enhances the method's accuracy and generalization ability. Experiment results on real-world dataset also support the feasibility of our method.

3 Problem Definition

To prepare for the fundamental concepts in graph mining, node embedding, user identity linkage and clarify the formal usage of symbols, following definitions are stated in advance.

Definition 1 (Social Network Graph). *An unweighted and undirected network is denoted as $G = \{V, E\}$, where V is the set of nodes and each node represents a user, E is the set of edges reflecting connections between nodes(i.e., the friendship relationship).*

Definition 2 (Node Embedding). *In a given network $G = \{V, E\}$, node embedding (a sub-task of network embedding) learns a projection function $\psi : V \mapsto \mathbb{R}^{|V| \times d}$, where $d \ll |V|$. For each node $v_i \in V$, $\psi(v_i) \in \mathbb{R}^d$ denotes its latent representation in the vector space.*

Definition 3 (n-th Order Neighbors). *The collection of all nodes which can be reached from the given root node $v_r \in G$ within exactly n hops, denoted as $C_r = \{v_i | hop(v_i, v_r) = n\}$.*

Definition 4 (User Identity Linkage). *Given two different networks, $G^S = \{V^S, E^S\}$ and $G^T = \{V^T, E^T\}$. The goal of User Identity Linkage (UIL) is to predict a pair-wise linkage between a user node v^s selected from the source network G^S and an unlabeled user node v^t in the target network G^T, which indicates they are the same user or entity (i.e., $v^s = v^t$).*

To better illustrate the UIL problem, a toy example is given in Fig. 1, where the corresponding same users (i.e., anchors) in two isolated networks are linked via Observed Anchor Links. Therefore, the task becomes how to predict the unknown dotted links based on known network structure information and these observed links.

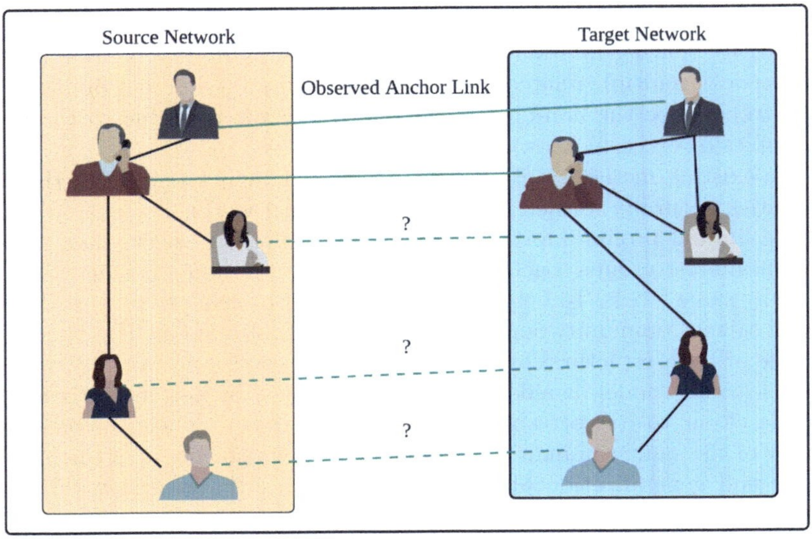

Fig. 1. User Identity Linkage: A toy example of the problem

4 Community Sensing User Identity Linkage

We proposed a novel embedding-based approach that considers and utilizes both individual similarity and community similarity, which consists of three major components:

1. Network Embedding
2. Community Clustering
3. Latent Space Mapping

A brief overview of the whole approach is illustrated in Fig. 2, where rounded blocks represent the core elements in each phase of our approach, green lines indicate the direction of the flow of structural information and blue lines show how core elements and algorithms are connected between different phases.

4.1 Network Embedding

First of all, as illustrated in Fig. 2, the data flow is divided into two parts. For data in regular euclidean space, we take typical examples like images (2-dimensional grid) and text (1-dimensional sequence), while for data in non-euclidean space, like the input social network stored in the form of a graph [18], it would be easier processed after transformed into an euclidean latent representation and more general data mining techniques could be applied on it.

The quality of the latent representation of each node in both source and target network is important to the results of the following clustering and mapping stages. Ideally, user nodes that have stronger connection, like sharing more

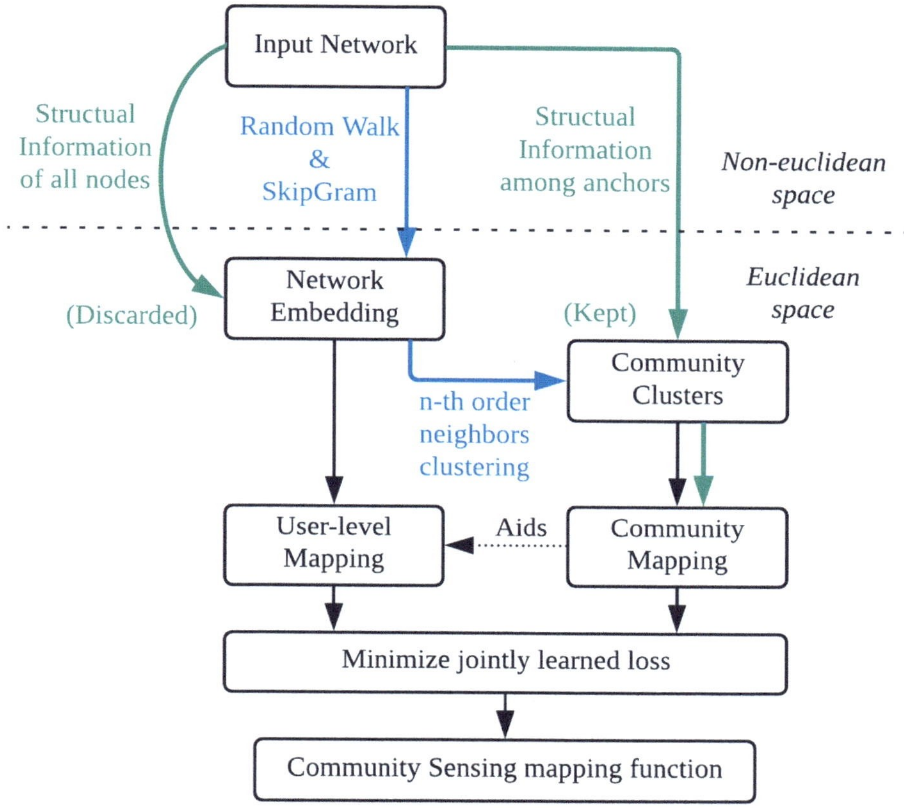

Fig. 2. A brief overview of CSUIL

common neighbors, or having shorter path between them should be closer to each other after they are projected into the latent space. To obtain the network embedding in good quality, an efficient model called DeepWalk [8] was adopted. DeepWalk mainly utilizes the truncated random walk and the SkipGram [6] model on network data.

In particular, a random walk generator is first applied to the network G, which will sample uniformly a random node $v_i \in G$ as the root of a random walk sequence W_{v_i}, then the generator will sample uniformly from the neighbors of the last node it visited until the maximum sequence length(l) is reached. The generated sequences could be thought of as short sentences, while the nodes within sequences are treated as words of a special kind of language, and therefore the original network could be treated as a document. We could then obtain the embedding of nodes as a byproduct when updating the weight matrix in the derived SkipGram model, which aims to maximize the co-occurrence probability of nodes that appear within a window size w near the center node v_j in the sequence W_{v_i}, that is to maximize the following log probability:

$$\max \quad \frac{1}{l}\sum_{i=1}^{l}\sum_{j=-w,j\neq 0}^{w}\log \Pr(v_{i+j}|v_i) \tag{1}$$

where $\Pr(v_{i+j}|v_i)$ is calculated with a hierarchical softmax function:

$$\Pr(v_{i+j}|v_i) = \frac{\exp(\boldsymbol{\psi}(v_{i+j})^{\mathrm{T}}\boldsymbol{\psi}(v_i))}{\sum_{m=1}^{l}\exp(\boldsymbol{\psi}(v_m)^{\mathrm{T}}\boldsymbol{\psi}(v_i))} \tag{2}$$

where $\boldsymbol{\psi}(v_i)$ is the embedding (i.e., vector representation) of node v_i that we need to update at each training step to maximize co-occurrence probability and finally being output to the next phase.

4.2 Community Clustering

In some supervised User Identity Linkage models such as PALE [5], it only focuses on learning the user level, pair-wise matching patterns between source and target network. However, these methods failed to consider the social communities naturally formed by people in the real world. Some drawbacks may exist under such settings that users with very limited profile information could be hard to distinguish from others and the model may fall into over-fitting of local pair-wise features when trained with small amount of labeled data [12]. More importantly, the knowledge contained in the structural relationship among anchor and non-anchor users in the original non-euclidean space is discarded after SkipGram is applied (illustrated by green lines in Fig. 2), thus later phases are not able to reuse such information.

Therefore, we made an assumption that compared to only considering the generated embedding or user-level similarity matching, the fact that which neighbors a user has in the original network, and which community a user belongs to could reveal more diffusible structural knowledge. Thus, we consider clustering the $n-th$ order neighbors of an anchor user to form their social community, the users in the same community have a closer relationship and higher similarity, which could be evaluated in some metrics including: the amount of common neighbors, or the minimum walk length between each other.

To utilize all the user information in a community, we reuse the structural information in the original network and derive a new embedding to represent this community by adopting the mean value of all community member embedding generated in Sect. 4.1 that are non-anchor nodes. The generated center node that represents a certain community cluster C_i is denoted as μ_i:

$$\psi(\mu_i) = \frac{\psi(v_r) + \sum_{v'\in C_i}\psi(v')}{N+1} \tag{3}$$

where v_r is the root user node, and N is the community size.

4.3 Latent Space Mapping

Let $\boldsymbol{z}^s = \psi(v^s)$ and $\boldsymbol{z}^t = \psi(v^t)$ be the node embedding generated in Sect. 4.1 and the final stage of CSUIL is Latent Space Mapping. In this phase, we try to find a mapping function Φ from the source network to the target network Φ: $\mathbb{R}^{|V^s|\times d} \mapsto \mathbb{R}^{|V^t|\times d}$, that will minimize the distance between the predicted embedding $\Phi(\boldsymbol{z}^s)$ and the true corresponding embedding $\boldsymbol{z}^t$ of $\boldsymbol{z}^s$ in the target network:

$$\min \quad \left\|\Phi\left(\boldsymbol{z}^s\right) - \boldsymbol{z}^t\right\|_F \tag{4}$$

We then train a novel two-inputs and two-outputs neural network model, which breaks down the whole task mentioned above into two simultaneously conducted parts: (1) minimize the distance between predicted and real user node (2) minimize the distance between predicted and real community center. The second sub-task will drive the mapping function to the direction that also exploits the relationship between community centers in both source and target networks to increase the generalization ability of the model on new unseen data.

Next, the design of the loss function could be one of the most critical parts of a machine learning model, a good loss function should reflect the regression error during training as well as the generalization error that guides network parameters for optimizing the model. Therefore, for the goal of above two sub-tasks, a new community sensing loss function is proposed as:

$$\begin{aligned} loss = (1-\gamma) & \sum_{(v^s,v^t)\in\{S,T\}} \left\|\Phi(\boldsymbol{z}^s;\theta) - \boldsymbol{z}^t\right\|_F \\ & + \gamma \sum_{\mu\in C} \left\|\Phi(\boldsymbol{\mu}^s;\theta) - \boldsymbol{\mu}^t\right\|_F \end{aligned} \tag{5}$$

where $\{S,T\}$ is the set of groundtruth anchor nodes pairs, C is the set of community clustering centers, F is the Frobenius norm, θ is the collection of all parameters need to be tuned in the model, and γ is the hyper-parameter of the weight coefficient of the community loss that could be co-optimized during the learning of the mapping function.

We finally employed a Multi-Layer Perceptron (MLP) model that does not require extensive feature selection or difficult parameter tuning to search for an optimized mapping function, while MLP model also keeps the flexibility of dealing with the non-linear relationships that may exist between the source and target network.

The overall algorithm design is shown in Algorithm 1.

5 Experiment

5.1 Data Preparation

A real-world social network dataset collected from Twitter and Foursquare [15] is used in this experiment, which was released in [3]. It was crawled from normal

Algorithm 1. CSUIL

Input: network $G(V, E)$, training anchor nodes $\{S, T\}$, test nodes $\{S', T'\}$, community clustering parameter n, community loss parameter γ
Output: mapping function Φ, matching result list R
foreach *node* $v_i \in G$ **do**
  Generate the embedding of v_i as $\boldsymbol{z}_i$.
end for
foreach *anchor node pair* $\{s_i, t_i\}$ *in* $\{S, T\}$ **do**
  Reuse the original network structure information, cluster the n-order neighbors of s_i and t_i respectively.
  Derive the community center $\boldsymbol{\mu}_i^s$ and $\boldsymbol{\mu}_i^t$ respectively.
end for
Train the MLP model by jointly minimize the node mapping loss $\left\|\Phi(\boldsymbol{z}^s; \theta) - \Phi(\boldsymbol{z}^t)\right\|_F$ and community loss $\left\|\Phi(\boldsymbol{\mu}^s; \theta) - \Phi(\boldsymbol{\mu}^t)\right\|_F$
foreach *test node* $s_i' \in S'$ **do**
  Add the predicted t_i' to result list R.
end for
$Evaluate(R, T')$ via metrics.

users' open-access profile pages, and all the sensitive personal information is removed under privacy concerns to form the final training and testing dataset. The ground truth of anchors links is obtained by crawling users' Twitter accounts from their Foursquare homepage. Table 1 lists the statistics of this dataset.

Table 1. Statistics of Twitter-Foursquare dataset

Network	#Users	#Relations	#Anchors
Twitter	5,220	164,919	1,609
Foursquare	5,315	76,972	

5.2 Evaluation Metrics

In this experiment, in a similar form to the paper [17], a metric called *Precision@k* was adopted, which is defined as following:

$$Precision@k = \frac{\sum_{i}^{n} TOP_k(\Phi(\boldsymbol{z}_i^s))}{N} \tag{6}$$

where $TOP_k(\Phi(\boldsymbol{z}_i^s))$ is a binary output function (0 or 1), for each predicted embedding $\Phi(\boldsymbol{z}_i^s)$ of $\boldsymbol{z}^s$ by our method, it tells whether the positive match $\boldsymbol{z}_i^t$ exists in the $top-k$ list or not, and N is the number of all testing nodes. In the context of UIL, as *Precision@k* is a metric of the true positive rate, it could be treated the same as *Recall@k*, and $F_1@k$.

5.3 Comparative Methods

We compare the proposed CSUIL with several existing embedding-based methods, and take them as the baseline of the link prediction task.

- **CSUIL**: The proposed method in this paper, it could explicitly exploit the individual as well as community features of a network, by jointly optimizing mapping function that concentrates on user-level and community-level similarity respectively.
- **IONE**: Proposed in [3] and adopted as a baseline result, Input-Output Network Embedding (IONE) is a network embedding and partial network alignment method. It takes follower-ship and followee-ship as input and output contexts and generates two relationship representations together with the user node representation.
- **INE**: Input Network Embedding (INE) is a simplified version of IONE, which only considers user node representation and input (i.e., follower-ship) representation for matching.

5.4 Results

The performance results are illustrated in Table 2 as well as Fig. 3. In the experiments, during the community clustering phase, the cluster size is set to be first-order neighbors for the simplicity to process. Then we examine the overall ability of the final model (with training rate = 90%) on the link prediction task. For CSUIL, we report the result in different settings of precision metrics k and community loss weight coefficient γ. For INE and INOE, we report results in the original paper's default setting.

From the experiment results, we could conclude that:

- Compared to the baseline model, INE and IONE, the best performance (when $\gamma = 0.8$, shown in Table 2 and Fig. 3(a)) of our approach has an improvement from about 6% to 21 % at most in different settings of precision metrics, which shows the feasibility of this approach.
- Figure 3(b) also illustrates that by changing the setting of community loss weight coefficient γ, the ability of the model to sense more positive matching in a larger searching space (i.e., higher k setting in precision), could be enhanced, which is an important improvement because many other papers only stress their performance at the $k = 30$ setting. However, adding too much weight to community loss may lead to a slight reduction of the ability to narrow the target to a finer scale (i.e., lower k in precision), compared with the $\gamma = 0$ setting, which means that the community similarity is not considered at all.

Table 2. Performance comparison between baselines

	Precision@k							
γ	P@1	P@5	P@9	P@13	P@17	P@21	P@25	P@30
INE	0.1108	0.2184	0.2975	0.3291	0.3703	0.4114	0.4304	0.4494
IONE	0.1899	0.3481	0.4494	0.4968	0.5253	0.5665	0.5854	0.6044
0.8	**0.2405**	**0.5190**	**0.6203**	**0.6835**	**0.7342**	**0.7722**	**0.7975**	**0.8165**

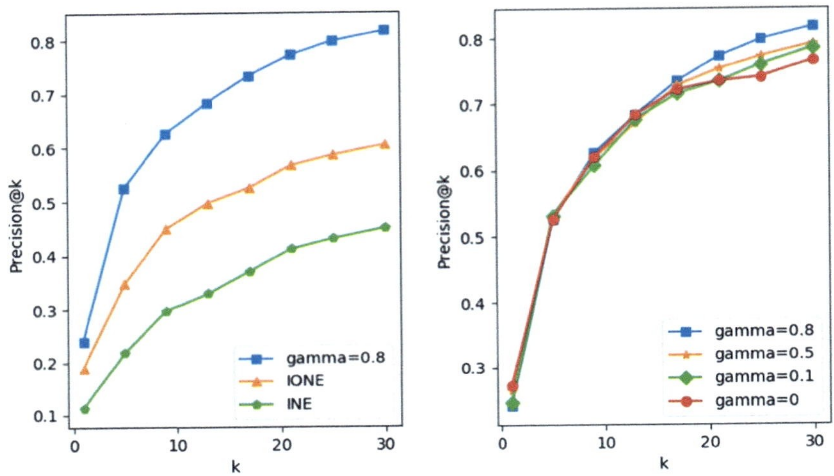

(a) Baselines (b) Precision on different γ

Fig. 3. Link prediction precision results. X-axis is the different value of k, for the top-k list being evaluated; Y-axis is the precision result in percentage.

6 Conclusion and Future Work

In this paper, we aim to study the UIL problem by reusing the discarded knowledge in the original Online Social Network after network embedding. Not limited to anchor same users across networks, we would also like the community formed by close users to have a positive match across networks. This is because some users may have limited profile and it could be hard to distinguish them from others. However, in the context of a community, users share common features, and they will be driven to the correct direction where a group of users with high similarity locates, even if some community members are known very little. This could also help to avoid over-fitting of the input data and increase the generalization ability of the method.

Therefore, we break down the main task into two related and simultaneously learned sub-tasks: User Mapping and Community Mapping, this is achieved by jointly optimizing the user loss and community loss in a single MLP model. Based on above theories, Community Sensing User Identity Linkage (CSUIL) is proposed. Results show that our approach outperforms current baseline models, and has the flexibility to adapt hyper-parameters for different needs or data input.

There are several directions that could be investigated in the future. First, a simplified clustering method was adopted here, some users that are not included in clusters due to fixed range threshold may also be strongly related to the community, new clustering technologies such as partition-based or hierarchical methods could be applied to improve the accuracy of community center generation. Second, in real-world social networks, the network is growing quickly, many new users come in everyday, and new following relationship is generated at real time, it could be helpful if we extend our approach to be able to work online to catch such dynamic mapping relation.

Acknowledgments. This work was funded by JSPS KAKENHI JP16H01836, JP16K12428, and industrial collaborators.

References

1. Cui, P., Wang, X., Pei, J., Zhu, W.: A survey on network embedding. IEEE Trans. Knowl. Data Eng. **31**(5), 833–852 (2018)
2. Kong, X., Zhang, J., Yu, P.S.: Inferring anchor links across multiple heterogeneous social networks. In: Proceedings of the 22nd ACM International Conference on Information & Knowledge Management, pp. 179–188. ACM (2013)
3. Liu, L., Cheung, W.K., Li, X., Liao, L.: Aligning users across social networks using network embedding. In: IJCAI, pp. 1774–1780 (2016)
4. Malhotra, A., Totti, L., Meira Jr, W., Kumaraguru, P., Almeida, V.: Studying user footprints in different online social networks. In: Proceedings of the 2012 International Conference on Advances in Social Networks Analysis and Mining (ASONAM 2012), pp. 1065–1070. IEEE Computer Society (2012)
5. Man, T., Shen, H., Liu, S., Jin, X., Cheng, X.: Predict anchor links across social networks via an embedding approach. IJCAI **16**, 1823–1829 (2016)
6. Mikolov, T., Sutskever, I., Chen, K., Corrado, G.S., Dean, J.: Distributed representations of words and phrases and their compositionality. In: Advances in Neural Information Processing Systems, pp. 3111–3119 (2013)
7. Mu, X., Zhu, F., Lim, E.P., Xiao, J., Wang, J., Zhou, Z.H.: User identity linkage by latent user space modelling. In: Proceedings of the 22nd ACM SIGKDD International Conference on Knowledge Discovery and Data Mining, pp. 1775–1784. ACM (2016)
8. Perozzi, B., Al-Rfou, R., Skiena, S.: Deepwalk: online learning of social representations. In: Proceedings of the 20th ACM SIGKDD International Conference on Knowledge Discovery and Data Mining, pp. 701–710. ACM (2014)
9. Ramsundar, B., Kearnes, S., Riley, P., Webster, D., Konerding, D., Pande, V.: Massively multitask networks for drug discovery. arXiv preprint arXiv:1502.02072 (2015)
10. Ren, S., He, K., Girshick, R., Sun, J.: Faster r-cnn: towards real-time object detection with region proposal networks. In: Advances in Neural Information Processing Systems, pp. 91–99 (2015)

11. Ribeiro, L.F., Saverese, P.H., Figueiredo, D.R.: struc2vec: learning node representations from structural identity. In: Proceedings of the 23rd ACM SIGKDD International Conference on Knowledge Discovery and Data Mining, pp. 385–394. ACM (2017)
12. Rozemberczki, B., Davies, R., Sarkar, R., Sutton, C.: Gemsec: Graph embedding with self clustering. arXiv preprint arXiv:1802.03997 (2018)
13. Shu, K., Wang, S., Tang, J., Zafarani, R., Liu, H.: User identity linkage across online social networks: a review. Acm Sigkdd Explor. Newsl. **18**(2), 5–17 (2017)
14. Zhang, J.: Social network fusion and mining: A survey. CoRR **abs/1804.09874** (2018). http://arxiv.org/abs/1804.09874
15. Zhang, J., Philip, S.Y.: Integrated anchor and social link predictions across social networks. In: IJCAI, pp. 2125–2132 (2015)
16. Zhang, Y., Yang, Q.: A survey on multi-task learning. arXiv preprint arXiv:1707.08114 (2017)
17. Zhou, F., Liu, L., Zhang, K., Trajcevski, G., Wu, J., Zhong, T.: Deeplink: a deep learning approach for user identity linkage. In: IEEE INFOCOM 2018-IEEE Conference on Computer Communications, pp. 1313–1321. IEEE (2018)
18. Zhou, J., Cui, G., Zhang, Z., Yang, C., Liu, Z., Sun, M.: Graph neural networks: a review of methods and applications. arXiv preprint arXiv:1812.08434 (2018)

Social Influence Prediction by a Community-Based Convolutional Neural Network

Shao Hsuan Tai, Hao-Shang Ma, and Jen-Wei Huang(✉)

Department of Electrical Engineering, Institute of Computer and Communication Engineer, National Cheng Kung University, Tainan, Taiwan
jwhuang@mail.ncku.edu.tw

Abstract. This is an extension from a selected paper from JSAI2019. Learning social influence between users on social networks has beextensively studied in a decade. Many models were proposed to model the microscopic diffusion process or to directly predict the final diffusion results. However, most of them need expensive Monte Carlo simulations to estimate diffusion results and some of them just predict the size of the spread via regression techniques, where people who will adopt the information becomes unknown. In this work, we regard the prediction of final influence diffusion results in a social network as a classification problem to avoid expensive simulations with knowing the final adopters. Furthermore, we propose a community-based convolutional neural network to capture the information of local structure with the aforementioned network. The proposed model is referred to as the Social Influence Learning on Community-based Convolutional Neural Network, SIL-CCNN. In the experiment, SIL-CCNN shows the promising results in both synthetic and real-world datasets. In addition, modeling local structure is indeed useful for the prediction of information diffusion.

Keywords: Social network analysis · Social influence prediction · Convolutional neural network

1 Introduction

Nowadays people tend to share their life, emotions, and opinions to others on social network websites. Online social networks play an important role in the diffusion of information [1]. Two representative diffusion models, Independent Cascade, IC, model and Linear Threshold, LT, model, were reformulated by Kempe [7]. These diffusion models would like to predict the diffusion of particular information and the subsequent activation of specific individuals following the propagation process within the network. However, there are several limitations of predicting the information diffusion using diffusion models. The influence probabilities between users and the active threshold of a user should be measured or learned from many personal features such as users' preferences and different relationships. In real social networks, the features are not easy to extract since

Y. Ohsawa et al. (Eds.): JSAI 2019, AISC 1128, pp. 203–214, 2020.
https://doi.org/10.1007/978-3-030-39878-1_19

the data sometimes is not complete. The relationships usually depend on the characteristic of network. In addition, to get the results of IC/LT model, we need to conduct a huge number of simulations.

In our opinion, the prediction of the information diffusion process and the final diffusion results models can be regarded as a classification problem. Given the information sources and the network structure, the individuals are classified into active or inactive classes in the final diffusion result. The active class is corresponding to the individuals being influenced successfully in the information diffusion process. On the other hand, the inactive class contains those individuals which are not influenced by other active ones. Some related works aim to predict the size of information spreads as a classification or regression problem [5,12]. Different with diffusion models, these methods usually do not learn the individuals who are actually influenced by the information. We would like to know exactly who are influenced and who are not.

To overcome above limitations and solve the classification problem, we propose an influence prediction model based on community-based convolutional neural network. First, we consider the influence propagation process in the past to learn the influence between individuals. The propagation traces record the information diffusion process from source nodes to influenced nodes. We use traces as features in the learning phase of neural networks. In addition, most of the diffusion models consider the network structure, i.e., the relations between individuals, as their features. The relations between individuals can help the model to simulate the influence spread. We also include the network structure to learn the influence of relations among individuals. We formulate a matrix to represent the relationships among all individuals within the social network. Elements in the relation matrix are weights representing the relations between individuals or binary values indicating whether individuals are connected.

There are some issues if we consider the entire network structure to predict the results. The relation matrix obtained from actual social networks tends to be huge and sparse. Each individual only has few relations to others. The information of the whole network structure may not be useful for the classification problem whereas the local network information of a single individual should be helpful [16]. We try to embed the local structure of an individual into our model to learn the local relations. The community structure in social networks represents a cluster of individuals sharing connections that are stronger than those with individuals outside the community. The individuals may have high probability to influence the others in the same community since the individuals in the same community are connected tightly. The information diffusion in a community should be faster than outside the community. Therefore, we propose to join the community structure instead of the entire network structure in this work. Moreover, we further consider using the convolutional neural networks to mine the local relations within the community structure. The idea is that convolutional neural networks are good to extract the effects of a small group of individuals around an individual by extracting valuable relations from the structure. The community-related information is extracted through the convolutional neural network.

Finally, the past traces and the community structure information are combined into our model as training features of the deep neural network. The proposed scheme is referred to as the Social Influence Learning on Community-based Convolutional Neural Network, SIL-CCNN. The remainder of the paper is organized as follows. Works related to our methodology is outlined in Sect. 2. Section 3 details the proposed methodology. Experiment results are presented in Sect. 4 and conclusions are drawn in Sect. 5.

2 Related Work

In this section, we will briefly introduce other related works on diffusion models and the prediction of information spread size.

2.1 Diffusion Models

Diffusion models can be used to identify social influence by tracking information propagating through a social network. Diffusion models can be categorized into two types, according to the forms of activation, say probability-like and threshold-like. Probability-like diffusion models, such as Independent Cascade, IC, model [7], predict activations based on the probability of successful influence which can be calculated from data. Threshold-like diffusion models, such as Linear Threshold, LT, model [7], predict activations by establishing a threshold of successful influence. In IC model, there is an influential probability on each edge. The influential probability represents the effect between nodes. If a node x has significant influence to another node y, the influential probability of the edge from x to y will be high. Since a node is active, it would "try" to influence its neighbors. Whether a node of its neighbors will be active or not is decided by the influential probability. Each decision of a successful influence will be decided independently. The diffusion process will follow the rules as mentioned above and repeat until converge. There are some other IC-based diffusion models considering more information such as the topic information [2], external influence [10], and negative influence [3]. Some LT-based models are proposed to consider mores features. Barbieri [2] proposed TIC and TLT models to consider topic-aware information. Chen [4] proposed Multi-state Open Opinion model to consider more information such as the negative influence and the time decay of information.

Nowadays, some diffusion models adopt the learning strategy to predict the activation since the technique of learning methodology has matured in these few years. Saito [11] first proposed a learning method for IC model. Wang [13] proposed a feature-enhanced approach, which considers not only temporal data in cascades but also additional features. Chou [6] proposed Multiple Factor-Aware Diffusion model, MFAD, that can consider many kinds of factors together. MFAD model adopts positive and unlabeled learning to train the classifiers for each individual.

2.2 Prediction of Information Spread Size

Different from diffusion models, the cascade prediction only focuses on predicting whether the information will become popular and widely spread. The problem is usually formulated as a classification or a regression problem to understand the size of potential influence of information.

Among the classification solutions of predicting information cascade, Cheng [5] proposed to use temporal and structural features for predicting the relative growth of a cascade size. In their observation, the key features for predicting the relative growth of a cascade size are temporal and structural features. Jenders analyze some user features, e.g., number of followers, tweet length, number of hashtags, number of mentions, etc., and latent features, e.g., sentiment and emotional divergence, to distinguish whether the tweets would become viral. Weng [14] proposed a model to predict the future popularity of a meme given its early spread patterns. They also demonstrated that features based on network structure, particularly communities, provide crucial insights into virality.

Our work differs from these related methods as we would like to predict the people who are actually influenced and who are not by given information. As for the prediction of information spread size, the methods only cares the final popularity but does not show whether each individual is influenced or not. We aims to learn the hidden influence propagation from the data instances and the network structure directly without the huge number of diffusion simulations.

3 Methodology

In this section, we first propose an ordinary deep neural network model for identifying traces of social influence. We first introduce the architecture of Social Influence Learning on Deep Neural Network, abbreviated as SIL-DNN. In SIL-DNN, we explain how traces of diffusions are used to feed the deep neural network. Second, a community detection algorithm is used to obtain the community structure. We adopt a convolutional neural network to extract the local structure information of a network from the communities. Then, we propose Social Influence Learning on Community-based Convolutional Neural Network, abbreviated as SIL-CCNN, which combines the SIL-DNN and a community-based convolutional neural network to predict the final influential results.

3.1 Social Influence Learning on Deep Neural Network

Given a social network $\mathbb{G} = (\mathbb{V}, \mathbb{E})$, the node set $\mathbb{V}$ corresponds to the individuals, and $\mathbb{E}$ is the edge set indicating the relationships between individuals. The traces record the information diffusion process from the source node to the influenced nodes. Each trace (u, v, x) indicates that the nodes v adopt the information x and is influenced by source nodes u, where u and v are sets of nodes in $\mathbb{V}$. There is an example shown in Fig. 1. The red user, v_2, provides an item and becomes an influence source. After the influence diffusion process, the green users, v_1 and v_4,

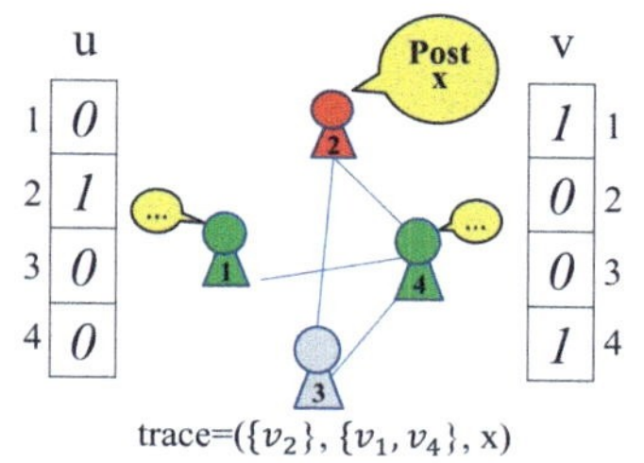

Fig. 1. The transformation from the data to a traces of influence diffusion

are the influenced users by the red node. The inputs are defined as the vectors of information sources. The outputs are defined as the vectors of influenced nodes.

The architecture of SIL-DNN is presented in Fig. 2. In SIL-DNN, each neuron in the input layer represents an individual in the social network. The number of neurons in the input layer is the same as the number of individuals in the social network $|V|$. The same setting is used in the output layer. For every trace, we put the vector of information sources as the input data of SIL-DNN and the output are the vector of influenced nodes. The input vector could be set as follows,

$$\begin{cases} y_i = 1, \text{if } i \in u & \text{for } (u, v, x) \\ y_i = 0, & \text{otherwise}, \end{cases} \tag{1}$$

where y_i represents an individual i at the input layer. $i \in u$ for (u, v, x) indicates all traces that transmit the information x from user i. The same setting is used for output traces.

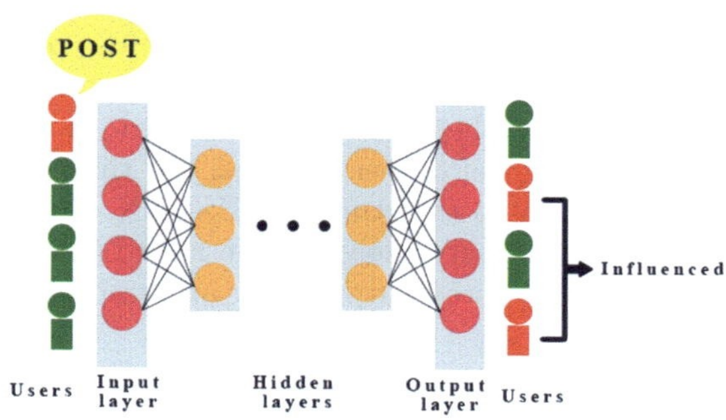

Fig. 2. Social influence learning on deep neural network architecture

In SIL-DNN, the number of neurons in the input layer and output layer is determined by the number of individuals in the network. SIL-DNN is trained by backpropagating derivatives of a cost function that measures the discrepancy between the target outputs and the actual outputs. We use a sigmoid function in the output layer and define the cross-entropy as the cost function. The cross-entropy equation is shown as follows,

$$Q\left(W^t\right) = -\sum_{i}^{I} t_i \log y_i, \tag{2}$$

where I is the elements number of input traces and W^t is the weight matrices of traces in SIL-DNN. Our objective is to minimize the loss between target t and the prediction y. For the optimization, we adopt Adaptive Moment estimation, Adam, to optimize the model parameters.

3.2 Social Influence Learning on Community-Based Convolutional Neural Network

Overall Architecture

In order to join the community structure to help us to predict the information diffusion results, we propose Social Influence Learning on Community-based Convolutional Neural Network, SIL-CCNN. The architecture of SIL-CCNN is presented in Fig. 3. First, we adopt the Speaker-listener Label Propagation algorithm, abbreviated as SLPA, to extract the community structure [15]. SLPA performs well both on the accuracy and efficiency especially for large scale and sparse social networks. In addition, SLPA is suitable for real world social networks since SLPA allows people to participate in multiple communities. Then, we design a community-based convolutional neural network to deal with community-related information by extracting features through a convolutional layer and a pooling layer. For the convolutional layer, the relation matrix containing the community information is defined to represent the local network structure of an individual. A filter is defined to extract local relations by convolution functions. In the pooling layer, the spatial size is reduced and pooling makes the convolutional neural network able to prevent overfitting. In SIL-CCNN, we use max-pooling as our pooling function. Max-pooling is particularly well suited to the separation of features that are sparse. Then, there is a merge layer to combine the output of the pooling layer and the input of diffusion traces. For the trace of information diffusion, we construct the same structure as SIL-DNN. The merge layer and the structure for learning traces are shown in the right of Fig. 3. Finally, a few hidden layers and one full-connected output layer are connected after the merge layer.

Community-Based Convolutional Neural Network

Based on traditional Convolutional Neural Network, we design a Community-based Convolutional Neural Network, CCNN, to deal with the locality information in social networks since the local network information of a single individual should be helpful for the influence prediction. The goal of CCNN is to identify the cases where small groups in the social network are mutually influential.

First, we need to extract the community information in the network and form the relation matrix of each community. A list of relation matrix RM for

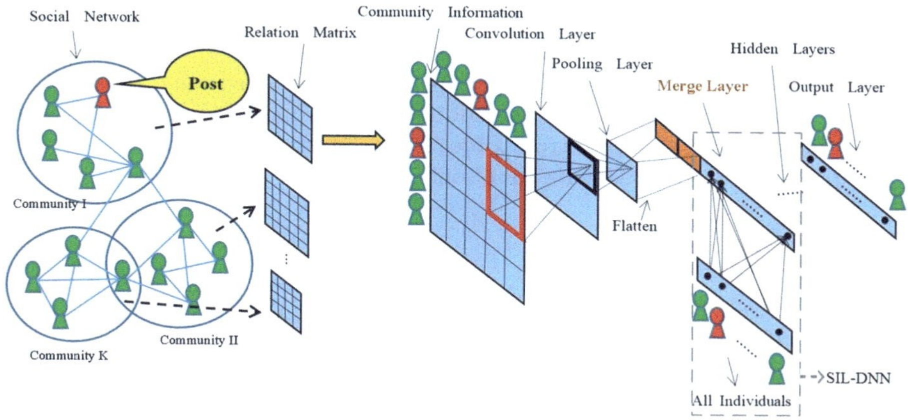

Fig. 3. Social influence learning on community-based convolutional neural network architecture

communities in $COMM$ can be formulated. Then, we design a community-based convolutional neural network to deal with community-related information by extracting features through a convolutional layer and a pooling layer. For the input of the convolution layer, we extract the relation matrices of communities to represent the local network information of the individual. The relation matrix RM of a community is defined as follows,

$$RM = [rm_{ij}]_{d \times d}, rm_{ij} = \begin{cases} w_{ij}, (v_i, v_j) \epsilon E \\ 0, otherwise, \end{cases} \tag{3}$$

where d is the number of nodes in the community. The element rm_{ij} in the relation matrix represents the weight on the edge between node v_i and node v_j in a community. In addition, in order to account for differences in the size of communities, the community information matrices are normalized to the size of the largest community and have zero-padding.

However, if we randomly assign the order of nodes in relation matrices and put the matrices into the convolutional neural network, the small extracted region in the convolution layer would be meaningless. Therefore, we design an arrangement strategy to determine the relations of nearby elements. The order of individuals in each relation matrix is determined by the ranking of the number of degrees. The individual having the largest degree will be arranged to the leftmost of x-axis and the topmost of y-axis. Therefore, each block in different relation matrices shares the same weight matrix in the convolutional neural network.

In the pooling layer, we use max-pooling as our pooling function. Max-pooling is particularly well suited to the separation of features that are sparse. After the pooling layer, SIL-CCNN constructs a merge layer to combine the output of the pooling layer and the input vector of SIL-DNN. Then, several hidden layers are trained in SIL-CCNN before the output layer. Finally, a few hidden layers and one full-connected output layer are connected after the merge layer.

As for the detailed setting of neural network, the kernel size of the convolutional layer is defined to 3 × 3. In addition, we define a sigmoid function in the output layer and a cross-entropy objective function. In the training step, our goal is to minimize the following equation:

$$Q\left(W^t, W^r\right) = -\sum_{i}^{I} t_i \log y_i, \tag{4}$$

where I is the number of inputs, W^t is the weight matrices of traces, and W^r represents the weight matrices of the structure relation in SIL-CCNN. The weight matrix contains the relations between the individuals and the local structure information. The equation above is shown for one individual i at output layer k ranging over all target labels. Our objective function aims to minimize the cross entropy between the target t and the prediction y. As for the optimization, we use the well-known backpropagation algorithm and Adam [8] to compute the parameters in SIL-CCNN.

4 Experiments

In this section, we introduce experiments aiming at evaluating predicting performance in social networks using some synthetic datasets and a real-world dataset. In the following, we describe the formulation of the synthetic dataset and how we crawl real-world dataset. We then outline the methods used to evaluate the performance of the proposed method and other related works. Finally, we present a discussion of the results.

4.1 Dataset

Synthetic Data. Using the Lancichinetti-Fortunato-Radicchi (LFR) benchmark [9], we generate two synthetic datasets, one with 1000 nodes (161,980 traces) and the other with 5000 nodes (2,666,674 traces). The parameters k and mu in LFR which are represented the average degree of nodes and mixing parameters. In 1000 nodes network, the k is set as 0.015 and the mu is set as 0.15. In 5000 nodes network, we set k as 0.005 and mu as 0.15. The adoption of a given node depends on the transmission probability on the edge. For the generation of diffusion data, we set the transmission probability uniformly between 0.1 and 1. We then choose a node at random to function as a source node and set it to be active. Nodes attempt to activate neighboring nodes based on the transmission probabilities. For each edge, the same item will only expose once (Table 1).

Table 1. Characteristics of three datasets

	Synthetic 1000	Synthetic 5000	Real Data
Users	1,000	5,000	20,453
Edges	15108	125,838	1,458,298
Avg. of degree	15.11	25.17	71.3
Number of communities	47	112	501
Avg. size of communities	21.27	44.64	40.82
Traces	161,980	2,666,674	3,782,305

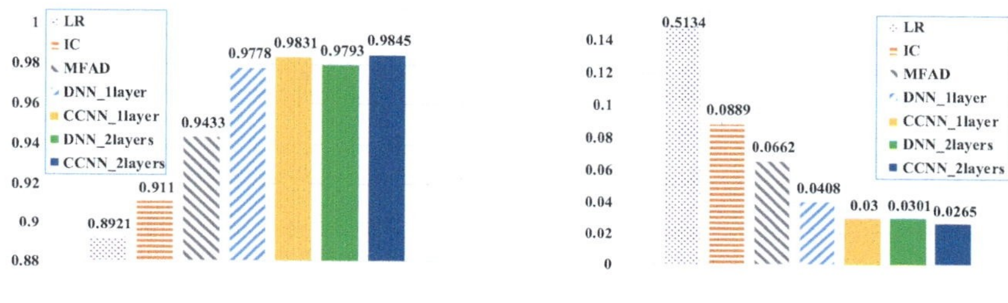

(a) Accuracy (b) MAE

Fig. 4. Accuracy and MAE on the synthetic dataset of 1000 nodes

Real Data. We crawl data from Twitter[1] for the period between September 2011 and May 2015. We use the experts in Healthcare Pundits[2] and Security[3] as the initial nodes, and crawl the followers of these experts to form a social network. The twitter data contains 20,453 users and 3,782,305 tweets. In this study, tweets were used as items, and the actions of sharing, replying, and liking are indications of the influence of a tweet.

4.2 Compared Methods

We include the following methods to predict the information diffusion results.

- **Logistic Regression** (LR). The in-degree and out-degree are included in a classifier for individual v.
- **Independent Cascade Model** (IC). IC is a conventional diffusion model in which the probability of transmission from individual u to v is the ratio of items individual v adopted items from individual u.
- **Multiple Factor-Aware Diffusion Model** (MFAD). MFAD is a diffusion model that can learn the social influence by multiple features [6].

[1] https://twitter.com.
[2] http://nursepractitionerdegree.org/top-50-health-care-pundits-worth-following-on-twitter.html.
[3] http://www.marblesecurity.com/2013/11/20/100-security-experts-follow-twitter/.

- **SIL-DNN and SIL-CCNN**. We evaluated the performance of SIL-DNN and SIL-CCNN using one hidden layer (DNN_1layer, $CCNN\_1layer$) and two hidden layers ($DNN\_2layers$, $CCNN\_2layers$) in the following experiments.

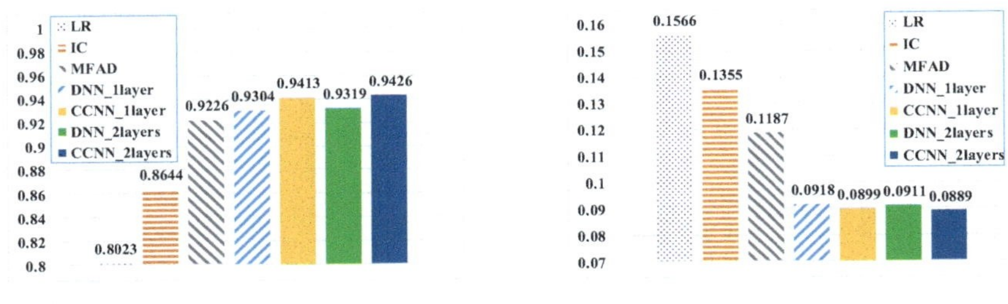

(a) Accuracy (b) MAE

Fig. 5. Accuracy and MAE on the synthetic dataset of 5000 nodes

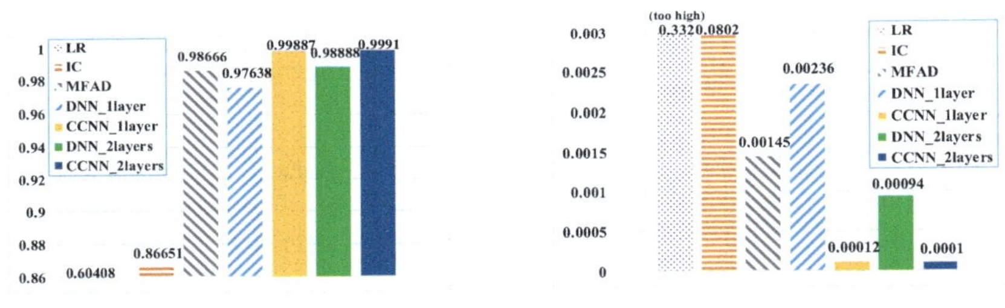

(a) Accuracy (b) MAE

Fig. 6. Accuracy and MAE on twitter dataset

4.3 Evaluated Metrics

We evaluate the performance of algorithms using the following two metrics.

- **Accuracy**. Accuracy is defined as the ratio of correct predicted answers over all answers in order to estimate the correctness of predictions.
- **Mean Absolute Error** (MAE). We computed the mean absolute error as follows: $MAE = \frac{\sum_{i=1}^{n}|y_i - x_i|}{n}$, where y_i is the truly adopted result of individual i, x_i is the estimated adoption probability of individual i, and n is the number of individuals in the network.

4.4 Results and Discussions

In this section, we compare the proposed SIL-DNN and SIL-CCNN with the other methods mentioned above. The comparison results on synthetic datasets are shown in Figs. 4 and 5. SIL-DNN and SIL-CCNN outperform the other three methods in both synthetic datasets. In addition, SIL-CCNN have higher accuracy and lower MAE than SIL-DNN using the same number of hidden layers. The results show that the information of the community structure actually helps the model to predict the influence results better. The proposed community-based CNN indeed extracts the local relations of individuals. Moreover, the performance of SIL-CCNN 2 layers is better than the SIL-CCNN 1 layer. Using more hidden layers also conducts a better result. We only adopt one and two hidden layers since there is no significant improvement when we add more hidden layers in this experiment. This phenomenon may be due to insufficient training data.

In the real twitter dataset, the results are shown in Fig. 6. Basically, SIL-DNN and SIL-CCNN still outperform the other three methods except for the DNN_1layer. We have examined the propagation results in these three datasets. We found that the influenced scale of two synthetic datasets is much larger than the Twitter dataset. The real Twitter data has fewer influenced nodes. This indicates that the diffusion results in synthetic datasets should be more difficult to predict. The experimental results show the proposed methods can learn the social influence of more complicated propagation process better. The superiority of SIL-CCNN over SIL-DNN shows that the performance can be improved by including the community information by the proposed community-based convolutional neural network.

5 Conclusions and Future Works

In this work, we proposed two neural network architectures, SIL-DNN and SIL-CCNN, to identify social influences based on the propagation of information in a social network. SIL-DNN uses only the traces of propagation to train the neural network. SIL-CCNN enables the extraction of network structural features through convolutional layers. The proposed framework makes it possible to obtain the diffusion results within the community. SIL-DNN and SIL-CCNN can predict the users who are actually influenced from the information without the Monte Carlo simulation. Experimental results demonstrate that SIL- DNN and SIL-CCNN both outperform existing methods.

For further improvement, we will design a strategy to extend the depth of SIL-CCNN in the future to overcome the problem with the insufficient number of traces. We also want to revise the structure of neural network to consider more features such as the content of items. An individual should be influenced by not only other's opinion but also the content of an item. The performance of SIL-CCNN could be improved if we could learn the content representations of items.

References

1. Bakshy, E., Rosenn, I., Marlow, C., Adamic, L.: The role of social networks in information diffusion. In: Proceedings of the 21st International Conference on World Wide Web, pp. 519–528 (2012)
2. Barbieri, N., Bonchi, F., Manco, G.: Topic-aware social influence propagation models. In: Proceedings of the 2012 IEEE 12th International Conference on Data Mining, pp. 81–90 (2012)
3. Chen, W., Collins, A., Cummings, R., Ke, T., Liu, Z., Rincon, D., Sun, X., Wang, Y., Wei, W., Yuan, Y.: Influence maximization in social networks when negative opinions may emerge and propagate. In: Proceedings of the 2011 SIAM International Conference on Data Mining, pp. 379–390 (2011)
4. Chen, Y.-C., Ma, H.-S., Huang, J.-W.: Multi-state open opinion model based on positive and negative social influences. In: Proceedings of IEEE/ACM International Conference on Advances in Social Networks Analysis and Mining, pp. 170–177 (2015)
5. Cheng, J., Adamic, L., Dow, P.A., Kleinberg, J.M., Leskovec, J.: Can cascades be predicted? In: Proceedings of the 23nd International Conference on World Wide Web, pp. 925–936 (2014)
6. Chou, C.-K., Chen, M.-S.: Multiple factors-aware diffusion in social networks. In: Proceedings of Pacific-Asia Conference on Knowledge Discovery and Data Mining, pp. 70–81 (2015)
7. Kempe, D., Kleinberg, J.M., Tardos, E.: Maximizing the spread of influence through a social network. In: Proceedings of the Ninth ACM SIGKDD International Conference on Knowledge Discovery and Data Mining, pp. 137–146 (2003)
8. Kingma, D., Adam, J.B.: A method for stochastic optimization. In: Proceedings of International Conference on Learning Representations (2015)
9. Lancichinetti, A., Fortunato, S.: Benchmarks for testing community detection algorithms on directed and weighted graphs with overlapping communities. Phys. Rev. E **80**, 016118 (2009)
10. Myers, S.A., Zhu, C., Leskovec, J.: Information diffusion and external influence in networks. In: Proceedings of the 18th ACM SIGKDD International Conference on Knowledge Discovery and Data Mining, pp. 33–41 (2012)
11. Saito, K., Nakano, R., Kimura, M.: Prediction of information diffusion probabilities for independent cascade model. In: Proceedings of Knowledge-Based and Intelligent Information and Engineering Systems, pp. 67–75 (2008)
12. Tsur, O., Rappoport, A.: What's in a hashtag?: content based prediction of the spread of ideas in microblogging communities. In: Proceedings of the Fifth ACM International Conference on Web Search and Data Mining, pp. 643–652 (2012)
13. Wang, L., Ermon, S., Hopcroft, J.E.: Feature-enhanced probabilistic models for diffusion network inference. In: Proceedings of the 2012 European Conference on Machine Learning and Knowledge Discovery in Databases, pp. 499–514 (2012)
14. Weng, L., Menczer, F., Ahn, Y.-Y.: Predicting successful memes using network and community structure. In: Proceedings of the Eighth International Conference on Weblogs and Social Media (2014)
15. Xie, J., Szymanski, B.K.: Towards linear time overlapping community detection in social networks. In: Proceedings of the 16th Pacific-Asia Conference on Advances in Knowledge Discovery and Data Mining, pp. 25–36 (2012)
16. Yang, M.-H., Chou, C.-K., Chen, M.-S.: Cluster cascades: infer multiple underlying networks using diffusion data. In: Proceedings of International Conference on Advances in Social Networks Analysis and Mining, pp. 281–284 (2014)

Segment Information Extraction from Financial Annual Reports Using Neural Network

Tomoki Ito(✉), Hiroki Sakaji, and Kiyoshi Izumi

Graduate School of Engineering, The University of Tokyo, Bunkyō, Japan
m2015titoh@socsim.org, {sakaji,izumi}@sys.t.u-tokyo.ac.jp

Abstract. This is an extension from a selected paper from JSAI2019. To extract business contents automatically from financial reports is an important problem in the financial area. Especially, segment names and their explanations are important contents that should be extracted. However, the methods for extracting these types of information from financial reports have not been established. In this study, we aim to develop a practical solution for extracting these types of information. To solve this problem, we developed a manually annotated dataset for the task of extracting the segment names and their explanations of each company from financial reports and then developed a recurrent neural network model to solve this task. Our method using the manually annotated dataset outperformed the baseline methods in the task of extracting segment names and their explanations of each company from annual financial reports. In addition, we experimentally demonstrated that our method can be available for this task even when we have a small training dataset. This work is the first work for applying a machine learning method to the task of extracting segment names and their explanations. The insights from this work should be valuable in the industrial area.

Keywords: Text mining · Financial documents · Neural network model

1 Introduction

With the development of information and communication technology, interest from investors on the technology of financial text mining has been increasing. When investors conduct investment activities, it is indispensable to gather performance information on listed companies. Among them, "financial summary reports" in which listed companies are published quarterly is one of the useful information sources for making investment decisions.

To analyze the financial summary reports, information on business segments name and segment explanation is important, because each company usually describes settlement information in business segment units, as shown in Fig. 1.

Y. Ohsawa et al. (Eds.): JSAI 2019, AISC 1128, pp. 215–226, 2020.
https://doi.org/10.1007/978-3-030-39878-1_20

However, it is known that some financial companies usually extract these segment information manually, and it makes costs. In addition, segment names can be changed irregularly, and this information should be extracted on every time of change. Therefore, automatic extraction of this information has a great demand in the financial area.

Several studies on extracting the important contents from financial reports have been conducted, [10,11,18]. However, the method for extracting the segment names and segment explanations has not been established.

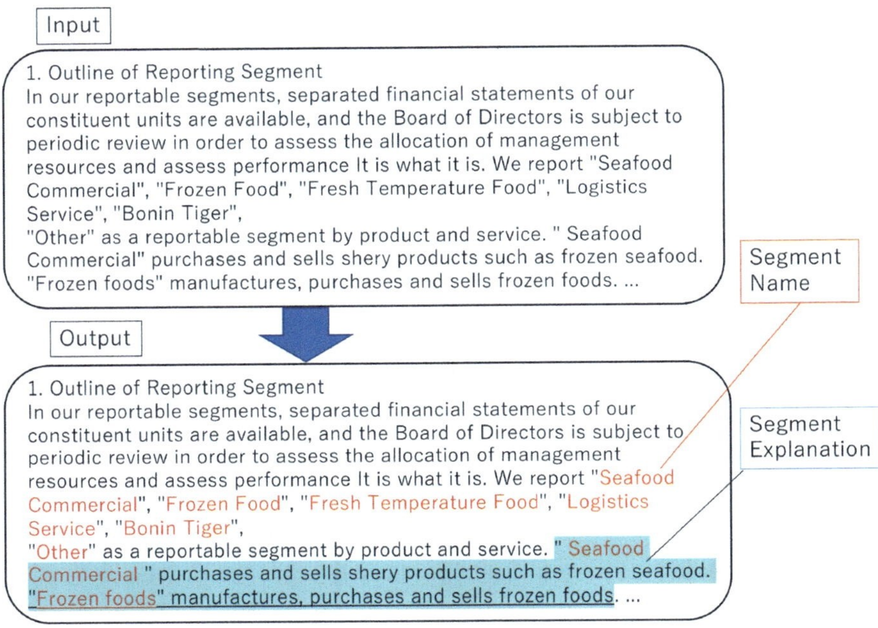

Fig. 1. Goal image: extraction of segment names and their explanations from a financial report

In this research, we aim to develop a practical method for extracting "business segment name" and "the explanation of the segment name" from the financial summary report. Using such information, we can know the business contents of companies and this type of information should be important for the decision making in the investment. To achieve our aim, we first developed manually annotated dataset including textual data for "overview of reporting segments" and "business segment name" for each financial summary reports. We then develop a method for extracting pairs of "business segment names" and their explanations using recurrent neural network model (RNN). This RNN extracts "business segment name" and start and end positions of the business segment explanation from financial summary reports using the annotated dataset. To demonstrate the practicality of our method, we demonstrate whether our appraoch can extract segment names using only small training dataset. This insight should be useful for the industrial area.

Our contributions are summarized as follows.

(1) We present a crucial task setting: extraction of segment names and segment extractions from financial reports, and created an annotated dataset for this task.
(2) We developed a practical method for extracting segment names and segment explanations from financial reports. This work is the first implementation for this task using a machine learning method, as far as we know.
(3) We demonstrated that we could extract segment names using only small training dataset, to some extent.

2 Task Setting

This section introduces the concrete task setting for this paper. Let $\mathbf{Q} = \{w_t^{\mathbf{Q}}\}_{t=1}^{N}$ be a document. Our task is divided into the following three tasks.

Task 1: Segment Name Extraction. This task aims to extract a set of segment names $\Omega^{\mathbf{Q}}$ from a document.$\mathbf{Q}$.

Task 2: Segment Explanation Extraction. This task aims to extract a set of segment explanations $\{\{w_t^{\mathbf{Q}}\}_{t=s}^{f}\}_{s \in S^{\mathbf{Q}}, f \in F^{\mathbf{Q}}}$ from a document.$\mathbf{Q}$ where $S^{\mathbf{Q}}$ means a set of start positions of a segment explanation in a document $\mathbf{Q}$, and $F^{\mathbf{Q}}$ means a set of start positions of a segment explanation in a document $\mathbf{Q}$,

Task 3: Pair of Segment Name and Its Explanation Extraction. This task aims to aims to extract a set of pairs of $\left\{(\phi^{\mathbf{Q}}, \{w_t^{\mathbf{Q}}\}_{t=s^{\phi^{\mathbf{Q}}}}^{f^{\phi^{\mathbf{Q}}}})\right\}_{\phi^{\mathbf{Q}} \in \Omega^{\mathbf{Q}}}$ where $\Omega^{\mathbf{Q}}$ means a set of segment names in a document.$\mathbf{Q}$, $\phi^{\mathbf{Q}}$ means a segment name, and $\{w_t^{\mathbf{Q}}\}_{t=s^{\phi^{\mathbf{Q}}}}^{f^{\phi^{\mathbf{Q}}}}$ means its explanation.

It should be noted that $|\Omega^{\mathbf{Q}}| and |S^{\mathbf{Q}}|, |F^{\mathbf{Q}}|$ are not constant values because the number of the segment differs between financial summary reports.

3 Extraction of the Segment Information Using RNN

This section introduces the proposed method for extracting segment names and their explanations from financial summary reports.

3.1 Structure of RNN

To solve the this task, we develop the RNN model as shown in Fig. 2. This RNN model is constructed using the idea in the pointer network model [20], which is known to be the powerful method for the information extraction.

Embedding. Given a comment $\mathbf{Q} = \{w_t^{\mathbf{Q}}\}_{t=1}^n$, this layer, first, converts all words in the comment to their respective word-level embeddings $\{\boldsymbol{e}_t^{\mathbf{Q}}\}_{t=1}^n$ where $\boldsymbol{e}_t^{\mathbf{Q}} \in \mathbb{R}^e$.

Segment Name Layer. This layer converts $\{\boldsymbol{e}_t^{\mathbf{Q}}\}_{t=1}^n$ to context representations $\{\boldsymbol{h}_t^{\mathbf{Q}}\}_{t=1}^n$ using a bi-directional gated recurrent unit , GRU [3]:

$$\boldsymbol{h}_t^{\mathbf{Q}} = \mathrm{GRU}(\boldsymbol{e}_t^{\mathbf{Q}}) \tag{1}$$

where $\boldsymbol{h}_t^{\mathbf{Q}} \in \mathbb{R}^e$. This layer, then, converts $\boldsymbol{h}_t^{\mathbf{Q}}$ to the segment name layer$\{y_t^{\mathbf{Q}}\}_{t=1}^n$:

$$\boldsymbol{a}_t^{\mathbf{Q}} = \boldsymbol{W}^O \boldsymbol{h}_t^{\mathbf{Q}} + \boldsymbol{b}, \tag{2}$$

$$y_t^{\mathbf{Q}} = \mathrm{argmax} a_t^{\mathbf{Q}} \tag{3}$$

where $y_t^{\mathbf{Q}}$ represents word $w_t^{\mathbf{Q}}$ is included in a segment name set Ω^Q ($y_t^{\mathbf{Q}} = 1$) or not ($y_t^{\mathbf{Q}} = 0$), and $\boldsymbol{W}^O \in \mathbb{R}^{2\times e}$ and $\boldsymbol{b} \in \mathbb{R}^2$ are the parameter values.

3.2 Segment Explanation Start Layer

This layer converts $\{\boldsymbol{e}_t^{\mathbf{Q}}\}_{t=1}^n$ to context representations $\{\boldsymbol{u}_t^{\mathbf{Q}}\}_{t=1}^n$ using a GRU:

$$\boldsymbol{u}_t^{\mathbf{Q}} = \mathrm{GRU}(\boldsymbol{e}_t^{\mathbf{Q}}) \tag{4}$$

where $\boldsymbol{u}_t^{\mathbf{Q}} \in \mathbb{R}^e$. This layer, then, converts $\boldsymbol{h'}_t^{\mathbf{Q}}$ to the segment explanation start point layer$\{y_t^{\mathbf{Q}}\}_{t=1}^n$:

$$\boldsymbol{c}_t^{\mathbf{Q}} = \boldsymbol{W}_s^O \boldsymbol{h}_t^{\mathbf{Q}} + \boldsymbol{b}_s, \tag{5}$$

$$s_t^{\mathbf{Q}} = \mathrm{argmax}\ \boldsymbol{c}_t^{\mathbf{Q}} \tag{6}$$

where $y_t^{\mathbf{Q}}$ represents word $w_t^{\mathbf{Q}}$ is included in a start point set $S^{\mathbf{Q}}$ ($y_t^{\mathbf{Q}} = 1$) or not ($y_t^{\mathbf{Q}} = 0$), and $\boldsymbol{W}_s^O \in \mathbb{R}^{2\times e}$ and $\boldsymbol{b}_s \in \mathbb{R}^2$ are the parameter values.

Segment Explanation Finish Layer. This layer converts $\{\boldsymbol{e}_t^{\mathbf{Q}}\}_{t=1}^n$ to context representations $\{\boldsymbol{g}_t^{\mathbf{Q}}\}_{t=1}^n$ using a GRU:

$$\boldsymbol{g}_t^{\mathbf{Q}} = \mathrm{GRU}(\boldsymbol{e}_t^{\mathbf{Q}}) \tag{7}$$

where $\boldsymbol{g}_t^{\mathbf{Q}} \in \mathbb{R}^e$. This layer, then, converts $\boldsymbol{g}_t^{\mathbf{Q}}$ to the segment explanation finish point layer$\{y_t^{\mathbf{Q}}\}_{t=1}^n$:

$$\boldsymbol{d}_t^{\mathbf{Q}} = \boldsymbol{W}_f^O \boldsymbol{h}_t^{\mathbf{Q}} + \boldsymbol{b}_f, \tag{8}$$

$$f_t^{\mathbf{Q}} = \mathrm{argmax}\ \boldsymbol{d}_t^{\mathbf{Q}} \tag{9}$$

where $f_t^{\mathbf{Q}}$ represents word $w_t^{\mathbf{Q}}$ is included in a finish point set $F^{\mathbf{Q}}$ ($f_t^{\mathbf{Q}} = 1$) or not ($f_t^{\mathbf{Q}} = 0$), and $\boldsymbol{W}_f^O \in \mathbb{R}^{2\times e}$ and $\boldsymbol{b}_f \in \mathbb{R}^2$ are the parameter values.

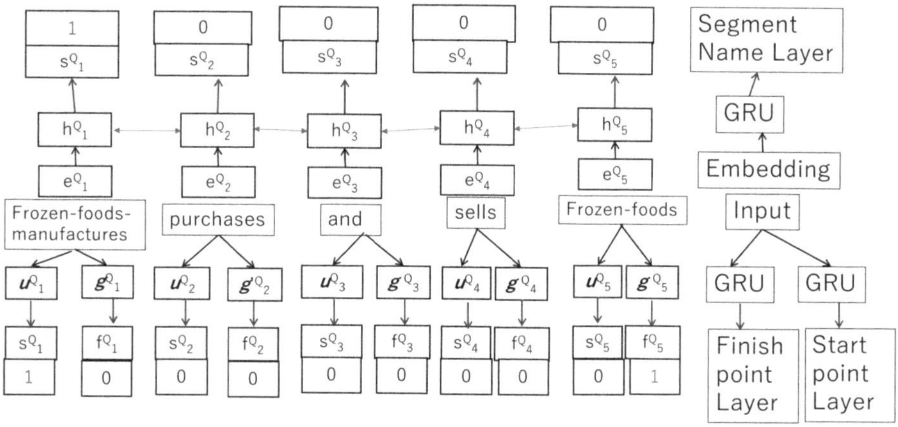

Fig. 2. RNN architecture

3.3 Learning

We can develop the RNN with a training dataset including $\{w_t^{\mathbf{Q}}\}_{t=1}^N$ and $\{\Omega^{\mathbf{Q}}, S^{\mathbf{Q}}, F^{\mathbf{Q}}\}$. We used the following L as a loss function

$$\begin{aligned} L = & CE(\{\boldsymbol{a}_t^{\mathbf{Q}}\}_{t=1}^N, \{w_t^{\mathbf{Q}} \in \Omega^{\mathbf{Q}}\}_{t=1}^N) \\ & + CE(\{\boldsymbol{c}_t^{\mathbf{Q}}\}_{t=1}^N, \{w_t^{\mathbf{Q}} \in S^{\mathbf{Q}}\}_{t=1}^N) \\ & + CE(\{\boldsymbol{d}_t^{\mathbf{Q}}\}_{t=1}^N, \{w_t^{\mathbf{Q}} \in F^{\mathbf{Q}}\}_{t=1}^N) \end{aligned}$$

where $CE(a, b)$ represents the softmac cross entropy between a and b.

3.4 Segment Information Extraction

Task 1. Segment Name Extraction. After the learning has finished, we can extract the a) segment names and b) pairs of segment names and their explanation in the following way. In extracting the segment names with the RNN, we extracted all the words $\Omega' = \{w_t^{\mathbf{Q}} | y_t^{\mathbf{Q}} = 1, 1 \leq t \leq n\}$ as the predicted segment names.

Task 2. Segment Explanation Extraction. In extracting the segment explanations with the RNN, we can extract the segment explanations by predicting the word positions $S' = \{t | s_t^{\mathbf{Q}} = 1, 1 \leq t \leq n\}$ and $F' = \{t | f_t^{\mathbf{Q}} = 1, 1 \leq t \leq n\}$ as the predicted start terms and finish terms.

Task 3. Pairs of Segment Names and Their Explanations Extraction. After extracting the Segment name and its explanation, we can extract the pairs pf segment name and its explanation considering whether each predicted segment explanation part includes the predicted segment name or not as shown in Algorithm 1.

Algorithm 1. Extraction of the Pairs of Segment Name and its Explanation

$P \leftarrow \phi$, $m \leftarrow 0$, $d \leftarrow \{\}$;
for $l \in S' + F'$ **do**
  if m = 0 **then** m $\leftarrow l$;
  else P $\leftarrow \{w_t^{\mathbf{Q}}\}_{t=m}^{l}$, m $\leftarrow 0$;
  end if
end for
for $\{w_t^{\mathbf{Q}}\}_{t=m}^{l} \in P$ **do**
  for $t \in [m, m+1, \cdots l]$ **do**
    if $y_t^{\mathbf{Q}} = 1$ **then** $d[w_t^{\mathbf{Q}}] = \{w_t^{\mathbf{Q}}\}_{t=m}^{l}$;
    end if
  end for
end for
return d: dictionary of segment name (key) and explanation (value);

4 Experimental Evaluation

This section introduces how we evaluated our method using a real textual dataset.

4.1 Dataset

To evaluate our method, we manually created a dataset including textual data for "overview of reporting segments", "business segment name", and "segment explanation" for each financial summary reports. We created dataset for 880 financial summary reports.

We then split the dataset into a training dataset and the remainder as a test dataset for this evaluation in a five variation as shown in Table 1.

Each training dataset was used for developing the RNN model, and each test dataset was used for evaluating the prediction performance.

To evaluate the practicality of our method, we evaluated our method in the case where the size of the training dataset is small and the size (i.e., dataset 1) of the training dataset is sufficiently large (i.e., dataset 5).

4.2 Evaluation Metrics

After that, we evaluated the results of our method using the Precision, Recall, and F_1 score following the SQuAD task [17].

4.3 Comparison Method

To evaluate the performance of our method, we compared the performance of our method with that of the baseline methods for each task.

Table 1. Dataset details

ID	1	2	3	4	5
(a) Training dataset					
Number reports	50	100	200	300	400
Segment names	116	250	543	836	1,134
Number of non-segment names	9,115	20,892	40,263	59,026	79,167
Number of start and finish positions	74	153	334	503	429
Number of the other positions	32,398	91,210	162,075	236,618	304,040
(b) Test dataset					
Number of reports	830	780	680	580	480
Number of segment names	2,351	2,217	1,924	1,631	1,333
Number of non-segment names	162,881	151,104	131,733	112,970	92,829
Number of start and finish positions	1,489	1,410	1,229	1,060	873
Number of the other positions	599,865	541,053	47,0188	395,645	328,223

Task 1. Segment Name Extraction. To evaluate our method for extracting segment names, we compared the results of our method with the results of extracting segment names using word embedding representations and the logistic regression model (baseline method).

Task 2. Segment Explanation Extraction. To evaluate our method for extracting segment explanations, we compared the results of our method with the results using the following baseline methods, namely, baseline (100), baseline (200), and baseline (300). In the baseline (100), baseline (200), and baseline (300) methods, we extracted the 100, 200, and 300 terms that existed after the predicted segment names as the explanation part.

Task 3. Pair of Segment Name and Its Explanation Extraction. To evaluate our method for extracting pairs of segment names and their explanations, we compared the results of our method with the results of the baseline (100), baseline (200), and baseline (300), where they were the same baselines used in Tasks 2.

4.4 Other Settings

Other experimental settings are summarized as follows: In developing RNN, we used the word embeddings calculated using the skip-gram method (window size = 5) [14] based on financial reports (between October 2002 and May 2018, 90,813 files). We set the dimensions of the RNNs' hidden and embedding vectors to 200, epoch to 40 with early stopping.

5 Results and Discussion

5.1 Result and Discussion

Task 1. Segment Name Extraction. Table 2 summarizes the results, showing that the proposed method outperformed the baseline method. In addition, the results demonstrated that the proposed method was practical because this method was able to extract segment names even with small training data. In addition, the fact that we could this task in high predictability using only small training dataset is specific. This practicality is considered to be caused by the consistent usage of words in the financial documents.

Table 2. Evaluation Result for Segment Name Extraction

ID	1	2	3	4	5
(a) Precision					
Baseline	0.123	0.524	0.624	0.603	0.593
RNN	**0.775**	**0.730**	**0.810**	**0.781**	**0.794**
(b) Recall					
Baseline	0.038	0.287	0.434	0.435	0.447
RNN	**0.935**	**0.964**	**0.952**	**0.960**	**0.975**
(c) F_1 score					
Baseline	0.058	0.371	0.512	0.505	0.510
RNN	**0.847**	**0.831**	**0.875**	**0.861**	**0.875**

Task 2. Segment Explanation Extraction. Table 3 summarizes the results, showing that the proposed method outperformed the baseline method. In addition, the results of baselines demonstrates that this extraction task can not be solved by a simple strategy, though the form of the financial report looks similar and therefore this extraction task looks easy. However, using neural network model, we can solve this problem to some extent without specific feature engineering.

Task 3. Pair of Segment Name and Its Explanation Extraction. Table 4 summarizes the results, showing that the proposed method outperformed the baseline method. It should be noted that in datasets 1–4, the precision score of the RNN in Task 3 was higher than the result in Task 2, although recall score was same. It indicates that to predict segment name and use this prediction result is helpful for extracting the segment explanation, especially in a situation where a size of traning dataset is small.

Table 3. Evaluation result for segment explanation extraction

ID	1	2	3	4	5
(a) Precision					
Baseline (100)	0.021	0.021	0.019	0.020	0.020
Baseline (200)	0.095	0.092	0.089	0.089	0.088
Baseline (300)	0.074	0.072	0.070	0.071	0.070
RNN	**0.864**	**0.856**	**0.879**	**0.872**	**0.875**
(b) Recall					
Baseline (100)	0.189	0.162	0.154	0.159	0.158
Baseline (200)	0.191	0.164	0.156	0.161	0.159
Baseline (300)	0.208	0.164	0.156	0.162	0.159
RNN	**0.556**	**0.586**	**0.663**	**0.639**	**0.639**
(c) F_1 score					
Baseline (100)	0.038	0.037	0.034	0.035	0.036
Baseline (200)	0.127	0.111	0.114	0.115	0.114
Baseline (300)	0.109	0.100	0.097	0.098	0.098
RNN	**0.676**	**0.695**	**0.756**	**0.737**	**0.738**

Table 4. Evaluation Result for Pairs of Segment Name and Segment Explanation

ID	1	2	3	4	5
(a) Precision					
Baseline (100)	0.021	0.021	0.019	0.020	0.020
Baseline (200)	0.095	0.092	0.089	0.089	0.088
Baseline (300)	0.074	0.072	0.070	0.071	0.070
RNN	**0.881**	**0.882**	**0.883**	**0.875**	**0.865**
(b) Recall					
Baseline (100)	0.189	0.162	0.154	0.159	0.158
Baseline (200)	0.191	0.164	0.156	0.161	0.159
Baseline (300)	0.208	0.164	0.156	0.162	0.159
RNN	**0.556**	**0.586**	**0.663**	**0.639**	**0.639**
(c) F_1 score					
Baseline (100)	0.074	0.037	0.034	0.035	0.036
Baseline (200)	0.208	0.111	0.114	0.115	0.114
Baseline (300)	0.109	0.100	0.097	0.098	0.098
RNN	**0.682**	**0.704**	**0.757**	**0.738**	**0.735**

5.2 Task Difficulty in Extraction of the Segment Explanation

To analyze the difficulty for Task 2, we investigated the prediction results of stat and finish positions by the RNN method, respectively.

Table 5 shows the results, showing that the detection of the end position was more difficult than that of the start point. Therefore, to improve the performance for this task, the improvement of the finish position detection part is an important issue.

Table 5. Prediction Result of the start and finish positions by the RNN model

ID	1	2	3	4	5
(a) Precision score					
Start position	0.894	0.884	0.887	0.895	0.904
End position	0.421	0.439	0.502	0.600	0.689
(b) Recall score					
Start position	0.639	0.670	0.758	0.714	0.743
End position	0.210	0.223	0.282	0.390	0.511
(c) F_1 score					
Start position	0.746	0.762	0.818	0.794	0.816
End position	0.280	0.296	0.362	0.473	0.587

6 Related Works

Several studies have been done for extracting important information from financial documents [1,7,10,11,13,16,18,19]. For example, in [19], the rule-based method for extracting up-date information form a news article was proposed. In [16], the method for extracting table contents from a financial document was proposed. In [13], method for extracting numerical information has been proposed. Moreover, several studies have been done for utilization of financial reports to a certain task [9,12]. In [12], the importance of financial reports for stock price prediction has been investigated, and this research demonstrated that financial reports were useful for the prediction of the stock price trends. In addition, in [9], the helpfulness of the annual reports for the detection of non-fraudulent firms has been demonstrated. However, the method for extracting business contents information from annual reports has hardly been established.

As for helpful technique for the task of extracting business contents from annual reports, techniques for information extraction can be presented [2,4,5,8,15]. In [2,4,8,15], utilize hand-crafted patterns from syntactic parsing in the information extraction. In [5], neural network based approach for extracting semantic relationships between entity pairs from plain text has been built.

Moreover, techniques for the Question Answering task [6,21,22] can be useful for the business contents extraction from annual reports. Especially, BERT [6] demonstrated that it has succeeded to outperform the performance of humans to some natural language understanding tasks. The application of these technique to our task can lead to the improvement of the extraction technique for extracting segment information.

7 Conclusion

In this study, we applied the RNN model to the task of extracting segment names and their explanations form financial reports. This work is the first implementation for these types of information using a machine learning method, as far as we know. We demonstrated that our method could extract segment names and their explanations with more higher F_1 scores than the baseline method. In addition, we experimentally demonstrated that we could extract segment names and their explanations with only small training dataset. This result should be the usual insight for the industrial area because this showed that our method was sufficiently practical.

In the future, first, the improvement of the extraction technology is necessary. Especially, to improve the performance in detecting the end positions of segment explanations is one of the main issues. We will apply our method to the other similar tasks, and finally, develop a text-visualization system that visualizes the financial report contents of each company in a user-friendly manner. Moreover, we will apply the proposed approach to other language resources such as Form 10-K, where Form 10-K is English annual financial reports.

Acknowledgment. This work was supported in part by JSPS KAKENHI Grant Number JP17J04768.

References

1. Alves, P., Rayson, P., Walker, M., Young, S.: Heterogeneous narrative content in annual reports published as pdf files: extraction, classification and incremental predictive ability. SSRN Electron. J. (2016)
2. Banko, M., Cafarella, M.J., Soderland, S., Broadhead, M., Etzioni, O.: Open information extraction from the web. In: Proceedings of the 20th International Joint Conference on Artifical Intelligence, pp. 2670–2676 (2007)
3. Cho, K., van Merriënboer, B., Gulcehre, C., Bougares, F., Schwenk, H., Bengio, Y.: Learning phrase representations using RNN encoder-decoder for statistical machine translation. In: Proceedings of the 2014 Conference on Empirical Methods in Natural Language Processing, pp. 1724–1734 (2014)
4. Corro, L.D., Gemulla, R.: ClausIE: clause-based open information extraction. In: In Proceedings of the 22nd International Conference on World Wide Web, pp. 355–366 (2013)
5. Cui, L., Wei, F., Zhou, M.: Neural open information extraction. In: Proceedings of the 56th Annual Meeting of the Association for Computational Linguistics, pp. 407–413 (2018)

6. Devlin, J., Chang, M.W., Lee, K., Toutanova, K.: Bert: pre-training of deep bidirectional transformers for language understanding. In: Proceedings of NAACL-HLT 2019, pp. 4171–4186 (2019)
7. EL-Haj, M., Rayson, P., Young, S., Walker, M.: Detecting document structure in a very large corpus of UK financial reports. In: Proceedings of The 9th Edition of the Language Resources and Evaluation Conference, pp. 26–31 (2014)
8. Etzioni, O., Fader, A., Christensen, J., Soderland, S., Mausam, M.: Open information extraction: the second generation. In: Proceedings of the Twenty-Second International Joint Conference on Artificial Intelligence, pp. 3–10 (2011)
9. Hajek, P., Henriques, R.: Mining corporate annual reports for intelligent detection of financial statement fraud - a comparative study of machine learning methods. Knowl.-Based Syst. **128**, 139–152 (2017)
10. Isonuma, M., Fujino, T., Mori, J., Matsuo, Y., Sakata, I.: Extractive summarization using multi-task learning with document classification. In: EMNLP (2017)
11. Kitamori, S., Sakai, H., Sakaji, H.: Extraction of sentences concerning business performance forecast and economic forecast from summaries of financial statements by deep learning. In: IEEE CIFEr (2017)
12. Lee, H., Surdeanu, M., MacCartney, B., Jurafsky, D.: On the importance of text analysis for stock price prediction. In: Proceedings of the Ninth International Conference on Language Resources and Evaluation (LREC-2014), pp. 1170–1175 (2014)
13. Madaan, A., Mittal, A., Mausam, Ramakrishnan, G., Sarawagi, S.: Numerical relation extraction with minimal supervision. In: Proceedings of Thirtieth AAAI Conference on Artificial Intelligence, pp. 2764–2771 (2016)
14. Mikolov, T., Sutskever, I., Chen, K., Corrado, G., Dean, J.: Distributed representations of words and phrases and their compositionality. In: NIPS (2013)
15. Niklaus, C., Cetto, M., Freitas, A., Handschuh, S.: A survey on open information extraction. In: Proceedings of the 27th International Conference on Computational Linguistics, pp. 3866–3878 (2018)
16. Pires, F.M., Abreu, S.: Automatic selection of table areas in documents for information extraction. In: EPIA (2013)
17. Rajpurkar, P., Zhang, J., Lopyrev, K., Liang, P.: SQuAD: 100,000+ questions for machine comprehension of text. In: Proceedings of the 2016 Conference on Empirical Methods in Natural Language Processing, pp. 2383–2392 (2016)
18. Sakaji, H., Murono, R., Sakai, H., Bennett, J., Izumi, K.: Discovery of rare causal knowledge from financial statement summaries. In: IEEE CIFEr (2017)
19. Sheikh, M., Conlon, S.: A rule-based system to extract financial information. J. Comput. Inf. Syst. **52**, 10–19 (2012)
20. Vinyals, O., Fortunato, M., Jaitly, N.: Pointer networks. In: NIPS (2015)
21. Wang, W., Yan, M., Wu, C.: Multi-granularity hierarchical attention fusion networks for reading comprehension and question answering. In: ACL (2018)
22. Wang, W., Yang, N., Wei, F., Chang, B., Zhou, M.: Gated self-matching networks for reading comprehension and question answering. In: ACL (2017)

One-Shot Learning Using Triplet Network with kNN Classifier

Mu Zhou[1,2](✉), Yusuke Tanimura[1,2], and Hidemoto Nakada[1,2]

[1] University of Tsukuba, Tsukuba, Japan
jassica.zm53@gmail.com
[2] Artifical Intelligence Research Center, National Institute of Advanced Industrial Science and Technology(AIST), Tsukuba, Japan

Abstract. This is an extension from a selected paper from JSAI2019. Humans have the ability to learn new things correctly without requiring large amount of data, while it is a challenging task in AI, which is called few-shot Learning or one-shot learning. Our key insight is using data augmentation technique to enlarge our dataset, then feeding them into a Triplet Network which is to collect same categories and separate the different. We have compared different augmentation methods, and we confirm that CVAE(Conditional VAE) can make sense as data augmentation method to slove one-shot classification problems.

Keywords: One-shot Learning · Data augmentation · Triplet network

1 Introduction

Humans can learn visual concepts quickly at a glance, whereas it is a challenge for machine to address, which is called few-shot learning. One-shot Learning or few-shot learning, which is to learn a concept from one or only a few training example, contrary to the normal Deep Learning problems [1–3] where huge amounts of labelled data is necessary.

Since deep networks need large size training data to achieve good performance, image augmentation is usually required to boost the performance of neural networks. Normal image augmentation methods create training images through different ways, such as random rotation, shifts, shear and flips. Besides, we can use generative model to create more images. All types of generative models aim at learning the true data distribution of the training set so as to generate new data points with some variations. There are two of the most commonly used and efficient approaches are Variational AutoEncoders (VAE) [9] and Generative Adversarial Networks (GAN) [12].

In our research, we use a Conditional VAE (CVAE) to augment the digits for the One-shot classification on training set, then we use a network called Triplet Network with a kNN classifier to identify the digits.

Y. Ohsawa et al. (Eds.): JSAI 2019, AISC 1128, pp. 227–235, 2020.
https://doi.org/10.1007/978-3-030-39878-1_21

2 Related Work

One of fundamental problem in computer vision is generative image modeling, which has lots of remarkable progress recently. VAE is one of generative model which formulates the problem of data generation as a bayesian model, the goal is to maximize the lower bound to learn the model. CVAE is one extension of VAE, which can generate the specific data with conditional label.

Several studies have investigated few-shot learning, one special type neural network is Siamese Networks [4]. The idea of the Siamese Network is based on distance metric learning which is to learn the distance metric from the input space of training data by a contrastive loss, then keep the samples belonging the same class close to each other and separate the dissimilar samples. The similar one is Triplet Network [7] which is composed of 3 parameter-shared convolutional neural networks.

Inspired by Siamese Networks and Triplet Networks, we improve the Triplet Network and use a triplet loss [8] in our work. The loss function is to minimize the distance between the data with same label and maximize the distance between the data with different label. Before we get the embeddings trained on networks, we do data augmentation on the training set. We use CVAE to generate the images which have only one sample. Then we make the prediction to the embedded query points by finding the nearest embedded support point by using k-Nearest Neighbor classifier.

3 Method

3.1 Conditional Variational Autoencoders

VAE learn a latent variable model for its input data. Instead of letting your neural network learn an arbitrary function, you are learning the parameters of a probability distribution modeling your data. If you sample points from this distribution, you can generate new input data samples [10].

CVAE is an extension of VAE, where we can generate some specific data. CVAE models lantent variables and data, both conditioned to some random variables.

In VAE, firstly, an encoder network turns the input samples X into two parameters z_μ and z_σ in a latent space, that describe the mean and the standard deviation of a latent normal distribution. Then, we randomly sample similar points z from the latent normal distribution that is assumed to generate the data, via $z = z_\mu + exp(z_\sigma) * \epsilon$, where ϵ is a random normal tensor. Finally, a decoder network maps these latent space points back to the original input data.

CVAE improve VAE by conditioning the encoder and decoder to another thing, a variable c. So the encoder is now conditioned to two variables X and c. The same with the decoder, it's now conditioned to two variables z and c. Hence, our generative model can generate specific data as we want. The difference between VAE and CVAE is shown in Figs. 1 and 2.

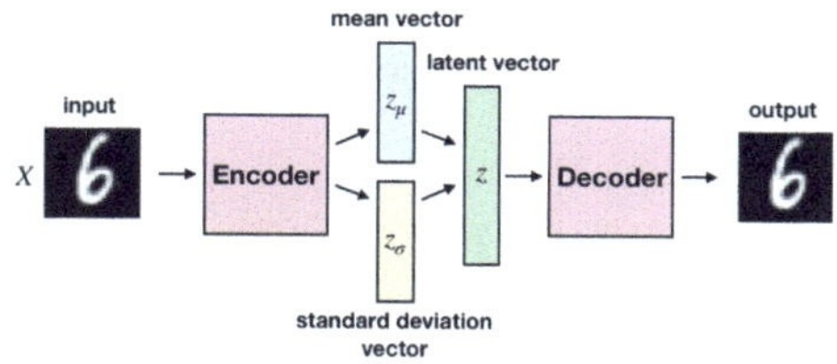

Fig. 1. Variational Autoencoders (VAE)

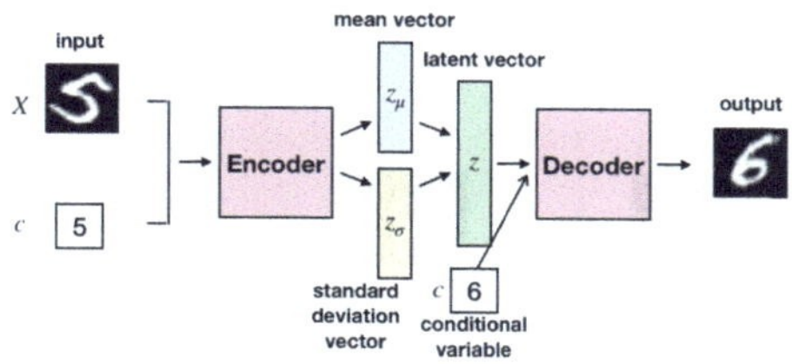

Fig. 2. Conditional Variational Autoencoder (CVAE)

3.2 Triplet Network

In this research, we use a triplet network to learn the distance metric from inputs of triplet images. The triplet network is a horizontal concatenation triplet with 3 identical Convolutional Neural Networks (with shared parameters), these ConvNets are trained using triplets of inputs. The input triplet $(\mathbf{x_a}, \mathbf{x_p}, \mathbf{x_n})$ is composed of an anchor instance $\mathbf{x_a}$, a positive instance $\mathbf{x_p}$ (same class as the anchor), and a negative instance $\mathbf{x_n}$ (different class from the anchor). The network is then trained to learn an embedding function *f(x)* called triplet loss. The model architecture is shown in Fig. 3.

Triplet Loss. Although we did not compare it to other loss function, we believe that the triplet loss is more suitable for this network, and triplet loss layer could improve the accuracy of ConvNets. A triplet loss is used to learn an embedding space for the images, such that embeddings of same class are close to each other, while those of different class are far away from each other. For the distance on the embedding space *d*, the loss of a triplet $(\mathbf{x_a}, \mathbf{x_p}, \mathbf{x_n})$ is:

$$L = max(d(x_a, x_p) - d(x_a, x_n) + \alpha, 0)$$

where α is a margin that is enforced between positive and negative pairs [8]. In our research, the triplet loss minimizes the distance between the anchor and the positive, both of which have the same identity, and maximizes the distance between the anchor and a negative of a different identity, as shown in Fig. 4.

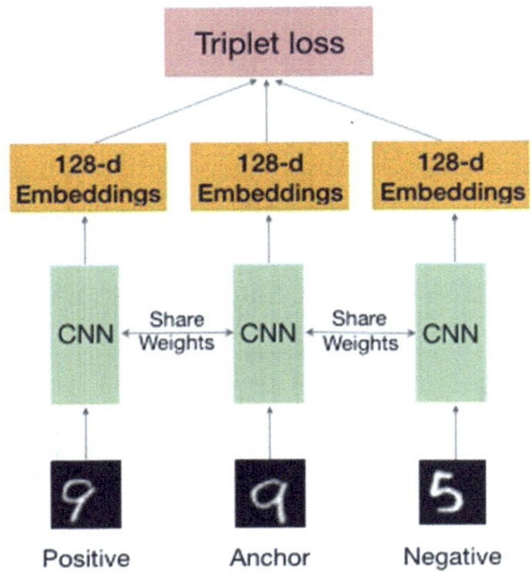

Fig. 3. Triplet network model.

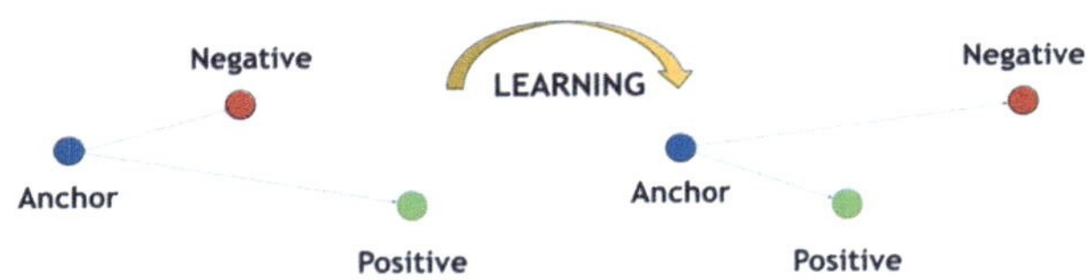

Fig. 4. Triplet loss function.

3.3 KNN Classifier

In our research, after we trained the dataset (both train and test dataset) on the Triplet Network, we obtained the embeddings of the data, each of which is a 128-dimensional feature vector. Then we used PCA (Principal component analysis) to reduce the dimension of the feature vectors. Since these vector embeddings are represented in shared vector space, we can calculate the similarity between the vectors by using the vector distance. Finally we used kNN classifier to calculate the distance between the test point and all the training points by giving the feature vector of labelled training and unlabelled test data. We gained the best choice of k and choose the corresponding classification that appears most frequently as the predictive class.

4 Experiment

4.1 Dataset

We generate our own dataset based on MNIST database which is a large database of handwritten digits. One is initial dataset and another one is augmented dataset.

Initial Dataset. To setup the training dataset, we chose whole digit images with label 0 to 4, while we randomly selected simple digit image with the label 5 to 9 from the MNIST dataset. This initial dataset was used for our comparison experiment.

Augmented Dataset. In addition to the initial dataset, we generated another training dataset by the technique of data augmentation and CVAE. Firstly, we enlarged the images with label 5 to label 9 by common augmentation technique, by which we generated our first augmented dataset. Due to the limitation of the digit images, (i.e. digit 9 may be recognized as digit 6 after the 180-degree rotation,) we did the random rotation operation with only 30 degrees combined with small random zoom and random shifts. Then, we fed these augmented images into our CVAE model, to generate more images. We learnt the low dimentional z for the style of digits from the digits 0–4, which means we want to generate digits 5–9 with the style of digits 0–4. Then we used the known z combined with our conditional variables c which are label 5 to 9, so we can enlarge the limited digits to different style. Figure 5 shows the generated images after we trained with CVAE using style of digits 0–4.

4.2 Triplet Selection

Input triplets for Triplet Network were generated in two ways. One kind of triplets was produced by the augmented dataset, while another one was created by the initial dataset which was not augmented. For the first type, we randomly selected 1 sample (used as the anchor instance) from the dataset, then chose another one (used as the positive instance) from the same label. Then we randomly obtained the other sample (used as the negative instance) from any other label. Finally, we concatenated them as a triplet pair. However, for the other type created by initial dataset, we used the same image as the positive instance to overcome the limitation of lack of samples.

4.3 Results

We evaluated the performance of our model on above two datasets (initial dataset and augmented dataset) in order to judge the effectivity of data augmentation. To estimate the performance on Triplet Network in comparison to other model,

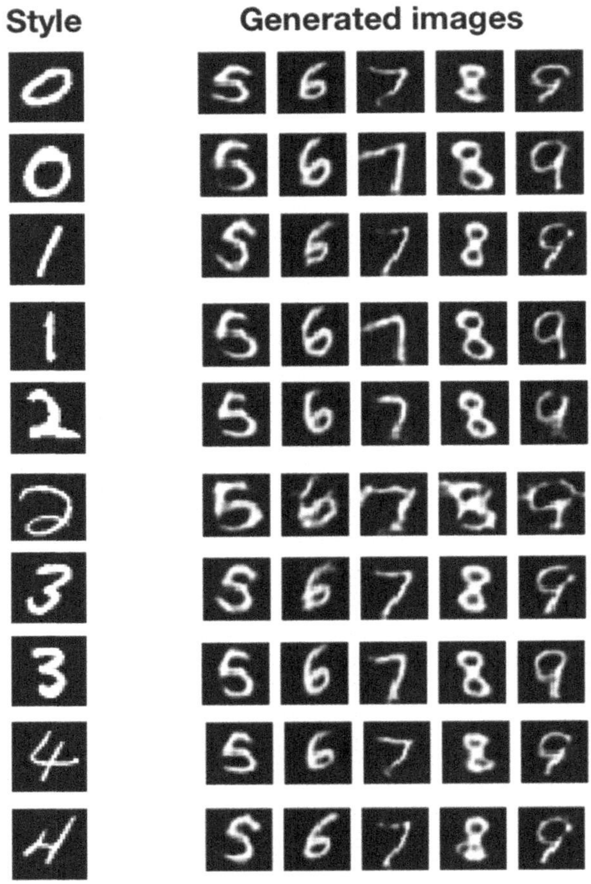

Fig. 5. The generated images of digits 5–9 with style of digits 0–4.

we applied the CNN model as is mentioned above on one-shot classification with the augmented dataset.

We obtained the embeddings of training points and test points from the Triplet Network. With these training points and test points, we evaluated the accuracy with different k ranging from 1 to 30, and selected the best choice of k. We predicted the label of test points with best k (k=11), and compared with the true label. The results are shown in Table 1, which present the accuracy of the test dataset with 1-shot classes (label 5 to label 9).

In our experiment on Triplet Network, the accuracy of the test dataset is 56.4% for 1-shot classes with CVAE augmented dataset, and 46.8% with common augmented dataset. This result suggests that data augmentation make sense in one-shot classification and can obtain better prediction result than without data augmentation. In addition, the Triplet Network gives a better performance than the CNN model in this experiment.

Figures 6 and 7 shows the result between actual label and predicted label among common augmented dataset and CVAE-augmented dataset. With regard to the accuracy of digit 9, it gets a lower score than the common augmented dataset and almost all the digit 9 are recognized as digit 4 in CVAE-augmented dataset. The reason we think may be that the handwritten digit 9 are significantly similar to handwritten digit 4. On the other hand, the reason why it worse on CVAE augmented dataset may be the generated images are not so clearly, some of them may be confused and noisy. We have tried to find out whether it depends on different style of handwritten digit 9, we choose specific digit-9 to experiment. We find the stroke of different digits may influence the accuracy, while in many cases, it cannot separate digit-4 and digit-9 as human. Figure 8 shows different score on different digits.

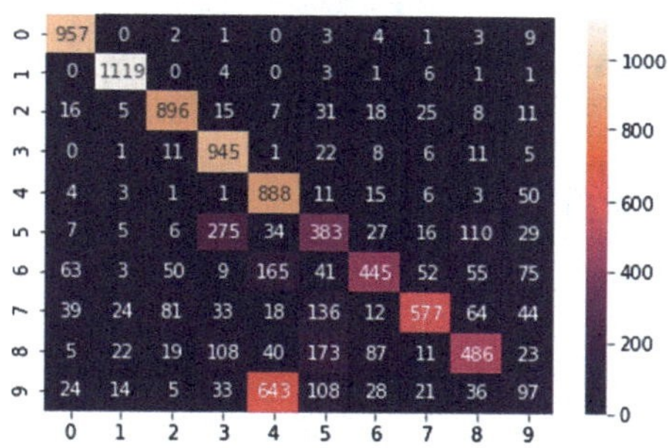

Fig. 6. The result on TripletNN with augmented dataset.

Table 1. Results of 1-shot classes.

Method	Accuracy					
	5	6	7	8	9	Average score
TripletNN(no Augmentation)	14%	18%	11%	6%	0%	9.8%
CNN(Augmentation)	25%	26%	16%	24%	13%	20.8%
TripletNN(Augmentation)	42%	56%	66%	56%	14%	46.8%
TripletNN+CVAE	**50%**	**74%**	**79%**	**69%**	**10%**	**56.4%**

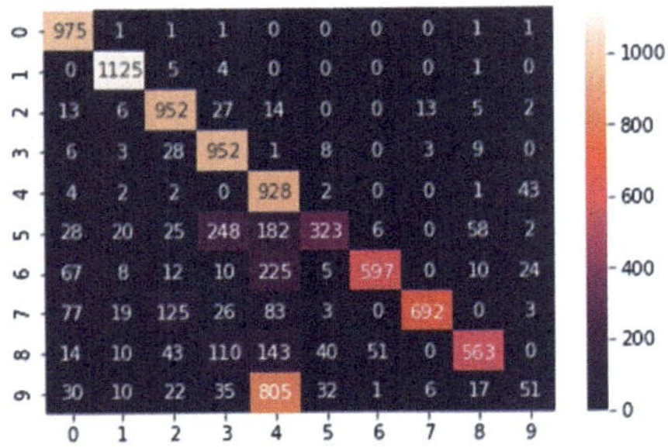

Fig. 7. The result on TripletNN with CVAE-augmented dataset.

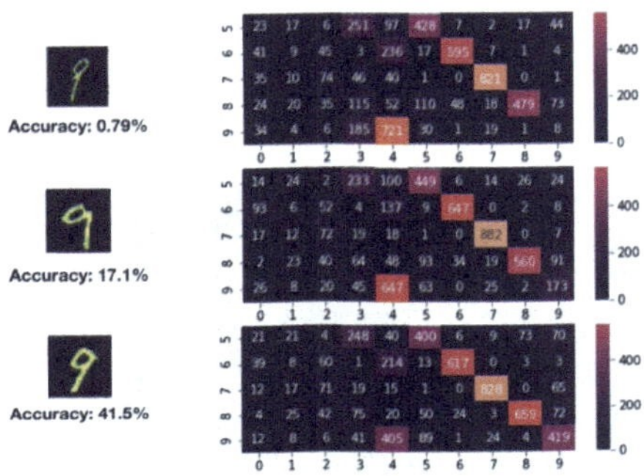

Fig. 8. The result on different digit 9.

5 Discussion

In this work, we describe a Triplet Network model which performs well on one-shot learning. We use kNN classifier to train on the training points and test points, which are the embeddings of the images trained on the Triplet Network. In this study we obtain great improvement by the effectiveness of data augmentation and CVAE. We find that in comparison experiment, almost all the data trained with 1 sample can not be recognized. However, we get better results after augmenting the data with common augmentation and CVAE augmentation. This study indicates that the benefits gained from data augmentation may work well on one-shot learning problem.

Although our experiment demonstrate a great improvement, this study still has several limitations. Most of digit 9 are recognized as digit 4 in the experiments. We try to find out the reason by the experiments on different digit 9, but we got the similar results. So we want to figure it out in future about how to seperate the difference between digit 9 and 4. In addition, due to some constraints, our experiments were unable to explore how our approaches work on other much larger and complex datasets. Therefore, given extra time and resources, we would do more experiments on other different datasets and we believe that our positive results here figure that our approach does indeed help to solve the one-shot learning problem. We would also interested in finding more effective approaches and investigating parallelism techniques to accelerate the training speed.

Acknowledgement. This paper is based on results obtained from a project commissioned by the New Energy and Industrial Technology Development Organization (NEDO). This work was supported by JSPS KAKENHI Grant Number JP16K00116.

References

1. Yu, D., Deng, L.: Automatic Speech Recognition: A Deep Learning Approach. Springer, London (2014)
2. He, K., Zhang, X., Ren, S., Sun, J.: Delving deep into rectifiers: surpassing human-level performance on imagenet classification. In: Computer Vision and Pattern Recognition (CVPR). arXiv:1502.01852 (2015)
3. Girshick, R.: Fast R-CNN. In: IEEE International Conference on Computer Vision (ICCV) (2015)
4. Koch, G., Zemel, R., Salakhutdinov, R.: Siamese neural networks for one-shot image recognition. In: ICML Deep Learning Workshop (2015)
5. Vinyals, O., Blundell, C., Lillicrap, T., Kavukcuoglu, K., Wierstra, D.: Matching networks for one shot learning. Adv. Neural Inf. Process. Syst. **29**, 3630–3638 (2016)
6. Snell, J., Swersky, K., Zemel, R.S.: Prototypical networks for few-shot learning. Adv. Neural Inf. Process. Syst. **30**, 4077–4087 (2017)
7. Hoffer, E., Ailon, N.Z.: Deep metric learning using triplet network. In: International Workshop on Similarity-Based Pattern Recognition (2015)
8. Schroff, F., Kalenichenko, D., Philbin, J.: Facenet: a unified embedding for face recognition and clustering. In: Proceedings of the IEEE Conference on Computer Vision and Pattern Recognition (2015)
9. Doersch, C.: Tutorial on Variational Autoencoders. arXiv:1606.05908 (2016)
10. Chollet, F.: Building Autoencoders in Keras (2016). https://blog.keras.io/building-autoencoders-in-keras.html
11. Kingma, D.P., Welling, M.: Auto-encoding variational bayes. In: International Conference on Learning Representations (ICLR) (2014)
12. Goodfellow, I., Pouget-Abadie, J., Mirza, M., Xu, B., Warde-Farley, D., Ozair, S., Bengio, Y.: Generative adversarial nets. Adv. Neural Inf. Process. Syst. **27**, 2672–2680 (2014)

Trees Detection on Google Street View Images Using Deep Learning and City Open Data

Lieu-Hen Chen[1(✉)], Hao-Ming Hung[1], Cheng-Yu Sun[1,3], Eric Hsiao-Kuang Wu[2], Toru Yamaguchi[3], Eri Sato-Shimokawara[3], and Hao Chen[1]

[1] National Chi Nan University, No. 303, Daxue Rd., Puli Township, Nantou County 545, Taiwan
lhchen@csie.ncnu.edu.tw

[2] National Central University, No. 300, Zhongda Rd., Zhongli Dist., Taoyuan City 320, Taiwan

[3] Tokyo Metropolitan University, 6-6 Asahigaoka, Hino-shi, Tokyo 191-0065, Japan

Abstract. This is an extension from a selected paper from JSAI2019. For almost every cities and towns, street trees play an important role in representing seasonal change of the street view. Nowadays, lots of countries start promoting open data. Among these data, very useful information related to street trees are well documented with free access by many city governments. At the same time, Google Street View provides the view of a certain surrounding by composing stitched images which are shot by specialized vehicles moving along streets and alleys. However, few research reports have been published on utilizing city open data for trees detection on Google Street View. Therefore, in this study, we aim to perform trees detection on Google Street View Images by utilizing Deep Learning technologies and city open data.

Keywords: Trees detection · Deep Learning · Open data

1 Motivation and Research Background

Street Trees are an indispensable component of great neighborhoods. Street trees play an important role for:

1. representing the style of a city, trees provide beauty and aesthetic appeal for the urban landscape.
2. improving environmental quality, especially for reducing air pollution.
3. strengthening urban amenity because trees provide spaces for rest and relaxation.

Therefore, street trees management is always an important issue for city and county governments. It is especially true for cities such as Tokyo and Washington DC where have beautiful cherry blossom seasons. Nowadays, many countries

Y. Ohsawa et al. (Eds.): JSAI 2019, AISC 1128, pp. 236–243, 2020.
https://doi.org/10.1007/978-3-030-39878-1_22

start promoting open data. The concept of open data is that certain data should be freely available to everyone to use and republish as they wish, without restrictions from copyright, patents or other mechanisms of control. Among the open data provided by some city governments, very useful information related to street trees are well documented with free access. For example, many well documented information of street trees can be accessed form the open database of Taipei and New York city, as shown in Figs. 1 and 2. These useful information includes: tree species, tree height, diameter at breast height, growth status, and position.

Fig. 1. Tree's data in Taipei

Fig. 2. Tree's data in New York

At the same time, Google Map has become the most common tool used for exploring street maps. There are many services available on Google Map. And

Google Street View is one of the feature derived from Google Map for improving users' perception of real world. It shows the view of a certain surrounding composed by stitched photographs shot by specialized vehicles moving along streets and alleys. However, these street images are static, no matter which season is now. Moreover, as shown in Fig. 3, it is still difficult to accurately locate trees on street images by using Google API because of the GPS position bias of moving vehicles and the image stitching algorithm. Therefore, in this study, we aim to develop a new approach which detects trees position on Google Street View Images by utilizing Deep Learning technologies and city open data. The remainder of this paper is organized as follows: in Sect. 2, the related works are introduced; in Sect. 3, our prototype system are explained; in Sect. 4, the current experiment results are shown; and finally in Sect. 5, the conclusion are discussed.

Fig. 3. The shifting problem of tree positions on street picture due to GPS errors

2 Related Works

2.1 Trees Detection

In comparison with the number of research works developed for detecting car and people, there are fewer papers published on solving street trees detection problem. In 2006, Wajid Ali proposed a tree detecting method by using the color and texture of trees as the basis for detection in a forest environment [1]. Later, Lu and Rasmussen proposed a tree trunk detection algorithm by using contrast templates in 2011 [2]. These studies have good results with limitations of vertical trees on grass fields. In addition, Harri Karrtinen et al. evaluated the quality, accuracy, and feasibility of automatic tree extraction methods, mainly based on laser scanner, which requires high precision and expensive instruments to collect and detect data from trees [3]. It is still very difficult to detect trees images because trees have very complex and various textures and shapes.

2.2 Machine Learning Technologies

Machine learning technologies have been widely applied in solving many problems such as object recognition and route planning, and have achieved many good results in recent years [4]. For example, in 1995, Cortes and Vapnik proposed the Support Vector Machine algorithm which separates the attribute space with a hyperplane to maximize the margin between different classes [5]. Later, in 1997, Dorigo, Gambardella used the Ant Colony Optimization approach to solve the Traveling Salesman Problem successfully [6].

3 System Implementation

Our system can be divided into the following four stages:

3.1 Using City Open Data to Find Isolated Trees' Positions, Then Collecting Their Pictures from Google Street View

At the early training stage, in our case, the size of overlapping between trees in the image is the smaller the better. In another word, when a tree is relatively far away from others, its street images will be more likely to be used as training data. Therefore, we first calculated the isolating factor for each tree recorded in the open database of New York. The isolating factor is defined as the minimum distance between the target tree with its near neighbor trees. Trees with high isolating factor were randomly selected to be used as training data. And the position information of these trees were used to retrieve suitable street tree images on Google Street View through Google APIs. In addition, clipping and resizing were also performed.

3.2 Adopting Deep Learning Approaches for Trees Image Segmentation

In our prototype system, we trained U-Net and Seg-Net for Deep Learning as shown in Fig. 4. Both U-Net and SegNet are kinds of Convolution Neural Networks. The U-Net was developed for biomedical image segmentation in 2015 [7]. SegNet was developed for autonomous driving applications to enable vehicles understand road scenes in 2016 [8]. For the research convenience, we try to reduce the network size and complexity by pre-categorizing trees images into several groups based on the tree species and seasons. Then each CNN is trained separately for different groups.

3.3 Selecting Appropriate Deep Learning Models for Tree Segmentation

By cross referencing the city open data, and the location/compass/time information of Google Street View images, appropriate Deep Learning models are then

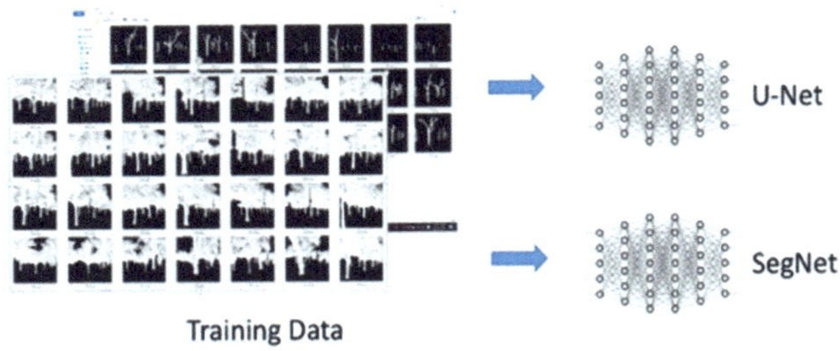

Fig. 4. System flow 1

used for image segmentation of tress. For example, as shown in Fig. 5, our system first determines that there should be a tree contained in the image, according to its GPS and compass information. Then we select the U-Net and Seg-Net which are trained by using deciduous trees images in Spring for image segmentation.

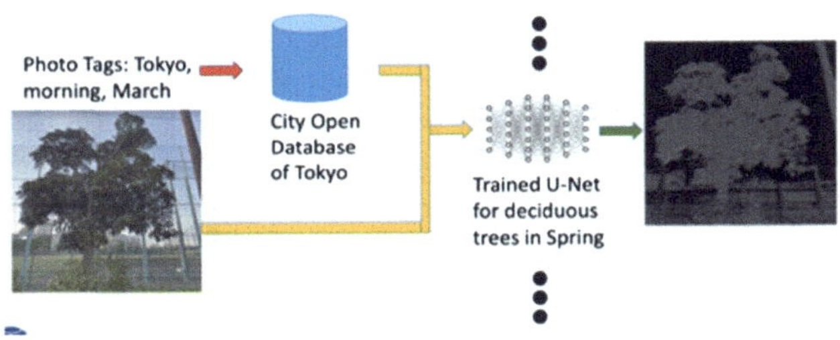

Fig. 5. System flow 2

3.4 Data Augmentation

There are different street tree species in different cities. For example, the common street tree species in New York include London planetree, Callery pear, Norway maple, and Honeylocust. On the other hand, the common street tree species in Taipei are Ficus microcarpa, Bischofia javanica, Cinnamomum camphora, and Koelreuteria elegans. Some of these trees may have very unique shapes and textures. Because shape is one of the key features for object recognition, we decided to add 17 new images of trees in Taipei with handmade labeling. The U-Net was retrained again by adding the above images to detect and segment street tress in Taipei.

4 Current Experiment Results

At the first stage, we used 100 street view pictures of New York for experiment, 40 labeled positive training data are used currently, 30 pictures for testing data,

Fig. 6. Hand-labeling tree images for Taipei

and 30 pictures for negative training data. Two segmentation results for tree images in New York and Taipei are shown in Figs. 7 and 8, respectively. It is noticeable that there are more noises in the results of Taipei. In addition, the densities of leaves are not clearly detected. After adding a few new images of trees in Taipei with handmade labeling (Fig. 6). The 2nd U-Net was retrained again to detect and segment street tress in Taipei. In comparison with the result shown in Fig. 6b, the better results are achieved as shown in Fig. 9. The dataset we collected so far is still insufficient for training Seg-Net to detect street trees very clearly. However, as shown in Fig. 10, Seg-Net can perform segmentation to multiple objects simultaneously. In Fig. 10, the blue regions are for sky, red regions for buildings, white regions for cars, black regions for road, and green regions for plants which are our target for further processing. This provides us a useful clue to filter out areas where may not contain trees and remove noises later.

Fig. 7. Segmentation for trees in New York

Fig. 8. Segmentation for trees in Taipei

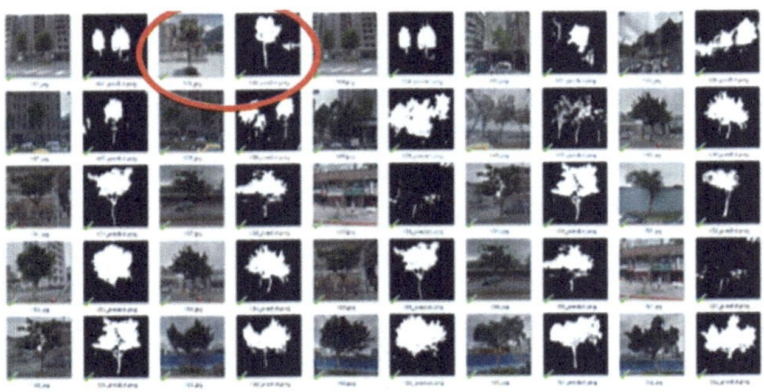

Fig. 9. Image segmentation results of street trees in Taipei

Fig. 10. A Seg-Net segmentation result for a street picture at Puli

5 Conclusion

In this research, by utilizing Deep Learning technologies and city open data, we developed a street trees detection method. First, based on the city open data, we located isolated trees and retrieved these street trees' images from Google Street View. Then we used these data for training U-Net and Seg-Net. The experiment result shows that our system has the potentials to:

1. provide an effective approach for automatically monitoring and managing street trees in smart cities.
2. serve as a useful tool for users to explore and share the great values of trees in cities and towns.

And we are currently collecting and labeling more images of common street trees in 4 seasons for training the Deep Learning models and fine tuning them. There are still some parts remained at the conceptual-level in our prototype system. In addition, more research efforts are required for extracting and separating the shapes of overlapping tree crowns. The following research works are currently ongoing:

1. Applying the trained Deep Learning models to detect trees in arbitrary video images from street cameras.
2. Performing the growth assessment of street trees by calculating the relative pixel coverage.
3. Detecting Trees with colorful leaves, flowers, and fruits in different seasons.

References

1. Ali, W.: Tree detection using color, and texture cues for autonomous navigation in forest environment. Umea University (2006)
2. Lu, Y., Rasmussen, C.: Tree trunk detection using contrast templates. In: 2011 18th IEEE International Conference on Image Processing, Brussels, pp. 1253–1256 (2011)
3. Batty, M.: Smart cities, big data. Environ. Plann. B Plann. Des. **39**, 191–193 (2012)
4. Liebig, T., Piatkowski, N., Bockermann, C., Morik, K.: Dynamic route planning with real-time traffic predictions (2015)
5. Cortes, C., Vapnik, V.: Support-Vector Networks (1995)
6. Dorigo, M., Gambardella, L.M.: Ant Colony System: A Cooperative Learning Approach to the Traveling Salesman Problem (1997)
7. Ronneberger, O., Fischer, P., Brox, T.: U-net: convolutional networks for biomedical image segmentation (2015)
8. Badrinarayanan, V., Kendall, A., Cipolla, R.: SegNet: a deep convolutional encoder-decoder architecture for image segmentation (2016)

Cyber Physics

Classification of Accessibility and Synergy in the Tokyo Area Multimodal Transportation Network

Aaron Bramson[1,2,3,4(✉)], Megumi Hori[1], Bingran Zha[1], and Hirohisa Inamoto[1]

[1] GA Technologies, Inc., Roppongi Grand Tower 40F, Roppongi 3-2-1, Minato-ku, Tokyo 106-6290, Japan
a_bramson@ga-tech.co.jp

[2] Laboratory for Symbolic Cognitive Development, RIKEN Center for Biosystems Dynamics Research, Minatojima-Minamimachi 6-7-3, Chuo-ku, Kobe 650-0047, Japan

[3] Department of General Economics, Ghent University, Tweekerkenstraat 2, 9000 Ghent, Belgium

[4] Department of Software and Information Systems, University of North Carolina Charlotte, 9201 University City Blvd., Charlotte, NC 28223, USA

Abstract. This is an extension from a selected paper from JSAI2019. Understanding accessibility in its many forms requires merging multimodal transportation network analysis with geospatial socioeconomic data and novel measures of their interaction. Building on previous research that scored and classified 500 random points within the Greater Tokyo Area, we now focus on the main urban and suburban regions of Tokyo and score all 95,118 locations in the 125 m inner radius hexagonal grid that covers it. We use the time-weighted (1) reachable area, (2) number of people, and (3) number of jobs, as well as (4) the distance to the furthest point reachable in 45 min to classify locations by their accessibility properties using unsupervised learning. We perform a similar clustering analysis on the synergy resulting from combining bus and train transport, as well as combining bus, train, and road transport. For both accessibility and synergy clusters we perform a statistical analysis of the factors determining cluster membership.

Keywords: Transportation networks · Accessibility · Synergy · Machine learning

1 Introduction

In order to better understand the geospatial patterns and determining factors of human mobility and accessibility we analyze the multimodal transportation network of the Tokyo Metro Area to classify locations by their accessibility profile. In the first case the accessibility profile is composed of four measures:

Y. Ohsawa et al. (Eds.): JSAI 2019, AISC 1128, pp. 247–266, 2020.
https://doi.org/10.1007/978-3-030-39878-1_23

reachability, *sociability*, *opportunity*, and *furthest reachable point* within 45 min. The first three measures are formally defined in Sect. 4, but essentially they are the inversely time-weighted number of hexes, people, and jobs reachable within 45 min. The values of those measures differ depending on the modes of transportation used; thus the multimodal aspect is crucial for understanding accessibility. In this paper we analyze the public transportation system (train+bus+walking) as well as the fully integrated network (adding the road network) to identify differences in accessibility via car usage. We then go on to perform a similar analysis on the mode synergy characteristics: how the use of multiple modes increases the accessibility above and beyond using any single mode. Finally, for both the accessibility and synergy groupings, we perform a statistical analysis to uncover relationships in the spatially embedded socioeconomic network structure features.

Most studies of transportation networks focus on one mode: typically train [1–3], road [4,5], or air [6] (see [7] for a review). Focusing on one mode fosters simpler structural analyses and makes comparison easier, but it fails to capture how people use a metropolitan transportation system like Tokyo's. Furthermore, in order to understand the relationship between geographical socioeconomic data and network structure/flows, the geographic and network data types need to be interfaced. As described in Sect. 2, our socioeconomic data comes from square grids of multiple scales that have been proportionally resampled into a regular hexagonal grid. In Sect. 3 we describe how we convert that hexagonal grid into a spatial network, how we construct each mode's transportation network, and how we weave all these networks together to create a single, fully integrated system.

Many details of the geographic and network data are identical or similar to previous work in the same region (for example, the method to resample the square grids into a hex grid is identical, the network data sources are identical, the first paper used reachability and sociability scores similar to those here, while the second used population and job data directly). Two previous treatments (in circulation) provide more detail on the data used and the network construction. This paper is an extension to a previous paper also using machine learning to discover clusters in accessibility scores, but here the scores are refined, the network is constrained, and we analyze the entire area instead of selected points.

2 Socioececonomic Data

Our socioeconomic data is stored in a hexagonal grid comprised of 54,127m^2 ($= 0.54\,\text{km}^2$, 125 m inner radius) hexagons. To build the hexagonal grid we use GoogleMap's coordinates of Tokyo Station (139.7649361E, 35.6812405N) as a fixed reference point and grow the hexes outward from there. Our population data obtained from [8] comes as a 250 m sided *square grid* aligned to the map using mesh coordinates from [9]. [10] reports the numbers of establishments and persons engaged/employed using a 500 m square grid, which is similarly aligned using mesh coordinates from [11]. In order to utilize this data together, we resample both datasets to our hex grid using the overlap proportions of the hexes and the respective square grids. This allows us to interpolate any geographic dataset into a homogenized foundation that also connects naturally to our geographically grounded transportation network.

The first step in our analysis is to reduce the analysis area from the entire Greater Tokyo Area (all of Tokyo, Kanagawa, Saitama, and Chiba prefectures) to just the main portion. Our previous analysis selected 500 random points from anywhere within those four prefectures, including several points on small islands, on remote mountains, and in rural farmland. From this we learned that the measurements in such areas are uninformative and uninteresting. In order to both focus on the area of genuine interest and reduce the computational load for the analysis, we restrict our coverage based on the time it takes to reach key areas of the city. Specifically, we include all hexes within any of the cutoff limits presented in Table 1. By doing so, the area covered was reduced by 61% (from 244,721 hexes to 95,118). However, the remaining 39% of the area covers 90.2% of the population (32,556,654) and 91.6% of the jobs (15,219,521).

Table 1. Cutoff thresholds for inclusion in the main area of Tokyo. Hexes satisfying any of these conditions are included in the analysis.

Subgraph center	Time threshold in minutes
Shibuya station	70
Tokyo station	60
Ikebukuro station	60
Yokohama station	45
Yokosuka station	30
Nishi Funabashi station	30

3 Network Construction

Our multimodal transportation network has five modal components: rail, bus, road, hex, and connecting links among them all. Although there are a variety of representations appropriate for transportation models, which to use depends on the research purpose. Because we are not including schedule data (we use daily average edge traversal times), we do not need a temporal network [12,13] structure. The train and bus networks are naturally seen as multigraphs with stations as nodes and a multiplicity of links connecting them [2,14]; however, such a representation requires additional machinery to incorporate transfer and waiting times. We opt for a flat (and undirected) network that expands these transfer and waiting aspects into additional edges of the network.

To construct the train network, we first take the set of physical stations and create nodes for them. Then we create a *route node* to represent the platform for each type of train (e.g., local, rapid, express) of every line that makes a stop at that station. These platform nodes are connected to their station nodes via an *access link* and to each other by *transfer links*. Finally, the route nodes at stations where a given train route makes consecutive stops are connected by a *route link*. In this way, route links of express trains directly connect the stops actually used by that route (leaving out passed intermediary stations), and different types of trains on the same tracks still require a transfer link (which captures the waiting time). For the train network, all transfer links have a traversal time of 5 min, while access links have a traversal time of 3 min. Traversal times for the route links are based on actual daily average traversal times [15]. The bus network is constructed using the same structure as the train network, using the same times for access and transfer edges (to capture wait times), and using actual daily average traversal times for the route links.

The road network is built using OpenStreetMap data for all roads larger than tertiary combined with unclassified roads (the 'road' category) [16]. First, a full network is made from all of the paths, then we simplify it by eliminating nodes with a degree of two and combining their endpoints; i.e., we eliminate non-intersection nodes. For approximate drive speeds we adopted a convention of 70 kph for major highways, 30 kph for other major roads, and 25 kph for minor roads based on speed limits from [17].

The hex network is simply the geographic hexagonal grid converted into a network by using the grid centers as nodes and connecting these nodes to neighboring grid spaces' nodes. Each hex link is approximately 250 m in length (give or take a few centimeters due to the curvature of the Earth), and we use a 5 kph walking speed to yield a 3 min traversal time. Movement on the hex grid covers walking from homes and offices to stations, bus stops, parking, taxi stands, etc., and also includes driving on smaller local access roads. Finally, any nodes (excluding train and bus *route* nodes) within 173 m of each other are connected with a walking link if not already connected. These edges are primarily to connect the distinct transportation modes together, so stations are linked to nearby hexes to ground it to the geography, as well as any bus stops and intersections located around the station. Nodes of the same mode are also sometimes connected; distinct train stations or bus stops that are nonetheless located next to each other, or (more often) street intersections for which the connecting link is below our size threshold.

In other work we used a directed version of our network in order to separately estimate and validate traffic flows going into and out of the city core. However, nearly the entire network is reciprocal, and we currently do not have heterogeneous (or time dependent) directional data for the transfer and/or access links. For the road network, genuine one-way roads exist; however, at the level of tertiary and higher, most of the directional links are just different directions of a divided road or connecting road segments among freeways. For our purposes here, we don't lose any accuracy in using an undirected version of the transportation networks. The numbers of nodes and edges for each component of our integrated multimodal transportation network appear in Table 2.

Table 2. Summary of basic network features for the fully integrated undirected network.

Transportation mode	Node count	Edge count
Train stations	1,259	—
Train transfer edges	—	16,965
Train access edges	—	4,567
Train routes	4,567	4,603
Bus stops	26,053	—
Bus transfer edges	—	236,784
Bus access edges	—	77,510
Bus routes	77,510	75,409
Road network	48,538	70,558
Hex network	95,118	282,469
Connecting links	—	172,096
Total	253,045	940,961

A flat network structure intuitively integrates the distinct modes of travel as well as connects them to our geographical foundation. It also does this while including waiting and transfer times natively as time-weighted edges. Such a structure makes incorporating heterogeneity and time-dependent values much easier while simultaneously fostering the use of simpler, more, and faster network measurement algorithms. One may consider this as a multi-layered graph: each mode comprises a layer, and the connecting edges act as inter-layer links. However, because the set of nodes is unique for each mode/layer, both the layered and flat conceptualizations of the network produce the same formal structure. We describe and analyze it as a flat network to avoid unnecessary complications.

4 Methods

In this section we describe the primary methods of our analysis. First, we provide definitions for our accessibility measures, then we define our network synergy measure based on those accessibility measures. We continue by describing the unsupervised machine learning methods used to cluster all 95,118 hexes by these accessibility and synergy features to uncover groups of areas with similar accessibility characteristics. Finally, we describe the statistical methods used to relate other features of the hexes to their accessibility and synergy groups.

4.1 Measures of Accessibility

Although the use of the term "accessibility" is varied across researchers [18, 19], we take it to be an umbrella concept that includes any assessment of the ability to reach/use surrounding resources, broadly construed. In this paper we capture four different aspects of accessibility: (1) reachable area, (2) number of people, and (3) number of jobs, as well as (4) the distance to the furthest reachable point. The first three use a similar formula to determine the time-weighted aggregate of key geosocioeconomic features, while the fourth method reflects the usefulness of transportation in terms of depth rather than breadth of the reach.

Let $S_i^{M|T}$ be the subnetwork centered at hex i and including all nodes and edges reachable from i within T minutes via transportation network M. For our analyses here we focus on analyzing two transportation networks: the hex+rail+bus network ("public") and the fully integrated network ("full") in order to highlight the accessibility effects of car usage. We also consider only subgraphs with a time horizon of $T = 45$ min based on previous results examining 20, 30, 45, and 60 min subnetworks.

Let A_j represent the area of geographic region j. For each hex node i, its *reachability* is the area of every hex j reachable from i within the time cutoff, weighted by the inverse of the time to reach j from i using transportation network M:

$$\text{reachability: } \mathcal{R}_i^{M|T} := \sum_{j \in S_i^{M|T}} \frac{A_j}{t_{ij}}. \tag{1}$$

We created a hexagonal grid covering the whole region, so in our case each j has identical area, and we can substitute 1 for the area to calculate reachability in terms of hexes (and multiply by the uniform hex area for comparison to other analyses). However, the measure in Eq. 1 is general enough to be used for square grids and irregularly shaped regions such as administrative units, Voronoi spaces, and land masses.

The calculations of *sociability* and *opportunity* follow a similar pattern as reachability, but where the hex area is replaced by the population (P_j) and economic opportunities (E_j) respectively. So, the resulting measure of sociability is

$$\text{sociability: } \mathcal{S}_i^{M|T} := \sum_{j \in S_i^{M|T}} \frac{P_j}{t_{ij}}. \tag{2}$$

In the current paper we use the total population of each hex; however variations of sociability can be constructed from subsets of the population (such as by sex, age, or wealth). Our economic data only covers full-time salaried employment engagements (aka "jobs"), so the particular variation of opportunity we utilized here is "salarayman opportunity":

$$\text{opportunity: } \mathcal{O}_i^{M|T} := \sum_{j \in S_i^{M|T}} \frac{E_j}{t_{ij}}. \tag{3}$$

Given access to additional data, variations that include part-time, seasonal, temporary, marginal, etc. kinds of jobs can also be created.

Our fourth measure of accessibility is the *furthest point* reachable from i within time T using transportation modes M. Using d_{ij} for the Haversine distance between hexes i and j:

$$\text{furthest point: } \mathcal{F}_i^{M|T} := \max_{j \in S_i^{M|T}} d_{ij}. \tag{4}$$

Beyond these measures we can define further accessibility measures using data for the number of establishments by type (i.e., the number of shops, restaurants, hospitals, schools, etc.), as well as parks, nature, recreation, sports, etc. to create fuller geographic accessibility profiles (future work).

4.2 Measures of Synergy

The common slogan for synergy is "The whole is more than the sum of its parts." We formalize this idea for modes of transportation by comparing the accessibility of modes used together compared to those same modes used separately. For each of the accessibility measures above we calculate two measures of synergy: one for the public transportation network and one for the fully integrated network. Let $\mathbf{X}$ stand for one of the accessibility measures, then we can measure public transportation synergy as:

$$\mathbf{X}\text{-synergy}_i^{\text{public}} := \frac{\mathbf{X}_i^{\text{public}|T}}{\max\left(\mathbf{X}_i^{\text{rail}|T}, \mathbf{X}_i^{\text{bus}|T}\right)}. \tag{5}$$

Adding the road network into the mix calculates the full network synergy:

$$\mathbf{X}\text{-synergy}_i^{\text{full}} := \frac{\mathbf{X}_i^{\text{full}|T}}{\max\left(\mathbf{X}_i^{\text{rail}|T}, \mathbf{X}_i^{\text{bus}|T}, \mathbf{X}_i^{\text{road}|T}\right)}. \tag{6}$$

Keep in mind that the rail, bus, and road subnetworks used here also include the underlying hex network portion and relevant connecting links. An alternative synergy calculation takes the accessibility measure on the union of the subnetworks in the denominator instead of the maximum value of the separate subnetworks. These are slightly different conceptions of synergy that imply different notions of mode choice; for now we only present the results based on the maximum value version.

4.3 K-Means Clustering

From our previous work we determined that no single clustering method was overall superior in clustering geosocioeconomic data; thus we concluded that an ensemble method that merged the result of multiple clusterings would be ideal. For this research we explored a variety of clustering methods and parameters, and we found that in many cases the clustering methods were too computationally demanding. As a result of these limitations, and based on our exploration, we have decided to present the results of k-means clustering for 9 groups and default parameters. All four clustering experiments here are run using the implementation from Python's Scikit-learn package [20]. The four experiments are the clustering of accessibility on the public and full networks, and the clustering of synergy of the public and full networks. Each experiment therefore creates clusters based on four measures (i.e., in 4D space).

4.4 Feature Importance

Although the patterns in the discovered groups are already insightful for revealing key features of the Tokyo Area transportation network, we also perform a statistical analysis of the underlying determinants of those groups. Our dependent variables are the groups found through k-means clustering. For our independent variables we have collected several features of each hex and its surroundings that are not directly determinants of accessibility, but plausibly play a role in the accessibility scores. These features include: (1) distance to the nearest station, (2) distance to the nearest bus stop, (3) distance to the nearest intersection, (4) distance to Tokyo station, (5) time to Tokyo station, (6–8) 1 km subnetwork max eigenvector centrality for each mode, (9) reachable people in 1 km, (10) reachable jobs within 1 km, (11) number of train nodes within 1 km, (12) number of bus nodes within 1 km, and (13) the number of intersections within 1 km.

Because our dependent variable is categorical (the groups), we are constrained in the statistical tests we can perform. We considered a multinomial logit/probit model, but decided to forego this method because the results do not indicate all features which may play a role in determining the category, only the features that suffice to produce those categories. For this reason we opted for a random forest analysis. Using this method allows us to relate the continuous explanatory variables to the groups and get a level of importance for each feature in assigning that category. Our analysis uses the random forest implementation from Scikit-Learn [20] with default parameters except 200 estimators.

Keep in mind that the groups are generated from four-dimensional data, the four different accessibility/synergy scores. We do not have a function that translates the four component scores (which are on highly different scales) into a single overall accessibility score. Below we use the mean of the four normalized values of each cluster's centroid to rank the outcomes for assigning colors to groups, but this rough ranking does not constitute an real aggregate score. Future work will take account of additional considerations to produce an aggregate accessibility function that rescales the components and weights them by importance. However, because of the difficulty in formulating and tuning such a score, finding similarly accessible locations is an excellent problem for unsupervised clustering analysis.

5 Results 1: Accessibility Clusters

Here we present the results of our clustering and feature importance analyses for the accessibility measures. The color in the plots below represents the group number, but (as described above) they are also ranked by the mean of the four accessibility measure values for the centroid of each cluster. By using a spectral color scheme we can roughly score areas by their aggregate accessibility score, but keep in mind that two groups could have similar aggregate values despite different accessibility profiles because one aspect of accessibility may trade off another in the mean value. We find several recognizable patterns in the groups found for both networks.

5.1 Public Transportation Accessibility

We start with the accessibility clusters in the public transportation network. The clusters identify locations with similar accessibility profiles; i.e., similar value in all four measures. Figure 1 shows the hexes belonging to each discovered cluster.

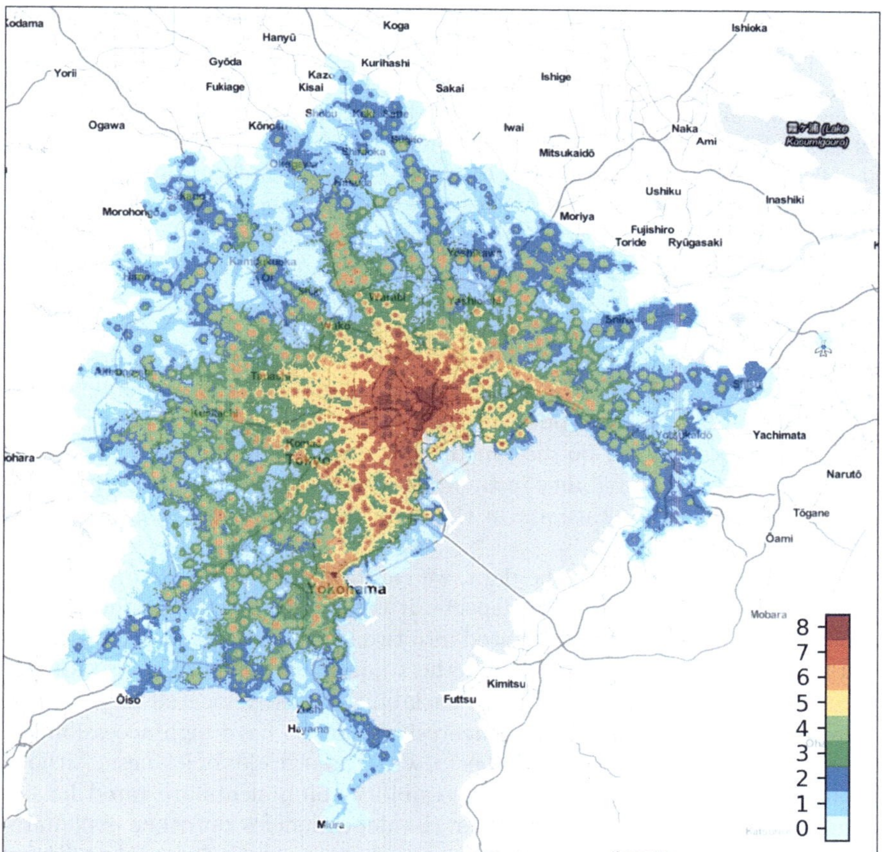

Fig. 1. Clusters in the public transportation (hex+train+bus) accessibility scores found by k-means clustering into 9 groups.

One recognizable pattern is the "pearl necklace" effect: lines radiating from the city dappled with blobs of warmer colors reflecting the locations of train stations. Even major and minor stations can be differentiated by the size of the warmer-colored blob. The second major pattern is decreasing average accessibility with greater distance from the core. We now look in more detail at the grouping results.

The city core is divided into two groups: those with major stations and large numbers of jobs are dark-red, and the surrounding areas that can easily reach these areas are bright-red. Note that in the station-dense core, these bright-red areas are also filled with nearby stations. Some stations at the fringes of the core are also bright red, transitioning into yellow (sometimes orange). Rather than being a single geographic entity, the core's group extends several kilometers, but only focused on major stations. Notice also that there is greater pull of higher values towards Yokohama than in other directions.

In this grouping, orange areas are not typically intermediate between red and yellow (as the color would indicate), but rather a high level of accessibility for a different reason. Orange hexes are predominantly suburban train stations, and beyond the 5 km ring they are surrounded by light green instead of yellow. Orange is uncommon near the core, with the exception of the main commuter lines to Chiba to the east. Thus the orange hexes play the role in the suburbs that dark-red hexes play in the core (stations); and similarly light-green hexes play the role of yellow: surrounding stations with an envelope of moderate accessibility. Although many of these light-green station surrounding areas transition into dark-green, there is a thicker spread of dark-green hexes permeating and connecting many wider areas around the core.

The next value is dark-blue, and aside from surrounding the stations in more remote residential areas, they also pick up a large number of bus stops in these areas (as do some dark-green hexes closer in). These bus stops are singular dots (darker hexes) often splayed out along a line. In some cases the contour of the underlying road can even be discerned. Although bus usage does not seem to offer much boost to accessibility in the train-centric Tokyo area, the appearance of bus stops (and train stations) in this way reflects a solid validation of the methods used.

Based on experience with the data, we can provide a plausible breakdown of the groups into their accessibility aspects. It is likely that the red zones are high in all aspects of accessibility, separated into two levels of magnitude. Orange represents areas with an especially high furthest point and moderate scores for other aspects. Greens likely score highly on sociability (mostly located in population centers), but are too far from their nearest station to have high accessibility in other aspects. Blues are low on all aspects, with three grades of lowness. Although we could look deeper into how the accessibility components are rated for each group, it is more insightful to uncover the dependencies on other explanatory features of the accessibility. We perform such an analysis below, but first we examine the accessibility clusters of the fully-integrated network.

5.2 Integrated Transportation Accessibility

Adding roads into the mix drastically changes some aspects of the results, but maintains some general patterns. For example, as can be seen in Fig. 2, it is still the case that higher values appear in the center of the core, and taper off on average with further distance. Although the train and bus networks are still present here, the dappled pattern has been replaced with a more diffusion-like pattern based on driving. That driving-dominated pattern also extends to reveal some details of the road network where specific intersections facilitate heavy traffic (especially along freeways).

As mentioned, the results here are more diffusion-like; the colors fan out from the center basically in the spectral order following natural and network-based contours. There are clumps of exceptions where expressways and train

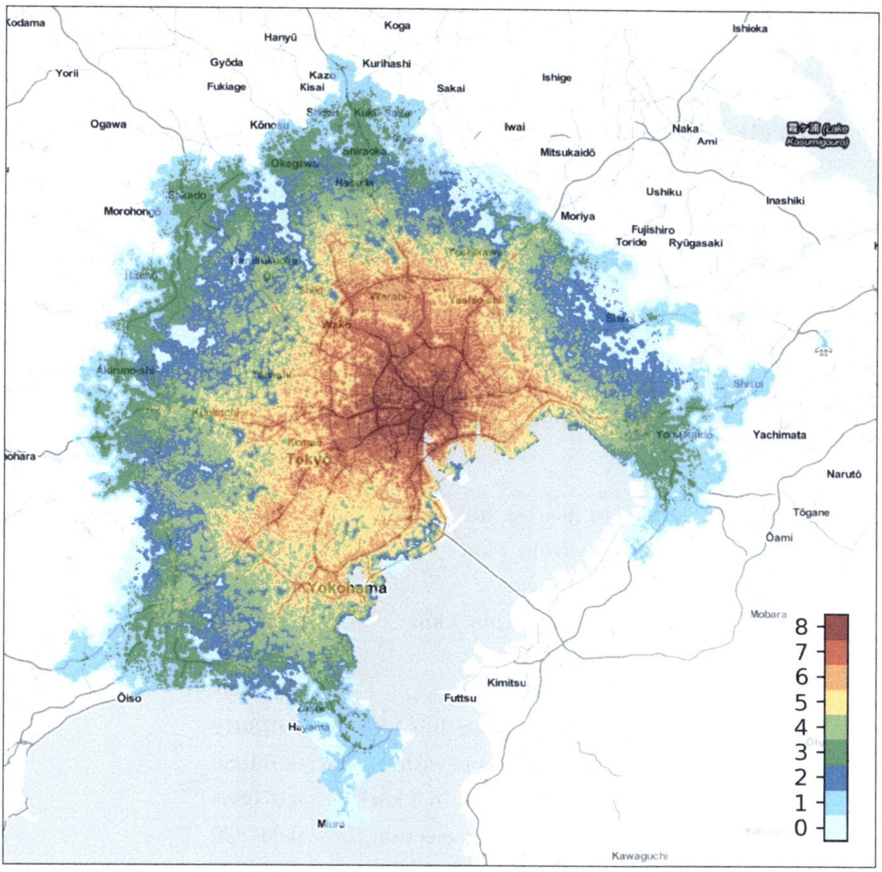

Fig. 2. Clusters in the fully integrated transportation accessibility scores found by k-means clustering into 9 groups.

stations allow more rapid access to further areas, but there is one more interesting exception to this pattern. Around the border of the included region from the west to the north there is a highway system; a ring that connects suburban regions together. The hexes around this highway system are consistently dark green. Other large expanses of dark green appear around similar highway systems to the east and southwest of Tokyo. The hexes belonging to this group surely have a similar profile because they are far from the center (hence jobs), but are able to travel great distances quickly (on the highway). For this case, most of the groups seem to merely correspond to levels of magnitude in the aspects of accessibility, but for these dark green hexes there is clearly a categorical difference.

5.3 Accessibility Determining Factors

In many ways, the results of our statistical tests confirm the qualitative description of the results above. Table 3 shows the importance values from the random forest analysis for our 13 included explanatory variables. Distance to the nearest station is the strongest factor in determining group membership, and distance to Tokyo is third, as could be seen from Fig. 1. Although maps of the local (within 1 km) population and number of jobs are not shown here, it is unsurprising that they would be strongly related to accessibility levels (and hence groups). Because sociability and opportunity are inversely weighted by travel time, even extending it to 45 min puts greatest weight on the closest people/jobs (i.e., those within 1 km). The time to Tokyo is strongly correlated with the distance, but on the public transportation network there is a great deal of spatial heterogeneity due to distances from the closest station, thus giving it a lower importance score.

Table 3. Importance level for each feature for the combined accessibility score for the public transportation network computed from a random forest analysis.

Feature	Importance
Distance to nearest station	0.155977
Population within 1 km	0.134759
Distance to Tokyo	0.111621
Number of jobs within 1 km	0.104383
Time to Tokyo	0.098885
Distance to nearest bus stop	0.076498
Number of bus stops within 1 km	0.073012
Number of intersections within 1 km	0.071985
Number of stations within 1 km	0.070885
Distance to nearest intersection	0.047830
Max eigenvector cent 1 km - train	0.018735
Max eigenvector cent 1 km - road	0.018490
Max eigenvector cent 1 km - bus	0.016941

The feature importance levels in Table 4 for the integrated network are similar, but we will point out some interesting differences. The first is how much more predictive the distance and time to Tokyo are (a combined 0.25.4% for public, to a combined 37.6% here). The second point is the change from the nearest station to the nearest intersection. Remember that trains and buses are still in this analysis, but distance to the nearest station is heavily downgraded. Despite the reputation for being a train-centric culture, the transportation system's accessibility features are dominated by the road network. We address this point in more detail through our analysis of network synergies below.

Table 4. Importance level for each feature for the combined accessibility score for the fully integrated network computed from a random forest analysis.

Feature	Importance
Time to Tokyo	0.209857
Distance to Tokyo	0.166580
Distance to nearest intersection	0.111304
Number of jobs within 1 km	0.088735
Population within 1 km	0.088435
Number of intersections within 1 km	0.080303
Number of bus stops within 1 km	0.066064
Distance to nearest station	0.051652
Distance to nearest bus stop	0.040103
Number of stations within 1 km	0.037628
Max eigenvector cent 1 km - train	0.022446
Max eigenvector cent 1 km - road	0.018631
Max eigenvector cent 1 km - bus	0.018263

6 Results 2: Network Synergy Clusters

Transportation network synergy occurs when using two modes of transportation together provides greater benefit than the combination of using either one of them separately. In this paper we are looking at synergy in the accessibility scores, meaning we are looking for areas where the accessibility of using (for example) both trains and buses together is greater than the accessibility of trains and buses by themselves.

6.1 Public Transportation Synergy

Figure 3 shows the hexes sorted into nine groups found by k-means clustering on the four synergy scores described above. These are synergy scores of the same accessibility measures as above, but we pick up different insights here. As before, the groups are ranked roughly from lowest synergy (blue) to highest synergy (red) using the mean value of the centroid's coordinates in the four-dimensional space. For the public transportation network, the greatest synergy value is slightly more than 7; meaning the value of that accessibility measure is seven times greater for the hex+train+bus network as whichever is greater between the hex+train or hex+bus networks. Locations with such high values occur where the hex is far from the station, but a bus runs through the hex to a nearby station.

The synergy map lacks the cohesive pattern found in the accessibility clusters, and they no longer clearly reflect the elements of the transportation network.

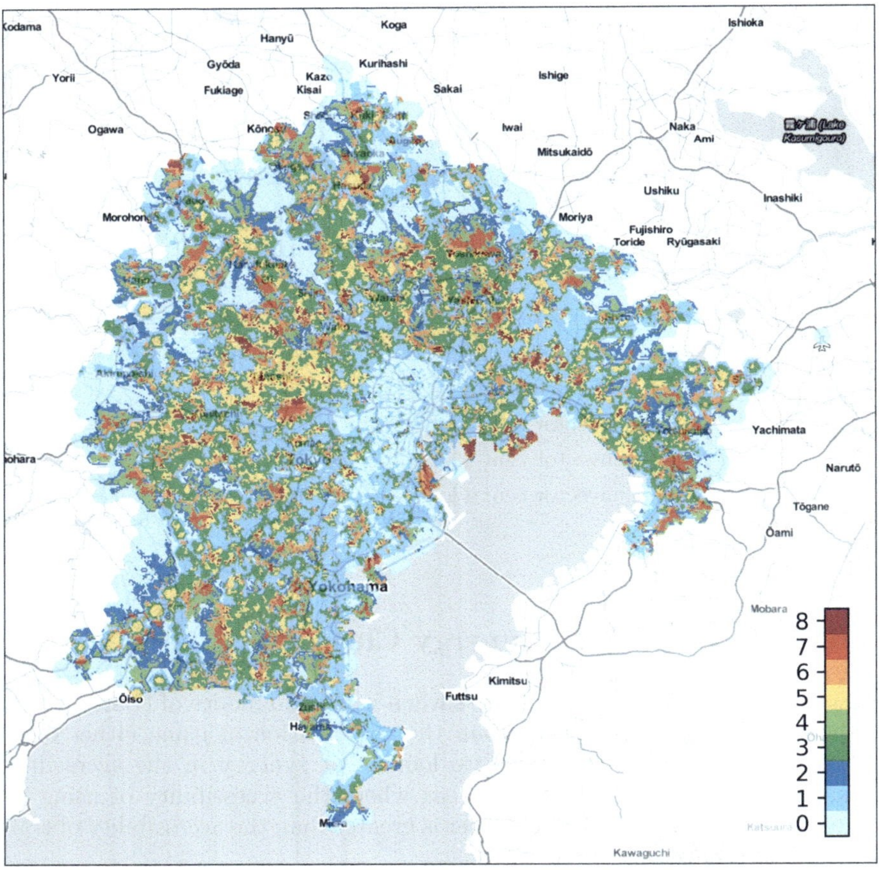

Fig. 3. Clusters in the public transportation synergy scores found by k-means clustering into 9 groups. The central core reveals little benefit of combining modes because any mode is convenient. However, many areas that are inconvenient by train alone, but well-serviced by station-bound buses, show up as red in this map.

However, there are some clear an interesting spatial patterns to examine. First, note that nearly the entire central core has no synergy benefit. This area is dense with trains, and to some degree buses. Trains are typically faster than buses, so what we are seeing is that in the core, trains dominate bus usage and there is no synergy. This is also true for various mountain tops and rivers and valleys where neither buses nor trains are useful. Because it is a ratio, similar values can be achieved where both the numerator and denominator are high, or both are low.

There is a slight station-oriented pattern here as well, and visible bus lines, but the pattern is more marbled than dappled. The core is unsynergistic, but the area just outside the core (especially from the west to the northeast) exhibits a high number of highly synergistic areas. The outer reaches of the area reveal the clearest train-based pattern, but there is an interesting reversal. Nearer the center, the hexes around stations show the lowest levels of synergy, but in the outskirts the train stations often produce high synergy. The reason for this is bus usage. In the station-dense inner areas people can generally walk directly from stations efficiently, but in the sparser areas buses become more important (and more useful). Thus areas near the center with low synergy may be the result of buses being generally unneeded rather than unavailable (or unused).

6.2 Integrated Transportation Synergy

Finally, we come to synergy in the fully integrated network. Recall my comment above regarding the dominance of car usage in what is generally regarded as a train-based transportation system. Figure 4 shows the accessibility synergy levels when the road network is added to the public transportation network. As before, synergy exists when a hex has greater accessibility using the integrated network than any of the three component networks alone. The large swaths of light and medium blue tell us that in most locations there is little to no benefit from integrated multimode transportation. For nearly all of these hexes the reason is that using the road network allows one to reach all the places the train or bus network can take you, plus a lot more.

The areas with the highest synergy (Narita airport and Mt Takao) are not directly reachable by the road network, and that gives them the greatest synergy. The patches of moderate values occur in locations like valleys, forests, and mountains where the roads are narrow and curvy, and there are few branching points to foster penetration into the nearby hexes. Trains running alongside these roads tend to be more efficient than driving, so the combination becomes superior to the roads in these cases.

6.3 Synergy Determining Factors

As before, we use a random forest method to determine the factors that contribute the most to group membership. Table 5 does so for the public transportation synergy groups. Distance to the nearest station is still the top factor (although in this case it is sometimes negatively and sometimes positively related, the random forest method itself does not reveal that detail). Distance

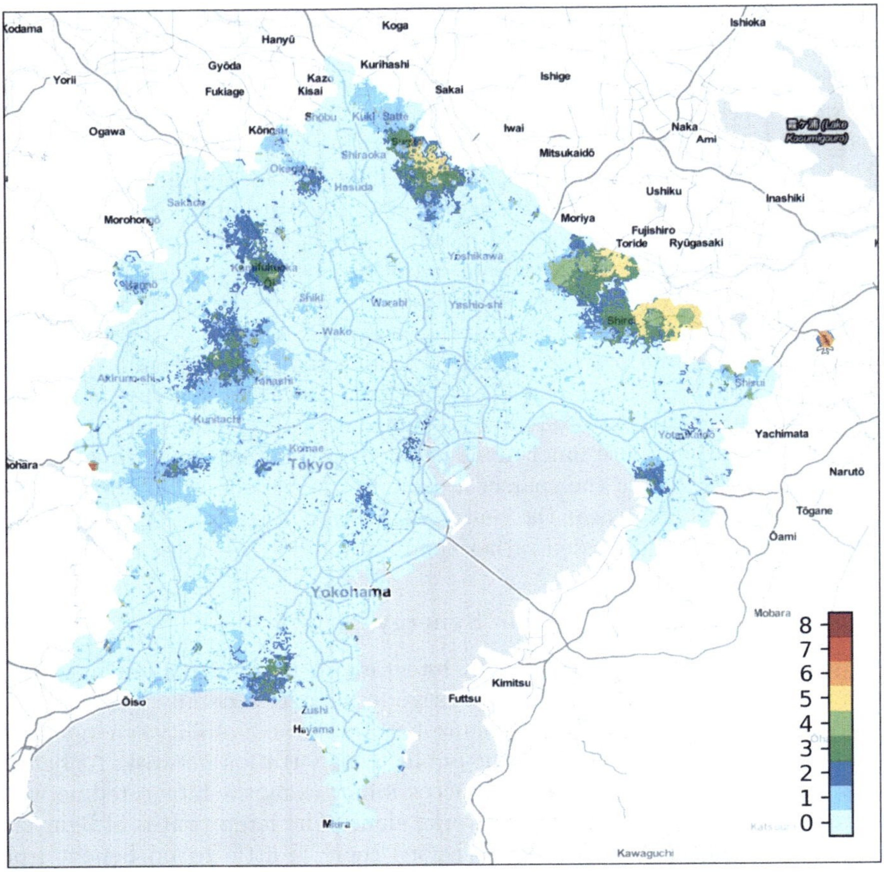

Fig. 4. Clusters in the full transportation synergy scores found by k-means clustering into 9 groups. Large areas of light blue correspond to areas with no synergy, nearly all of which are car-dominated.

and time to Tokyo both inform us of the obvious spatial pattern, although here the values are lower rather than highest in the core, we are only explaining group membership with these factors.

Local population and jobs are still important for group membership. This was clear for accessibility groups because these variables are closely related to the accessibility scores, but this is less obvious for synergies. Because synergies are ratios of accessibility values, the synergy values can be similar even where the accessibility values are not. This difference produces the highly differing patterns in the maps, but (at least among these 13 explanatory variables) these two variables are nearly as important for synergy as for accessibility.

Table 5. Importance level for each feature for the combined synergy score for the public transportation network computed from a random forest analysis.

Feature	Importance
Distance to nearest station	0.148403
Distance to Tokyo	0.106564
Population within 1 km	0.104020
Number of jobs within 1 km	0.097425
Time to Tokyo	0.091698
Distance to nearest bus stop	0.089041
Number of bus stops within 1 km	0.086959
Number of intersections within 1 km	0.077219
Number of stations within 1 km	0.074835
Distance to nearest intersection	0.063061
Max eigenvector cent 1 km - road	0.025207
Max eigenvector cent 1 km - train	0.018488
Max eigenvector cent 1 km - bus	0.017082

Table 6 reports the importance values for the synergy groups in the fully integrated network. One thing to notice is the dramatic drop in the importance level of the highest factors. Rather then being sufficiently explained by the top 3–6 variables, the values are more widely spread across 8 or 9. Distance and time to Tokyo station are still on top because most groups only have members at the fringes of the region.

Referring back to Fig. 4, note the medium and dark blue specks among the sea of light blue. These are areas with similar (and low) levels of synergy. There are also larger areas that share the same group membership. For some locations it may be the case that a train offers a particular useful express route that is faster than any road route, but that taking a car (e.g., taxi) to/from the origin and destination stations adds a bit more accessibility compared to walking. This reasonably explains the dark blue surrounding Chiba station to the east of the core as well as the dark-blue around Shinagawa station filling the south part of the core. What is curious about the isolated locations is that being one hex away can sufficiently differentiate the synergy score to put is in a separate group from all of its neighbors. This can be explored more with different clustering parameters as well as directly mapping all the individual accessibility and synergy scores (which is beyond the scope of the current paper).

Table 6. Importance level for each feature for the combined synergy score for the fully integrated network computed from a random forest analysis.

Feature	Importance
Distance to Tokyo	0.139507
Time to Tokyo	0.126022
Number of jobs within 1 km	0.097276
Population within 1 km	0.094745
Distance to nearest intersection	0.090485
Number of bus stops within 1 km	0.088033
Distance to nearest station	0.085011
Number of intersections within 1 km	0.081462
Distance to nearest bus stop	0.072312
Number of stations within 1 km	0.054618
Max eigenvector cent 1 km - train	0.028015
Max eigenvector cent 1 km - road	0.021694
Max eigenvector cent 1 km - bus	0.020817

7 Conclusions

When interpreting the results here and above, we need to understand that there is a border effect in our network, especially for the road network, and especially for reachability. The effect exists because in most cases the edge of the included region is not a natural boundary. Another way to interpret our area reduction is to say that the region contains more than 90% of the people and jobs, but it only includes 40% of the land area. Many of the removed areas can only be reached by car, so the accessibility of the road network should be even stronger than it is near the region boundaries. This artificial depression likely affects group membership as well. The solution to this bias is not obvious, so here we merely point it out.

In no cases were standard network measures (such as eigenvector centrality, degree, etc.) strongly related to accessibility measures. In fact, for the public transportation network analysis (that doesn't use roads) having a high-eigenvector centrality node within 1 km is less of a determinant than even the distance to the nearest intersection. Measures like accessibility require the fusion of socioeconomic data, geographic measures, and network analysis; but traditional network analysis is not the right set of tools to do this. The characteristics related to importance in social networks do not translate well into transportation networks (despite several efforts to apply them [1–4,18,21]). There are novel approaches to capturing and measuring these spatially explicit networks ([22,23]), and we look forward to continuing to contribute to this body of research.

From this analysis we uncovered patterns in accessibility for the Tokyo area for two cases: public transportation and all modes of transportation. We found several patterns with intuitive explanations that work to validate the data and methods used. We then leverage that face validity to uncover additional, and less obvious, accessibility characteristics. Because creating a tuned function to aggregate the various components of accessibility requires unavailable data and would be use-dependent, we instead reveal the patterns through the use of k-means clustering among four disparate accessibility factors. Because these groups are categorical, we use the random forest method (instead of regression methods) to tie geographical and network-based features to the group membership.

This research is ongoing; and future work aims to both overcome some limitations of our network construction and also to expand the aspects included in determining accessibility. In most cases the limiting factor is a lack of data for the Japanese transportation system and related socioeconomic features [24]. We continue to gather and generate additional data to incorporate an increasing number of considerations into our analyses. We look forward to continued fruitful research and collaboration.

References

1. Derrible, S., Kennedy, C.: Transp. Res. Rec. **2112**(1), 17 (2009)
2. Barthélemy, M.: Phys. Rep. **499**(1–3), 1 (2011)
3. Derrible, S.: PloS One **7**(7), e40575 (2012)
4. Crucitti, P., Latora, V., Porta, S.: Chaos: Interdiscip. J. Nonlinear Sci. **16**(1), 015113 (2006)
5. Ren, Y., Ercsey-Ravasz, M., Wang, P., González, M.C., Toroczkai, Z.: Nature Commun. **5**, 5347 (2014)
6. Guimera, R., Mossa, S., Turtschi, A., Amaral, L.N.: Proc. Nat. Acad. Sci. **102**(22), 7794 (2005)
7. Derrible, S., Kennedy, C.: Transp. Rev. **31**(4), 495 (2011)
8. Official Statistics of Japan: Statistical data/Census/2015/5th layer mesh/(250m mesh)/Part 1 Population and aggregation (2015). www.e-stat.go.jp. Accessed 12 Dec 2018
9. Association for Promotion of Infrastructure Geospatial Information Distribution. By prefecture 250m Mesh (2015). www.geospatial.jp. Accessed 12 Dec 2018
10. Official Statistics of Japan: Economic census for business frame, tabulation of establishments, results for Japan (2014). www.e-stat.go.jp. Accessed 12 Jan 2012
11. Association for Promotion of Infrastructure Geospatial Information Distribution. By prefecture 500m Mesh (2015). www.geospatial.jp. Accessed 4 Apr 2018
12. Bramson, A., Vandermarliere, B.: Sci. Rep. **6**, 34052 (2016)
13. Masuda, N., Holme, P.: Temporal Network Epidemiology. Springer (2017)
14. Goczyłla, K., Cielatkowski, J.: Eur. J. Oper. Res. **87**(2), 214 (1995)
15. Ekitan: Train and bus schedule data (2019). www.ekitan.co.jp
16. OpenStreetMap Contributors. Planet dump (2019). https://planet.osm.org, https://www.openstreetmap.org
17. Japan Traffic Safety Association: Rules of the road (2017)
18. Levinson, D.: PloS One **7**(1), e29721 (2012)
19. Biazzo, I., Monechi, B., Loreto, V.: arXiv preprint arXiv:1810.03017 (2018)

20. Pedregosa, F., Varoquaux, G., Gramfort, A., Michel, V., Thirion, B., Grisel, O., Blondel, M., Prettenhofer, P., Weiss, R., Dubourg, V., Vanderplas, J., Passos, A., Cournapeau, D., Brucher, M., Perrot, M., Duchesnay, E.: J. Mach. Learn. Res. **12**, 2825 (2011)
21. Chorus, P., Bertolini, L.: J. Transp. Land Use **4**(1), 45 (2011)
22. Austwick, M.Z., O'Brien, O., Strano, E., Viana, M.: PloS One **8**(9), e74685 (2013)
23. Kivelä, M., Arenas, A., Barthelemy, M., Gleeson, J.P., Moreno, Y., Porter, M.A.: J. Complex Netw. **2**(3), 203 (2014)
24. Kawasaki, S.: IATSS Res. **39**(1), 1 (2015)

Modelling Naturalistic Work Stress Using Spectral HRV Representations and Deep Learning

Juan Lorenzo Hagad[1(✉)], Ken-ichi Fukui[2], and Masayuki Numao[2]

[1] Graduate School of Information Science and Technology, Osaka University, Suita, Japan
hagad@ai.sanken.osaka-u.ac.jp

[2] Institute of Scientific and Industrial Research, Osaka University, Suita, Japan

Abstract. This is an extension from a selected paper from JSAI2019. With the proliferation of wearable devices and the inflow of new health data, artificial intelligence is expected to revolutionize the field of wellness and health management by providing potential tools for analyzing harmful conditions like prolonged stress. Currently, one of the standard measurements used by medical practitioners to measure stress is heart rate variability (HRV), a set of numerical indices that reflect autonomic balance. However, recent advances in machine learning have shown that learned features tend to outperform hand-crafted features. In this work we propose a more expressive intermediate data representation based on Lomb-Scargle periodograms combined with the feature learning capabilities of deep learning. Using stress data from naturalistic work activities, we tested different shallow and deep learning architectures and show that significant improvements can be achieved compared to traditional HRV indices. Results show that models trained on our spectral-temporal representation significantly outperform models trained on traditional HRV indices for predicting naturalistic work stress.

Keywords: Physiological data · Health data · Artificial intelligence · Spectral analysis of HRV · Spectral-temporal representations

1 Introduction

Findings from a large number of clinical studies [7, 10] have linked cardio-vascular disorders to psychological stressors. In particular, long-term exposure to stressful environments, such as those commonly found in offices, has been shown to be a major risk factor for major depressive disorder (MDD), a disorder that affects an estimated 350 million people across the world [11]. In response to this, we have witnessed a surge in the development of ambient intelligences for the purpose of mental health and stress monitoring [1].

Y. Ohsawa et al. (Eds.): JSAI 2019, AISC 1128, pp. 267–277, 2020.
https://doi.org/10.1007/978-3-030-39878-1_24

1.1 HRV Stress Analysis

Historically, heart rate variability (HRV) spectrograms were used by medical practitioners as the primary tool to measure cardiac neural regulation externally. This was brought out by findings that mental and physical states have certain effects on cardiac rhythm that could easily and unintrusively be measured with electrocardiographs (ECG) [8]. In the past few decades, studies have established the connection between HRV and the sympatho-vagal stress response as being visible in both the time and frequency domains. Through experiments and empirical anlaysis, these works have identified indices which show high correlations with physical and mental stress, particulary with the high and low frequency ranges [12].

Despite our growing understanding of physiological signals, even now there is no widely accepted standard for stress evaluation based on HRV features. In recent surveys [9,16] it has been found that neurobiological evidence links some components of HRV to activity in cortical regions related to stress appraisal as well as other brain mechanisms. As it stands, it seems that one of the main challenges is identifying which components of HRV may be relevant to evaluating stress.

1.2 Deep Learning and Heart Rate Analysis

Traditionally, affective analysis such as those done for stress have mostly focused towards using visual or auditory measures [17], where deep learning has been found to be effective. In recent years, others works in the field of ECG analysis for medical applications have also pushed for end-to-end applications of deep learning.

In one of the most prominent works [14] they used a 34-layer residual convolutional neural network (CNN) to analyze raw ECG sequences in order to identify 14 different types of arrhythmia. Using a very large dataset of over 60,000 recordings they were able to train a model that performed better on average than actual cardiologists recruited for the experiment. Their work demonstrates how deep learning is able to identify structural patterns in heart rate data.

In the current work, we attempt to exclude morphological patterns and instead focus on frequency-related features of heart beats. Towards this end we introduce the use of short-time Lomb-Scargle (LS) spectrograms to train CNNs.

2 Methodology

This work aims to use spectral HRV data representations together with deep learning models to predict mental stress. Specifically, we use a sliding window Lomb-Scargle approach to build a spectrogram of heart rate in lieu of the traditional short-term Fourier transform (STFT). Next, we use pre-built deep models designed for image recognition and compared them with more simplified architectures and train them to recognize stress. In the following sections we describe the experiments and tools used in this research.

2.1 Dataset

For the following experiments we use the same work stress dataset introduced in a previous work [5]. It consists of ECG recordings of subjects performing naturalistic desk work activities in front of a PC. The physiological recording device used was a wearable ECG sensor by Imec[1]. This allowed unconstrained and continuous measurement of participants' heart rate. The sensor is worn with leads placed diagonally over the heart as shown in Fig. 1. RR beats were then extracted from the raw ECG signal using a QRS detection toolkit [4], which were then used to build the HRV spectrograms.

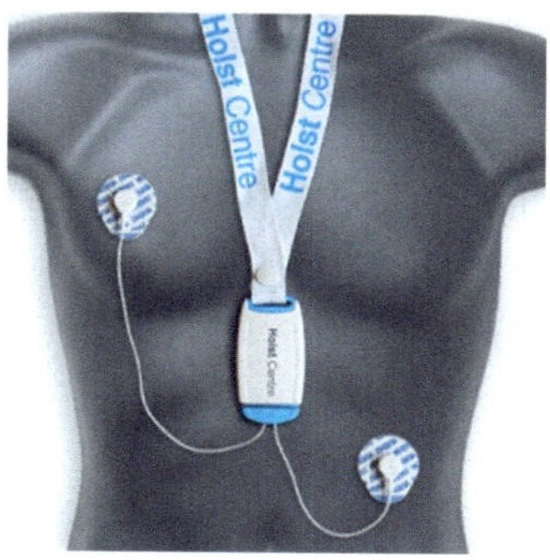

Fig. 1. The 2-lead wearable ECG sensor used in the experiments.

The dataset was gathered from four subjects each contributing five separate hour-long sessions annotated with activity and stress labels as shown in Table 1. This provided us with a total of 20 h of ECG data that was used for training and testing.

Table 1. Stress annotations distribution

	Tasks	Very low	Low	High	Very high
Total	196	46	78	53	18
Percentage		23.47%	39.80%	27.04%	9.18%

2.2 Signal Processing

Previous works on spectral anlaysis of HRV applied the fast fourier transform (FFT) [19] to produce periodograms from which the power of different frequency bands can be measured and compared. However, one limitation of the FFT algorithm is that it assumes that all samples used for analysis are evenly spaced in time. This poses a fundamental issue for heart rate since the RR intervals used

[1] https://www.imec-int.com/nl/expertise/wearables.

to represent detected heart beats are not evenly sampled. A study [2] has shown that data resampling methods used to align RR interval data for FFT analysis can lead to artifacting and higher errors in the final PSD estimation. A better alternative has been put forward in the form of the Lomb-Scargle periodogram, a method based on FFT that does not require evenly sampled data.

Lomb-Scargle Periodogram. Periodograms are some of the most basic tools used for spectral analysis, the most common of which is the discrete Fourier transform (DFT). The DFT can be defined for any sampled dataset, $X(t_i), i = 1, 2, ..., N_0$ as,

$$FT_X(\omega) = \sum_{j=1}^{N_0} X(t_j) \exp(-i\omega t_j). \tag{1}$$

The periodogram can then be defined as,

$$\begin{aligned} P_X(\omega) &= \frac{1}{N_0} |FT_x(\omega)|^2 \\ &= \frac{1}{N_0} \left| \sum_{j=1}^{N_0} X(t_j) \exp(-i\omega t_j) \right|^2 \\ &= \frac{1}{N_0} \left[\left(\sum_j X_j \cos \omega t_j \right)^2 + \left(\sum_j X_j \sin \omega t_j \right)^2 \right] \end{aligned}, \tag{2}$$

[3] which while sufficient for calculating an accurate estimation of the FFT of most signals was found to be sensitive to noisy signals and other artifacts [18]. In practice, it is necessary to apply additional spectral smoothing equations, such as *spectral window functions*, to reduce the variances of the spectrum. Most of these methods are designed to be applied on evenly sampled data. Though some may also be applied to the unevenly sampled case, such as in RR intervals, it has been found that it can lead to higher error and increased sensitivity to signal noise [13]. This can also be mitigated by the addition of more data for calculating the spectra as a way to taper off the signal-to-noise ratio, but this may not be an option for data-scarce or time-intensive domains. A reasonable alternative is the Lomb-Scargle periodogram [15], a notable method designed for use in calculating frequency spectrograms from unevenly sampled data streams such as heart beats. The equation is based on the FFT and is wholly similar but addresses some of the noise sensitivity and *spectral leakage* issues of the standard FFT equation. Here, the periodogram is defined as,

$$P_X(\omega) = \frac{1}{2} \left\{ \frac{\left[\sum_j X_j \cos \omega (t_j - \tau) \right]^2}{\sum_j \cos^2 \omega (t_j - \tau)} \right\} + \frac{\left[\sum_j X_j \sin \omega (t_j - \tau) \right]^2}{\sum_j \sin^2 \omega (t_j - \tau)} \tag{3}$$

with τ defined as,

$$\tan(2\omega\tau) = \left(\sum_j \sin 2\omega t_j\right) \Big/ \left(\sum_j \cos 2\omega t_j\right). \tag{4}$$

This modified equation, outputs similar values to the original FFT in most cases, however it has a simpler statistical behavior and is equivalent to the reduction of the sum of squares in least-squares fitting of sin waves to data [15]. It is also time-translation invariant and reduces to a similar representation as DFT in cases where data is evenly spaced. This makes it ideal for HRV analysis and is the method used in this research. The next issue is how to tackle the non-stationary nature of heart rate. For this, we construct spectrograms from the periodograms.

Spectrogram Analysis. Spectrograms are visual representations of frequency spectra over time. They are tools that are used extensively in applications that deal with audio and radio signals. Recently, they have also seen frequent application in the analysis of physiological signals such as brainwaves and heart rate. The most common method used for building spectrograms is the short-time Fourier transform (STFT). Put simply it is built by applying the DFT over a pre-defined window of data and sliding the windowed function over the data series using a fixed amount of overlap. Formally, for any series $x(n)$ we can defined the STFT as,

$$X_w(mS, \omega) = \sum_{n=-\infty}^{\infty} x(n)w(n - mS)\exp(-j\omega n) \tag{5}$$

for analysis window w, frame index m and step size S [20]. The internal equation $\sum_{n=-\infty}^{\infty} x(n)w(n - mS)\exp(-j\omega n)$ represents the Fourier transform of $x_w(mS, \omega)$ and is stored in the matrix $\hat{X}$ whose columns are Fourier periodograms $P_X(\omega)$. To adapt the spectrogram to the unevenly sampled heart rate data, we replace this equation with the Lomb-Scargle equation,

$$\begin{aligned} X_w(mS, \omega) = \frac{1}{2} \left\{ \frac{\left[\sum_j x(n)\cos\omega(t_j - \tau)\right]^2}{\sum_j \cos^2\omega(t_j - \tau)} \right. \\ \left. + \frac{\left[\sum_j x(n)\sin\omega(t_j - \tau)\right]^2}{\sum_j \sin^2\omega(t_j - \tau)} \right\} \end{aligned} \tag{6}$$

In this way we mitigate the sensitivity of the standard STFT while maintaining its ability to capture the nonstationary patterns in the temporal data. In effect, we end up with a very high dimensional representation that could potentially have more expressive and appropriate features for detecting stress. The next step is to use machine learning to discover these features and compare the performance difference with models based on traditional HRV features. Due to the 2D spatial nature of data representation, convolutional neural networks should provide an efficient learning architecture.

3 Experiments and Validation

3.1 Dataset and Models

For the following experiments, we applied a sliding window method to the stress dataset in order to segment the data. This method also lends itself to real-time processing in the future. While deciding the appropriate window length, we observed that shorter segments tended to be more stable but not very expressive. Meanwhile, longer segments were more expressive but tended to overlap across samples with different labels. Through empirical testing, we found that 505-beat windows ($\tilde{6}$ mins) showed the best balance between stability and expressiveness while staying above the 5-minute minimum length recommended for accurate short-term HRV analysis (See Fig. 2).

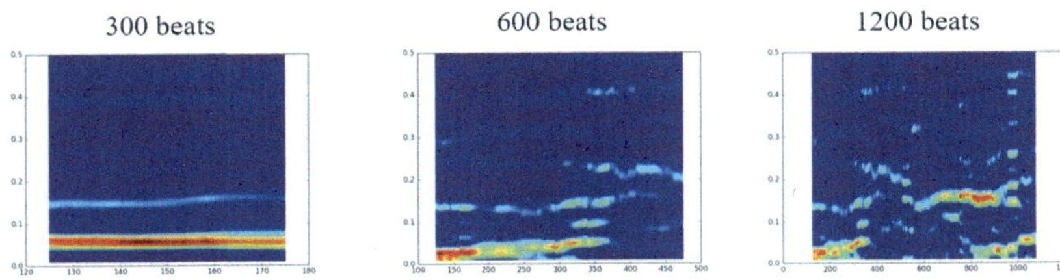

Fig. 2. Image samples of extracted ECG spectrograms at different window widths.

We also applied a modal labelling scheme for window segments which overlapped between multiple labels. A threshold of 70% was set in order to mark stable segments which could be labelled using the modal value of the contained samples and separate them "transitional" segments which were considered to be more ambiguous. The final experiment dataset was comprised of 1,500 windowed samples which were either labelled based on the 4-level stress annotation or as transitions.

For modeling, we trained perceptron and autoencoder models using the stress HRV data to serve as baselines and compared these to convolutional neural network (CNN) models trained on our spectral HRV data representation. The convolutional layers were topped with a 128-unit dense layer to match the ones used for the non-deep HRV stress models. For the CNN models, we tested both standard deep learning models typically used for image classification tasks as well as shallower CNN architectures to verify the hierarchical feature requirements of the dataset. Performance was validated using stratified 10-fold cross-validation.

3.2 Results and Discussion

First, we analyze the performance results for models trained on the standard HRV features, which were 14 time and frequency domain features, and compared them to the deep ResNet [6] model trained on the LS spectrum data. As seen in Fig. 3, the baseline HRV-MLP model achieved 52% test accuracy which was improved by up to 60% by applying denoising autoencoders for more efficient latent feature selection. These results are well-above the 20% random classification on this 5-class dataset.

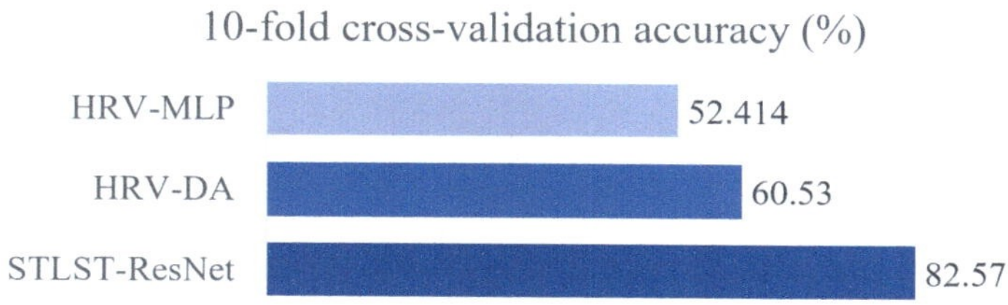

Fig. 3. Comparison of test performance results between models trained on standard HRV features (MLP and DA) and the ResNet model trained on STLST heart rate spectrum data.

Deep Spectral Model. For the spectral models, ResNet50 was able to achieve over 80% testing accuracy, which is well above both of the traditional HRV baselines. These significant gains could be explained by the effective combination of the more expressive spectral data with the feature learning ability of deep learning algorithms. The traditional indices used in traditional HRV analysis help humans understand physical conditions, but they may also obfuscate some intricate features that may be necessary for deeper affective analysis. By allowing the model to learn from a spatial-spectral representation, we are able to train a more effective model. However, this comes at the cost of increased complexity since the standard 50-residual unit model requires over 25 million parameters. This not only leads to higher data requirements, but may also lead to more challenges in interpretation and optimization.

Shallow Spectral Model. To help alleviate the problem of increased complexity, we attempted to train a more compact model. For these tests, the goal was to reduce the depth and overall complexity of our model without compromising performance. This ultimately culminated in a model featuring 5 intermediate 128-filter CNN layers with BatchNorm for regularization. The relatively shallow model also meant residual units were unnecessary and training was much faster than the deeper model. It should also be noted that tests were also performed on shallower models below 5 layers, but this typically lead to larger performance variations and more noticeable overfitting. Figure 4 shows a comparison of the learning process of both the deep and shallow models. Despite using less than 1 million parameters, as opposed to the over 25 million of ResNet50, the shallow model is able to match the performance of the much larger ResNet model while also avoiding overfitting. Actual training time could also reduced by up to 5x which may be useful in situations where occasional retraining is necessary.

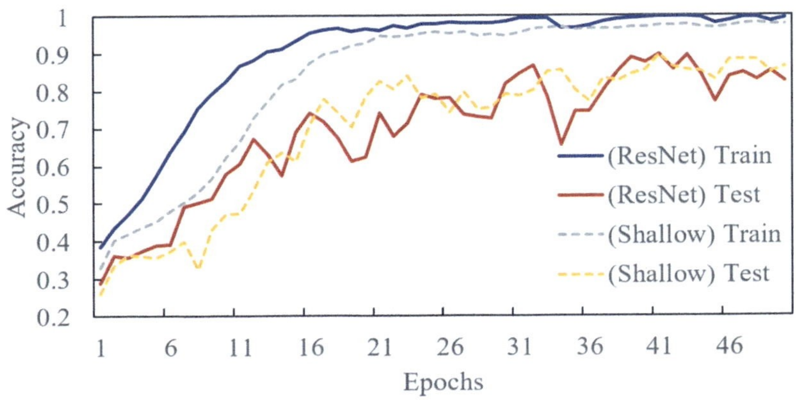

Fig. 4. Learning and test performance of ResNet50 and the shallower 5-layer CNN model.

3.3 ROC Analysis

Next, we wish to analyze how well the model is able to discriminate between the different stress classes. For this we look at the receiver operating characteristic (ROC) curves for each of the classes, using a one versus all methodology. As seen in the in Fig. 5, using the deep ResNet model has a high rate of discrimination for the edge labels (i.e., very low and very high). Meanwhile, it is slightly more challenging to correctly distinguish the median labels (low and high). This could indicate that there is more ambiguity with regards to the physiological signals shown for moderate stress levels. Given that the stress levels were expected to be ordinal, it would be easier to separate classes on either side of the spectrum rather than in the middle. This may be a sign that the features discovered by the model follow the same intended ordinality as the stress labels.

Alternatively, this could also be due to labelling inconsistencies common to psychological labels. Simply, it is easier to write accurate labels for extreme emotions rather than moderate ones. Future work should focus on exploring this further in order to determine a more efficient annotation methodology, or possibly even correct or omit inconsistent labels.

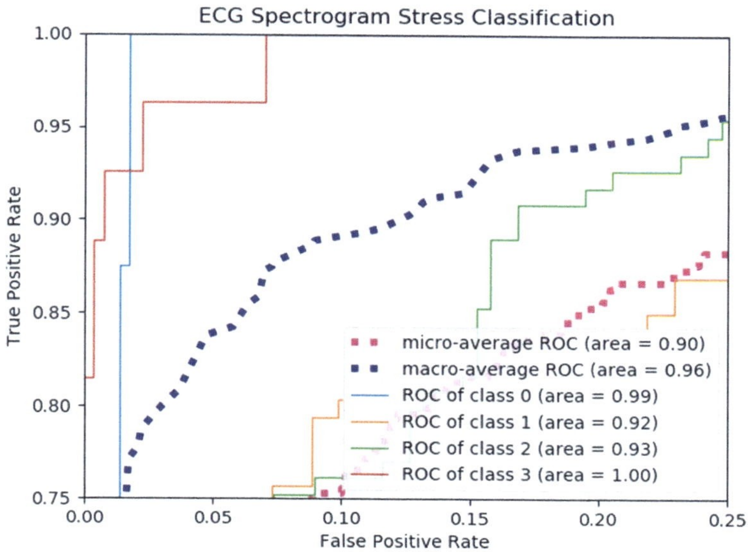

Fig. 5. ROC for the classification results of the ResNet50 model.

4 Summary and Future Work

In this work we trained a deep model of work stress using a spectral representation of heart rate extracted using short-time Lomb-Scargle transform. The experiments showed that the deep models trained on the spectral representation could easily outperform those trained on the traditional HRV features for detecting naturalistic stress levels. Additionally, we show that even relatively shallow models featuring at least 5 convolutional layers can match the performance of ResNet50, this could indicate that at least some shallow hierarchical features may be important for effectively modelling naturalistic stress from heart rate.

In future work we plan to investigate further the specific features learned by our model through feature activation analysis. We also plan to train and compare the performance of the model on other affective computing datasets for emotion. Finally, it would benefit the affective computing community greatly if we can develop optimal representations that can enable the aggregation of data using different devices and dataset to build a single model that contains all the richest features for HRV analysis.

References

1. Acampora, G., Cook, D., Rashidi, P., Vasilakos, A.: A survey on ambient intelligence in healthcare. Proc. IEEE **101**(12), 2470–2494 (2013)
2. Clifford, G.D., Tarassenko, L.: Quantifying errors in spectral estimates of hrv due to beat replacement and resampling. IEEE Trans. Biomed. Eng. **52**(4), 630–638 (2005)

3. Deeming, T.J.: Fourier analysis with unequally-spaced data. Astrophys. Space Sci. **36**(1), 137–158 (1975)
4. Goldberger, A.L., Amaral, L.A.N., Glass, L., Hausdorff, J.M., Ivanov, P.C., Mark, R.G., Mietus, J.E., Moody, G.B., Peng, C.K., Stanley, H.E.: PhysioBank, PhysioToolkit, and PhysioNet: components of a new research resource for complex physiologic signals. Circulation **101**(23), e215–e220 (2000)
5. Hagad, J.L., Fukui, K.I., Numao, M.: A hierarchical model of authentic work stress using physiological signals and stress coping profiles. In: Proceedings of 30th Annual Conference of the Japanese Society for Artificial Intelligence, pp. 3E34–3E34. The Japanese Society for Artificial Intelligence (2016)
6. He, K., Zhang, X., Ren, S., Sun, J.: Deep residual learning for image recognition. In: Proceedings of the IEEE Conference on Computer Vision and Pattern Recognition, pp. 770–778 (2016)
7. Iso, H., Date, C., Yamamoto, A., Toyoshima, H., Tanabe, N., Kikuchi, S., Kondo, T., Watanabe, Y., Wada, Y., Ishibashi, T., Suzuki, H., Koizumi, A., Inaba, Y., Tamakoshi, A., Ohno, Y.: Perceived mental stress and mortality from cardiovascular disease among Japanese men and women: the Japan collaborative cohort study for evaluation of cancer risk sponsored by monbusho (JACC study). Circulation **106**(10), 1229–1236 (2002)
8. Kamath, M.V., Ghista, D.N., Fallen, E.L., Fitchett, D., Miller, D., McKelvie, R.: Heart rate variability power spectrogram as a potential noninvasive signature of cardiac regulatory system response, mechanisms, and disorders. Heart Vessels **3**(1), 33–41 (1987)
9. Kim, H.G., Cheon, E.J., Bai, D.S., Lee, Y.H., Koo, B.H.: Stress and heart rate variability: a meta-analysis and review of the literature. Psychiatry Invest. **15**(3), 235 (2018)
10. Kowata, E., Hozawa, A., Kakizaki, M., Tomata, Y., Nagai, M., Sugawara, Y., Kuriyama, S., Tsuji, I.: Perceived stress and cardiovascular disease mortality. The Ohsaki Cohort Study. Nihon Koshu Eisei Zasshi **59**(2), 82–91 (2012)
11. Marcus, M., Yasamy, M., van Ommeren, M., Chisholm, D.: Depression: a global public health concern. World Health Organization, Technical report (2012)
12. Melillo, P., Bracale, M., Pecchia, L.: Nonlinear heart rate variability features for real-life stress detection. case study: students under stress due to university examination. Biomed. Eng. Online **10**, 96+ (2011). http://dx.doi.org/10.1186/1475-925x-10-96
13. Moody, G.: Spectral analysis of heart rate without resampling. In: Proceedings Computers in Cardiology 1993, pp. 715–718, September 1993
14. Rajpurkar, P., Hannun, A.Y., Haghpanahi, M., Bourn, C., Ng, A.Y.: Cardiologist-level arrhythmia detection with convolutional neural networks. arXiv preprint arXiv:1707.01836 (2017)
15. Scargle, J.D.: Studies in astronomical time series analysis. ii-statistical aspects of spectral analysis of unevenly spaced data. Astrophys. J. **263**, 835–853 (1982)
16. Thayer, J.F., Åhs, F., Fredrikson, M., Sollers III, J.J., Wager, T.D.: A meta-analysis of heart rate variability and neuroimaging studies: implications for heart rate variability as a marker of stress and health. Neurosci. Biobehav. Rev. **36**(2), 747–756 (2012)
17. Tzirakis, P., Trigeorgis, G., Nicolaou, M.A., Schuller, B.W., Zafeiriou, S.: End-to-end multimodal emotion recognition using deep neural networks. IEEE J. Sel. Top. Sig. Process. **11**(8), 1301–1309 (2017)
18. VanderPlas, J.T.: Understanding the lomb-scargle periodogram. Astrophys. J. Suppl. Series **236**(1), 16 (2018)

19. Welch, P.: The use of fast fourier transform for the estimation of power spectra: a method based on time averaging over short, modified periodograms. IEEE Trans. Audio Electroacoust. **15**(2), 70–73 (1967)
20. Zhu, X., Beauregard, G.T., Wyse, L.L.: Real-time signal estimation from modified short-time fourier transform magnitude spectra. IEEE Trans. Audio Speech Lang. Process. **15**(5), 1645–1653 (2007)

Privacy-Preserving Resident Monitoring System with Ultra Low-Resolution Imaging and the Examination of Its Ease of Installation

Shogo Murakami[1(✉)], Takumi Kimura[2], and Ikuko Eguchi Yairi[1]

[1] Graduate School of Science and Technology, Sophia University, 7-1 Kioicho, Chiyoda-ku, Tokyo 102-8554, Japan
murakami@yairilab.net

[2] Sophia University, 7-1 Kioicho, Chiyoda-ku, Tokyo 102-8554, Japan

Abstract. This is an extension from a selected paper from JSAI2019. Monitoring systems using infrared array sensors allow monitoring of residents while protecting their privacy. However, since such a sensor is vulnerable to subtle movements, accuracy of posture classification is low, and limits the locations and methods available for installation. This study proposes a posture classification method with higher accuracy. Over 93% accuracy was achieved in posture classification by color conversion of infrared array sensor images and successfully decreased loss due to displacement by DCNN. Additionally, this research considers methods to create artificially simulated data for postural-behavioral study. To check the validity of this method, postures of 3 subjects were examined using a classifier with studied simulation data. Finally, simulation environments with different sensor altitudes and angles were created to examine the ease of installation for the proposed method. As a result, the experiments showed that accuracy was highest at approximately 90% when the sensor was located 50 cm below the height of the target and when the tilt angle was within $\pm 2°$.

Keywords: Monitoring system · Ultra-low resolution image · Machine learning

1 Introduction

Behavior analysis is one of the oldest yet effective methods in uncovering changes that can help improve quality of life. Although such methods have been thought as too technical and beyond the reach of the average man, the spread of artificial intelligence and internet has made behavior analysis much more accessible. For example, smartphones and smartwatches: we use both devices mindlessly, but just by carrying them around they collect physical and behavioral information to analyze and output the most suitable advice for the user [1–3]. Another

Y. Ohsawa et al. (Eds.): JSAI 2019, AISC 1128, pp. 278–290, 2020.
https://doi.org/10.1007/978-3-030-39878-1_25

widespread use of behavior analysis is in home-monitoring services to detect abnormalities and emergencies. In most cases, these devices are installed by a third party on behalf of users, and the ones being monitored spend their daily lives experiencing no merit at all. In this ever-growing digital age, being recorded every minute of the day rises privacy concerns. Since users should feel safe and unviolated in their own home, just enough data to detect abnormal conditions is necessary. To eliminate both physical stresses and behavioral limitations for the user, monitoring systems using infrared sensors were proposed [4–11]. In these systems, only temperature data are used in human detection such that installation is simple and that privacy leakage risks are at a minimum. These IR monitoring devices can be largely grouped into those using single infra-red beam sensors and those using infrared array sensors. Single infrared beam sensors are exactly as stated-single rayed, and infrared array sensors are simply multiple criss-crossed infrared rays. Systems using infrared array sensors obtain spatial information in 4 dimensions along with received light quantity and use these data to make posture estimations of the people within the given region [8–12]. Current concerns are that the accuracy of infrared array sensors is sensitive to changes in background temperature, installation location, and installation angle. Studies so far have recorded data only in optimal conditions, so effects of these subtle condition variances are unknown. If these effects are not investigated, use of these devices would be highly limited for both possible installation locations and installation methods. Another task of importance is to consider how to obtain learning data to increase classification accuracy. Investigating participants under different conditions—room size and temperatures, body type, and posture—would need extensive preparation and would lead to inaccuracies. Collecting data from actual participants in this way would be unrealistic due to the time and high cost involved. To resolve the aforementioned concerns, this paper assesses the classification accuracy by the single 8×8 infrared array sensor, proposes the methods of artificially creating simulated data to study postures for machine learning, and analyzes the effects caused by the angle of the installed infrared array sensor.

2 Indoor Monitoring System Using an Ultra-low-resolution Infrared Array Sensor

2.1 Posture Classification and Experimentation System

To examine posture classification accuracy, a data collection device including an infra-red array sensor was developed. A diagram of the device is shown in Fig. 1. This device composes of a Raspberry Pi3 Model B mounted with a Grid-EYE (AMG8833) sensor [13]. The Grid-EYE will output an 8×8 pixel image data of the surface temperature for objects detected in the observation space. Temperatures between $0\,^{\circ}\mathrm{C}$ $-80\,^{\circ}\mathrm{C}$ can be detected with a step increment of $0.25\,^{\circ}\mathrm{C}$. Deep Convolution Neural Network (DCNN), is an effective method for high accuracy image recognition. It is used in this study to examine the obtained infrared image data for use in the production of a posture classifier. This network contains five layers: an input layer, two convolutional layers, a fully connected layer, and an output layer. The convolution layers were treated using convolution and max pooling [14]. To prevent over-learning, Dropout was implemented and to update parameter weighting to optimal levels, a method called Adam was used [15].

Fig. 1. An action recognition device equipped with a 8×8 infrared array sensor.

2.2 Posture Classification Using DCNN

A posture classification experiment was conducted on 3 subjects to check the operation of the posture classification system and to evaluate the performance of the DCNN. The first two experiments were conducted in a 9.5m2 Japanese-style room at roughly 13 °C room temperature. Here, the data of a male subject of age 24 and height of 170 cm, and a female subject of age 20 and a height of 160 cm was obtained. The experiment for the third subject was done in a 20m2 room with a room temperature of roughly 11 °C. The subject was male, of age 22, and was 170 cm in height. For all three scenarios, the sensor was placed 140 cm from ground level such that the entire body of each subject could be observed. The subjects were stationed 1–3 m away from the sensor and were told to stand, sit, or lie down within the view of the sensor. A total of 14983 frames worth of data were obtained. 10% of the above data were randomly selected and studied by the classifier, then were used to classify the remaining 90%. It was confirmed that accuracy plateaus near 100 epochs, so epoch numbers were set at 100. Posture classification results and accuracy evaluations are shown in Table 1. Classification results are a sum of all subjects, and evaluations were conducted on these results. Considering practical use, a resident monitoring system needs to be able to detect instantaneous dangerous incidents such as slips and therefore it is desirable for the F-measure to be above 90%. Experimental results showed an F-measure of roughly 87%. Observing categorization results, incorrect categorizations were prominent in similar postures such as "standing" to "seated," "seated" to "horizontal."

Table 1. Users' action classification results of 8×8 infrared array sensor images by DCNN.

Posture classification results

Recog-nition	Answer Stand	Sit	Lie
Stand	3582	266	15
Sit	498	5243	435
Lie	1852	590	4171

Posture classification evaluations

	Stand	Sit	Lie
Accuracy	0.9272	0.8492	0.8433
Recall ratio	0.8402	0.8596	0.9026
F-measure	0.8816	0.8544	0.8719

2.3 Effects of Data Conversion on Classification Accuracy

Hitherto, values from the 8×8 IR images were directly inputted into the DCNN. Considering how similar postures were incorrectly categorized multiple times, it could be hypothesized that with no processing, accurate feature extraction is not possible. DCNN is capable of handling multiple channel inputs, including color images. Making use of this characteristic, the 8×8 IR sensor values were converted to 8×8 pixel color images, then inputted to the DCNN. Ito et.al conducted a study on car camera images and deep learning. They took converted greyscale data to color and added edge processing and succeeded in significantly increasing accuracy [16]. Rachmadi and Punama used color spaces such as RGB, HSV, CIE Lab, and CIE XYZ to detect car colors with high recognition accuracy [17]. Yuriy et.al used learning data with multiple combined color spaces to create an efficient face recognition system [18]. As mentioned, many researchers are succeeding in increasing categorization accuracy by converting their data to color. Since all studies mentioned only used high-resolution image data, effects of this process on ultra-low-resolution image data needs investigation. By applying processing with various color spaces, the most effective process for this research will be chosen. Evaluation of Color Spaces. In this paper, 8×8 grayscale image data were color converted to evaluate categorization accuracy. Representative color spaces determined by the CIE (Commision Interon acznationale de l' Eclairage)—RGB, CIE XYZ, CIE L*a*b*—were used. For conversions to RGB color space, temperature data were mapped to color space such that lower temperature values are shown as blue and higher values are shown as red. The images were then divided into three separate images, each representative of R, G, and B, then fed into the DCNN. In doing this, the input layer was switched from channel 1 to 3 such that the three images could be processed. For conversions to color spaces other than RGB, Open CV (Open Source Computer Vision Library), an open source library developed by Willow Garage Inc., was used. Real data used in the previous chapter were converted to color space, 10% were randomly selected to be learned by classifier, then the remaining 90% were categorized and evaluated. This process was repeated 100 times to assess classification accuracy. Results are shown in Tables 2, 3, and 4, and the state of the classifier during learning is shown in Fig. 2. The results show that regardless of sensor height, categorizations in color space had significantly increased accuracy than in grayscale and that although by only a minute difference, RGB color space had the highest validity amongst the color spaces. As for learning situations, learning rates varied for grayscale but for color spaces RGB, XYZ, and L*a*b*, learning rates increased in a steady manner and repetitions necessary for plateau decreased significantly. From these results it was confirmed that the most effective color space for this research is RGB. Evaluation of Bit Depth. It was confirmed that accuracy was maximized when temperature data was converted to RGB. Other than the fact that colored data is easier for man to visualize, the core nature of the data is unchanged from the original. Nevertheless, the precision yielded by the classifier had increased. This can be hypothesized to be because of the net increase in information. RGB color space is generally measured in the range of 24 bits (8 bits per color), but depending on circumstances, can be expressed with more or less bits. The number bits used is called "bit depth." The relationship between bit depth, information amount, and precision was investigated. Data presented in the previous chapter were used again, and an improved version of the RGB conversion algorithm was used in converting RGB data to bits. 7 different bit patterns—3bit, 6bit, 9bit, 12bit, 24bit, 48bit, and 96bit—were prepared, 10% of this data were randomly selected to be studied by the machine, and the rest were classified and evaluated. This process was repeated 100 times, then accuracy was evaluated. Results are shown in Fig. 3. The results show that classification accuracy directly increased as bit size increased. Once bit size reached 24, the slope of accuracy plateaued. From this, it was concluded that for data dealt with in this research, accuracy follows bit depth until 24bit, where maximum precision is reached.

Table 2. Action classification result of 8×8 RGB converted images of infrared array sensor.

Posture classification results

Recog-nition	Answer Stand	Sit	Lie
Stand	3929	137	20
Sit	217	5620	113
Lie	121	336	4490

Posture classification evaluations

	Stand	Sit	Lie
Accuracy	0.9615	0.9445	0.9076
Recall ratio	0.9207	0.9223	0.9712
F-measure	0.9407	0.9333	0.9383

Table 3. Action classification result of 8×8 XYZ converted images of infrared array sensor.

Posture classification results

Recog-nition	Answer Stand	Sit	Lie
Stand	3898	221	21
Sit	350	5603	235
Lie	19	269	4366

Posture classification evaluations

	Stand	Sit	Lie
Accuracy	0.9415	0.9054	0.9381
Recall ratio	0.9135	0.9195	0.9446
F-measure	0.9273	0.9124	0.9413

Henceforth, all grayscale data are first converted to 24bit RGB color space prior to DCNN input.

3 Development of Learning Data-Generating Applications and Evaluation of Posture Classifications

3.1 Production of Pseudo-data Simulator for Data Learning

Methods of creating learning data sans real participants were considered. Infrared images used in this research are indoor temperature distributions compacted into 8×8 pixels. It is hypothesized that by having a human model appear within the viewing range of the sensor, it is possible to create images similar to those created through experimentation without the need for high precision human models and infrared light propagation simulations. A simulator that outputs pseudo-data for data learning was developed using an engine with preinstalled libraries of physical functions. In creating this, Unity, a cross-platform game engine developed by Unity Technologies was used. The main reason why a game engine was selected is because it allows freedom in many settings such as human model height, model physique, sensor tilt, sensor height, and room size.

Table 4. Action classification result of 8×8 L*a*b* converted images of infrared array sensor.

Posture classification results

Recognition	Answer: Stand	Answer: Sit	Answer: Lie
Stand	3958	279	6
Sit	272	5333	243
Lie	37	481	4374

Posture classification evaluations

	Stand	Sit	Lie
Accuracy	0.9328	0.9119	0.8941
Recall ratio	0.9275	0.8752	0.9461
F-measure	0.9302	0.8932	0.9193

Most cases where data collection is conducted from simulators require high reproducibility for human models [19,20] but since this research only uses low resolution devices, a simple 3D model with no unnecessary parts was used [21]. The original sensor uses 64 elements that receives infrared light to detect temperature. To replicate this setting, the simulator uses an embedded sensor that uses ray tracing from the point of observation. Ray tracing not only helps in locating the human model but accurately recreates the physical phenomenon in which the further the sensor is, the lower the precision of measured apparent temperature as well. The human model used is composed of several parts including a head, arms, and a body. Temperature distributions are set for each part separately since concealed body parts and revealed parts vary in temperature. Temperature distributions for the room and for the walls are preset as well. After the simulator is turned on, the sensor emits a traceable ray of light. When the

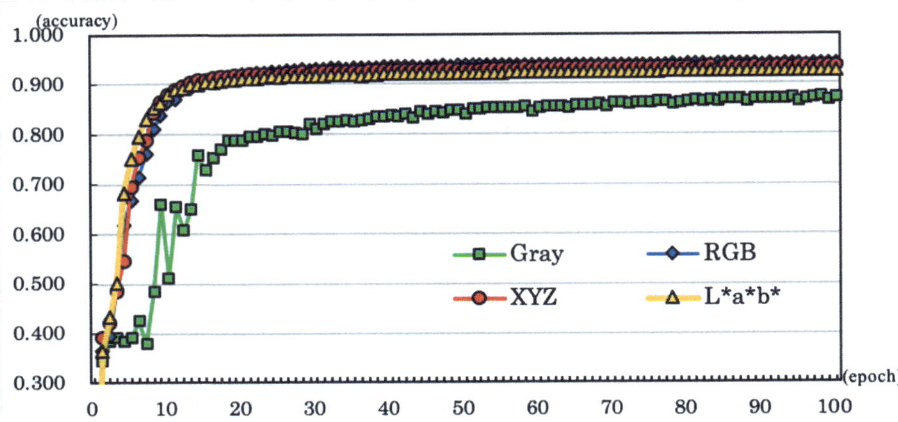

Fig. 2. Accuracy for various color space.

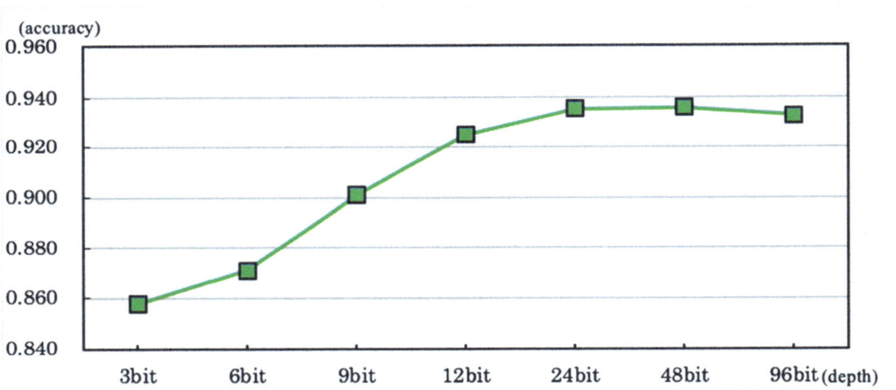

Fig. 3. Accuracy for various bit depth.

ray reaches the human model, it measures and records that temperature If it fails to reach the human model, the room temperature is recorded. The human model has three types of postures namely, standing, seated or horizontal.

3.2 Evaluation of Data Learning Using Pseudo-Data

Actual IR image data from participants and pseudo-image data from the simulator are shown and compared in Fig. 4. Production of simulator data made use of participant data and real space data from Sect. 2.3. Figure 4 is color-coded to show red portions as high-temperature regions and blue portions as low-temperature regions. It can be observed that the human portion of both data are almost indistinguishable and that reproducibility is high.

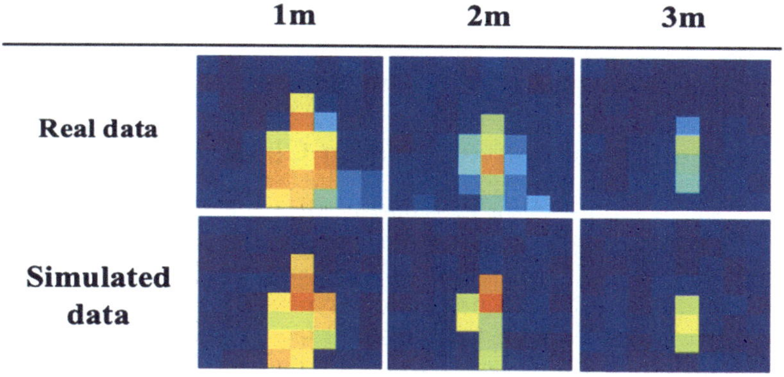

Fig. 4. Comparison between real data from the sensing device and simulated data from the learning data generator.

One area of concern, though, is the background portion where real data has a gradient of higher temperatures surrounding the human model as for the pseudo-data, there is a clear division between the model and background. After feeding the DCNN pseudo-data to learn, real environment data were classified. The same data from Sect. 2.3 were used. Data collection for pseudo-data was conducted by placing a human model in random locations 1–3 m away from the sensor with one of the three aforementioned postures. Room temperatures and sizes were set to correspond with real data. The human model settings were made to represent subjects from the real experiment, and the learning data generator was run for 30 min, outputting 50000 sample data for each simulated subject. Finally, the posture classifier was used to analyze the obtained data. Classification results and evaluation results are shown in Table 5. Classification results are shown as a sum of all models, and evaluations have been carried out after summation. The results show that approximately 70–80% accuracy is achieved for most cases. Although this is an improvement, it is yet too low for practical use. Placing a finger on miscategorizations, it could be seen that adjacent states— standing and sitting, sitting and laying down— states with similar temperature distributions— had relatively high rates of miscategorization at 22.5% and 27.6% respectively. The causes for this issue are as previously stated—that temperatures surrounding the human models are higher than expected, and that the background images occasionally have random high temperature noise. Processing methods that could reduce these errors were considered; eliminating background temperatures from real data. First, background data between frames were examined, then an algorithm that detects and eliminates back-ground data from temperature differences between frames was devised. This algorithm effectively decreased heat sources such as lighting, and left only data from the human model. Background elimination was conducted on both pseudo-data for learning and on real data for categorization assessments. Table 6 summaries the categorization results and evaluations. After applying processing, F-values for all categories in-creased from 70% to above 90%, indicating a successful increase in classification ability. This lead to a conclusion that in using pseudo-data to categorize real data, background elimination is highly effective in raising accuracy.

4 Effects of Data Conversions on Classification Accuracy

Heretofore, the monitor's sensor was consistently set at 140 cm from the ground and consistently succeeded in scoring above 90% in accuracy. In real conditions, though, height need not be always steady, and at what height accuracy begins to decline needs to be investigated. Another possible variable factor includes changes in sensor tilt resulting from external factors such as the sensor being bumped during use, or the sensor being shaken during an earthquake. To assess the monitoring device's performance, tests with differing installation conditions were conducted.

Table 5. Action classification results of the real data by learning with the simulated data.

Posture classification results

Recognition	Answer: Stand	Sit	Lie
Stand	3306	537	216
Sit	959	5129	1274
Lie	5	427	3130

Posture classification evaluations

	Stand	Sit	Lie
Accuracy	0.8144	0.6966	0.8787
Recall ratio	0.7742	0.8417	0.6774
F-measure	0.7938	0.7623	0.7650

Table 6. Effects of background image removal on action classification of the real data by learning with the simulated data.

Posture classification results

Recognition	Answer: Stand	Sit	Lie
Stand	3929	137	20
Sit	217	5620	113
Lie	121	336	4490

Posture classification evaluations

	Stand	Sit	Lie
Accuracy	0.9615	0.9445	0.9076
Recall ratio	0.9207	0.9223	0.9712
F-measure	0.9407	0.9333	0.9383

4.1 Evaluations on Installation Height

Using the learning data generator, sensor height was varied from 50–170 cm with a step increment of 10 cm, and human model height was set at either 170 cm, 160 cm, or 150 cm. 10% of each case were selected randomly then fed into the classifier for learning. The remaining 90% were categorized and evaluated. These steps were repeated 100 times. For each condition, an assessment of classification accuracy was performed. The results are shown in Fig. 5. By averaging the classification accuracy of these three postures, it was concluded that accuracy reaches its maximum when the sensor is installed 50 cm below user height.

4.2 Evaluations of Sensor Angle

In the previous section, effects of height variances on classifier reliability were evaluated. Next, effects of sensor angle were investigated. Model height was set at 170 cm, and sensor height was set at 120 cm (50 cm below model height) where accuracy should be at its maximum. Tilt range was set from $-5°$ to $5°$ with increments of $1°$. As for learning data, previously established data with no tilt were used. Results are summarized in Fig. 6. The results showed that when the sensor is tilted $2°$ upwards, the F-measure is maintained at over 90%, but when the angle reaches $5°$, the F-measure decreased to roughly 86%. When the sensor is tilted downwards, F-measure is sustained at high values until $-1°$. However, a tilt of over $-2°$ drastically decreases the F-measure until under 80% at $-5°$. Since this research takes privacy preservation seriously, images used are ultra-low-resolution. Therefore, the posture classification methods adopted rely heavily on temperature distributions per row rather than pixels per image. The one-row differences caused by sensor tilt gave an extensive impact on classification results. It was concluded that height is less of an issue than tilt, and that variations in tilt cause a dynamic decrease in classification accuracy.

5 Discussion

To investigate effects of external factors on classification accuracy, installation height and tilt were analyzed in Sect. 4. It was uncovered that accuracy peaks when the sensor is installed 50 cm under model height, and that sensor tilt has an unneglectable impact on accuracy. Currently, the most urgent concern is how to deal with installation in abnormal conditions since adaptability of the monitoring device is extremely low.

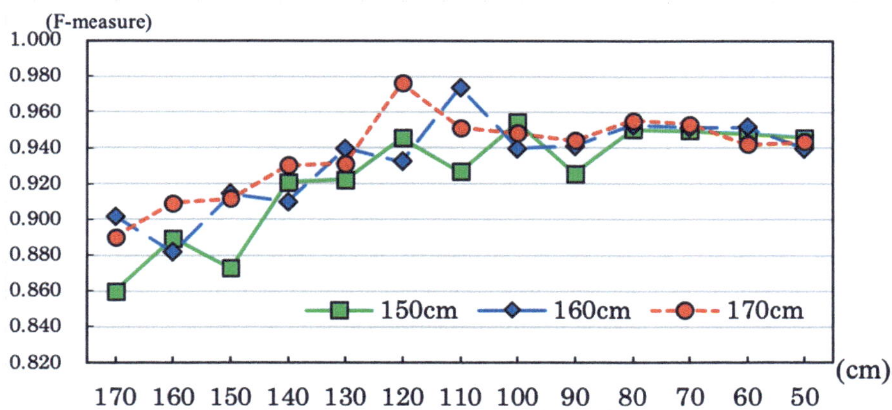

Fig. 5. Accuracy for various sensor altitudes.

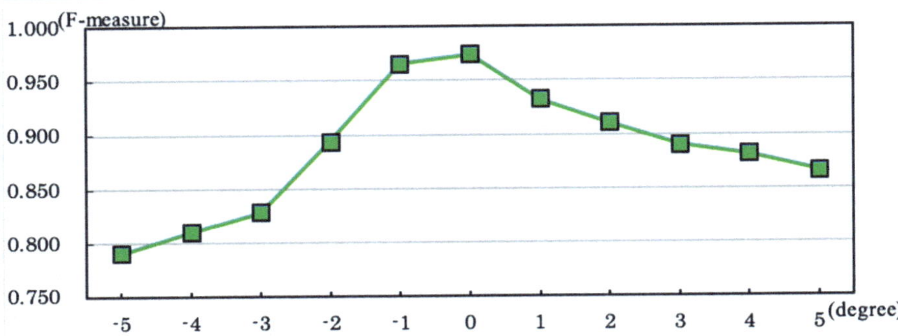

Fig. 6. Accuracy for vertical sensor angle.

Taking into consideration the above conclusions, commercial use is contemplated. If the surface of installation is perfectly perpendicular, the user could easily measure and install the monitor 50 cm below their own height. Sensor installation would only need several screws on the top, bottom, and sides of the monitor, and even an unexperienced user could complete their own installation with only a simple guide to help them. If the surface of installation is bumpy, tilted, or blocked, further consideration is necessary. One possible solution is to collect surface data from the user upon purchase, to in-put this angled data into the classifier, and to make a specialized monitoring device that covers for tilt. This method would yield a very accurate and useful device, but needs heavy planning beforehand and would take extensive costs and time. If cost and time are of no issue, though, this shows that with just the use of a pseudo-data generator, monitors applicable to any household could be created.

6 Conclusion

This paper justified that an infrared array sensor resident monitoring system using an infrared array sensor image with 8×8 pixels would output over 90% accuracy for posture action classification of the target. Noise analysis was performed on tilt and it was concluded that approximately 90% accuracy was sustained for angles within $\pm 2°$ by extension of the classifier. In addition, a high accuracy classifier was successfully constructed via posture learning data simulation and comparison to real data. As for further research, improvements in pseudo-data generators and learning algorithms will be pursued for application in real environments.

We would like to show our gratitude to all the supporters for their participated and assistance. This research was conducted under the support of 23rd and 24th year of Heisei period research grant from the Support Center for Advanced Telecommunications Technology Research, Grant-in-Aid for Scientific

Research B(17H01946), and Grant-in-Aid for challenging Exploratory Research (16K12537).

References

1. Reeder, B., David, A.: Health at hand: a systematic review of smart watch uses for health and wellness. J. Biomed. Inform. **63**, 269–276 (2016)
2. Yi, W., Sarkar, O., Mathavan, S., Saniie, J.: Wearable sensor data fusion for remote health assessment and fall detection. In: IEEE International Conference on Electro Information Technology (EIT), Milwaukee, WI, USA, pp. 303–307 (2014)
3. Nakazawa, J., Sasaki, W., Obuchi, M., Egashira, K., Nishiyama, Y., Okoshi, T., Yonezawa, T., Tokuda, H.: A platform for mutual watch-over among the elderly using PAN and gamification. Inst. Electron. Inf. Commun. Eng. J. D **J101–D**(2), 306–319 (2018)
4. Ogoshi, Y., Ogoshi, S., Hirose, S., Kimura, H.: A study on the recognition of human activities by datamining from infrared sensor information. Trans. Inst. Electron. Inf. Commun. Engs. D-II **J85–D–II**(5), 959–964 (2002)
5. Shinagawa, Y., Kishimoto, T., Ohta, S.: Development of an algorithm for automatic emergency calls using non-response intervals of infrared sensors. Kawasaki Med. Welf. Soc. J. **15**(2), 553–563 (2006)
6. Sixsmith, A., Johnson, N.: A smart sensor to detect the falls of the elderly. IEEE Pervasive Comput. **3**(2), 42–47 (2004)
7. Kemper, J., Hauschildt, D.: Passive infrared localization with a probability hypothesis density filter. In: 2010 7th Workshop of Positioning, Navigation and Communication (WPNC), Dresden, Germany, pp. 68–76 (2010)
8. Okada, R., Yairi, I.: An indoor human behavior gathering system toward future support for visually impaired people. In: Proceedings of the 15th International ACM SIGACCESS Conference on Computers and Accessibility, Washington, no. 36, October 2013
9. Takagi, Y., Takahashi, N., Otsuka, S.: Watching elderly people using temperature sensors. In: Proceedings of Forum on Data Engineering and Information Management 2016 (DEIM 2016), Fukuoka, P6-1 (2016)
10. Hevesi, P., Wille, S., Pirkl, G., Wehn, N., Lukowicz, P.: Monitoring household activities and user location with a cheap, unobtrusive thermal sensor array. In: Proceedings of the 2014 ACM International Joint Conference on Pervasive and Ubiquitous Computing (UbiComp 2014), Seattle, Washington, pp. 141–145 (2014)
11. Kawashima, T., Kawanishi, Y., Ide, I., Murase, H., Deguchi, D., Aizawa, T., Kawade, M.: Action recognition from extremely low-resolution thermal image sequence. In: IEEE International Conference on Advanced Video and Signal Based Surveillance, Italy (2017)
12. Kusukame, K., Yoneda, A., Shikii, S., Silawan, N., Nosaka, K., Kubo, H.: Contactless estimation method of thermal sensation using infrared thermography. Panasonic Tech. J. **63**(2), 118–122 (2017)
13. Omron Japan news release. https://www.omron.co.jp/press/2013/05/e0529.html. Accessed 22 Mar 2018
14. Yang, J.B., Nguyen, M.N., San, P.P., Li, X.L., Krishnaswamy, S.: Deep convolutional neural networks on multichannel time series for human activity recognition. In: Proceedings of the 24th International Joint Conference on Artificial Intelligence (IJCAI), Buenos Aires, Argentina, pp. 3995–4001 (2015)

15. Kingma, D., Ba, J.: Adam: a method for stochastic optimization. In: International Conference on Learning Representations (2015)
16. Ito, T., Takata, Y., bin Mohammad Sofian, M.H.H.: Crossing Pedestrian Detection Using Deep Learning by On-board Camera (2018)
17. Rachmadi, R.F., Purnama, I.K.E.: Vehicle color recognition using convolutional neural network, CoRR, abs/1510.07391 (2015)
18. Kurylyak, Y., Paliy, I., Sachenko, A., Chohra, A., Madani, K.: Face Detection on Grayscale and Color Images using Combined Cascade of Classifiers (2009)
19. Sun, B., Saenko, K.: From virtual to reality: fast adaptation of virtual object detectors to real domains. Proceedings of British Machine Vision Conference (BMVC). BMVA Press (2014)
20. Peng, X., Sun, B., Ali, K., Saenko, K.: Learning deep object detectors from 3D models. In: Proceedings of the 2015 IEEE International Conference on Computer Vision (ICCV), Santiago, Chile, pp. 1278–1286 (2015)
21. UNITY-CHAN! OFFICIAL WEBSITE, Data Set.: box-unity-chan. http://unity-chan.com/. Accessed 21 July 2017

Identity Verification Using Face Recognition Improved by Managing Check-in Behavior of Event Attendees

Akitoshi Okumura[1(✉)], Susumu Handa[2], Takamichi Hoshino[2], Naoki Tokunaga[2], and Masami Kanda[2]

[1] Information-technology Promotion Agency, 2-28-8 Honkomagome, Bunkyo-ku, Tokyo 113-0021, Japan
ak-okumu@ipa.go.jp

[2] NEC Solution Innovators, Ltd., 2-6-1 Kitamikata, Takatsu-ku, Kawasaki 213-8511, Japan

Abstract. This is an extension from a selected paper from JSAI2019. This paper proposes an identity-verification system using continuous face recognition improved by managing check-in behavior of event attendees such as facial directions and eye contact (eyes are open or closed). We previously developed Ticket ID system for identifying the purchaser and hold-er of a ticket. This system carries out face recognition after attendant's check-in using their membership cards. The average face-recognition accuracy was 90%, and the average time for identity verification from check-in to admission was 7 seconds per person. The system was proven effective for preventing illegal resale by verifying attendees of large concerts. The problem with this system is regarding face-recognition accuracy. This can be mitigated by securing clear facial photos because face recognition fails when unclear facial photos are obtained, i.e., when event attendees have their eyes closed, are not looking directly forward, or have their faces covered with hair or items such as facemasks and mufflers. In this paper, we propose a system for securing facial photos of attendees directly facing a camera by leading them to scan their check-in codes on a code-reader placed close to the camera just before executing face recognition. The system also takes two photos of attendees with this one camera after an interval of about 0.5 s to obtain facial photos with their eyes open. The system achieved 93% face-recognition accuracy with an average time of 2.7 seconds per person for identity verification when it was used for verifying 1,547 attendees of a concert of a popular music singer. The system made it possible to complete identity verification with higher accuracy with shorter average time than Ticket ID system.

Keywords: Face recognition · Illegal ticket resale · Identity verification

Y. Ohsawa et al. (Eds.): JSAI 2019, AISC 1128, pp. 291–304, 2020.
https://doi.org/10.1007/978-3-030-39878-1_26

1 Introduction

Identity verification is required in an increasing number of situations. Let us take an example of a case in which many people are admitted to an event. It used to be that in such cases, having a document, such as a ticket or an attendance certificate, checked was sufficient to gain entry; the need for personal authentication was not seriously considered due to the limited amount of time for admitting all participants. Many events with high ticket prices had designated seating, so it was not necessary to assume that some tickets may have been counterfeit. However, the advent of Internet auctions in recent years has made it easier to buy and sell tickets at the individual level. This has resulted in an increase in illegal ticket scalping, i.e., tickets being purchased for resale purposes. Equity in ticket purchasing is required not only by ticket purchasers but also by event organizers and performers [1]. Consequently, event organizers have had to deal with complaints about malicious acts by undesignated individuals who take advantage of fans by buying and selling tickets on the Internet. In many cases, therefore, any ticket buying and selling outside the normal sales channels is prohibited. Ticket-sales terms now often stipulate that tickets are invalid when people apply for them using a pseudonym or false name and/or false address or when they have been resold on an Internet auction or through a scalper. Illegally resold tickets have in fact been invalidated at amusement parks and concert halls [2]. Verification has therefore become a more important social issue than ever before. The problem in verifying ticket holders is how to simultaneously verify identities efficiently and prevent individuals from impersonating others at a large-scale event at which tens of thousands of people participate. To solve this problem, we previously developed Ticket ID system that identifies the purchaser and holder of a ticket by using face-recognition software [3]. Since the system was proven effective for preventing illegal resale by verifying attendees at large concerts of popular music singers and groups, they have been used at more than 100 concerts. However, it is necessary to improve face-recognition accuracy because face recognition fails when unclear facial photos are obtained, i.e., when event attendees have their eyes closed, not looking directly for-ward, or have their faces covered with hair or items such as facemasks and mufflers. We propose an identity-verification system for attendees of large-scale events using continuous face recognition improved by check-in behavior of event attendees such as facial directions and eye contact (eyes are open or closed).

2 Personal Authentication

The following episodes denote personal authentication using a countersign, pill-box and tattoo during the Edo period (between 1603 and 1868) in Japanese history, The Akou Roushi (The Forty-seven Ronin) gave the countersign yama (mountain) and kawa (river) when they fought in the darkness to avenge their master's honor [4]. Those who were not ready with these countersigns were to be treated as enemies. A retired lord of the domain of Mito, Mitsukuni Tokunaga fought wrongdoing while masquerading as a commoner. He revealed his identity to the evildoer by having one of his two servants show them a small pillbox bearing the symbol of the Tokugawa clan [5]. A magistrate, Kinshiro Tohyama fought to catch wrongdoers while masquerading as an ordinary citizen and baring a distinctive cherry blossom tattoo without identifying himself. He bared his tattoos again at his court to convince the wrongdoers that the magistrate himself was a witness to their crimes [6]. In today's modern society, personal authentication can be divided into three methods: (1) knowledge certification using information only the person in question knows, such as a password or personal identification number, (2) possession certification, i.e., possessions such as an ID card or driver's license, and (3) biometric authentication by confirming a person's fingerprints face, etc. Although knowledge and possession certification are already widely used, such as at bank terminals and in e-commerce, both methods can be transferable if the individuals purchasing tickets and those entering the event match. Therefore, neither can be effective means to prevent individuals from impersonating others.

Many anticipate that biometric authentication can be a means of solving these problems [7]. One advantage of biometric authentication is that there is no risk of biological information being lost or forgotten. Also, biometric authentication can be considered a means to prevent individuals from impersonating others because it uses person-specific biological information. Biometric authentication verifies identity by matching pre-registered biometric information and collation information obtained through a sensor. For example, both the vein authentication procedure used in financial institutions [8] and fingerprint authentication procedure used in national and local governments [9] require dedicated biometric information sensors. Biometric authentication, which requires dedicated biometric information sensors, is not acceptable for verifying the identity of the purchaser and holder of a ticket because it is not practical for ordinary people to have their biological information registered in advance at home and checked at event sites on the day of the event with such sensors. On the other hand, a sensor for face recognition is a normal camera that ordinary people can easy handle. Face recognition has been put into limited practical use for verifying identity such as entrance and exit control for rooms, immigration control, and reception control [10, 11]. They are aimed at enhancing security, that is to say, preventing individuals from impersonating others. Their purpose is not improving efficiency of identity verification, which is required for checking a large number of people. There has been no reports on a system that is successful in increasing the throughput of identity verification, that is to say, the number of people who can be verified per unit time, by using face recognition. No conventional identity verification system using electric tickets and personal authentication was able to verify identities efficiently as well as prevent individuals from impersonating others at a large-scale of event in which many people participate.

3 Ticket ID System Using Face Recognition

3.1 Outline of Ticket ID System

Thorough verification for preventing individuals from impersonating others is in a trade-off relationship with efficient verification. The problem in verifying ticket holders is how to simultaneously verify identities efficiently and prevent individuals from impersonating others at a large-scale event in which tens of thousands of people participate. The solution should be suitable within practical operation costs for various sized events held in various environments including open air. As a practical solution combining efficiency, scalability, and portability for a large-scale event, we developed Ticket ID system, which consists of two sub-systems, a one-stop face recognition system (one-stop system) and a check-in system [3]. The one-stop system uses the high-speed and high-precision commercial face recognition product NeoFace [12]. The one-stop system is implemented in a commercially available tablet terminal, and the recognition result is displayed with regard to the facial-photo information of 100,000 people within about 0.5 s. The check-in system supports identity verification of attendees. A venue attendant checks in by placing his/her membership card on the card reader and initiates face recognition by the taking of his/her photos. Figure 1 shows the ticket-verification procedure from the first step of ticket application to the last step of admission supported with our one-stop face recognition system.

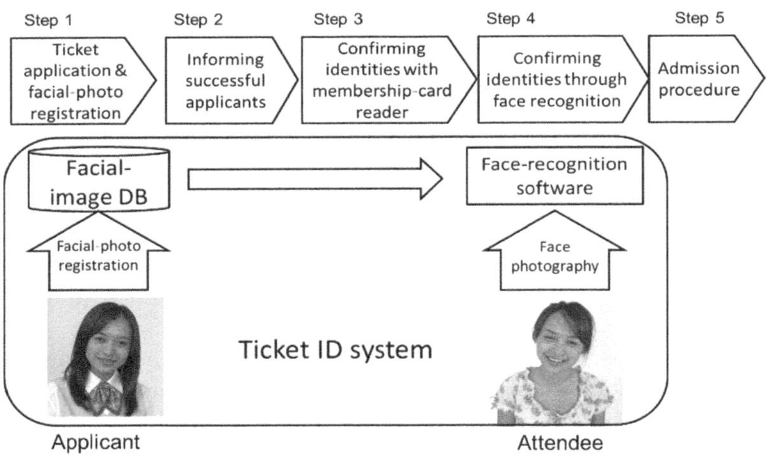

Fig. 1. Ticket-verification procedure with Ticket ID system

The following steps make up the ticket-verification procedure from ticket application to admission [3]:

Step 1: Tickets to popular events are often sold on a lottery basis at fan clubs or other organizations where membership is registered. Individuals applying for

tickets register their membership information as well as their facial photos. In the same way for an ordinary ID photo, the registered facial photo is a clearly visible frontal photo taken against a plain background. The face must not be obstructed by a hat, sunglasses, facemask, muffler, or long hair.
Step 2: Event organizers notify ticket winners, i.e., successful applicants that have been selected.
Step 3: On the day of the event, venue attendants receive membership cards from attendees, and use a card reader to verify that attendants entering the venue are successful applicants at the event venue, as shown in Fig. 2.
Step 4: The attendants use the one-stop system to confirm that the photo taken at the time of application and the collation photo show the same person. The attendants explain the verification through face recognition to the attendees and instruct them where to stand in front of the terminal. Then, they execute the face-recognition process using the terminal to confirm the attendees are those who applied for the tickets.
Step 5: The admission procedure is carried out in accordance with the face-authentication results.

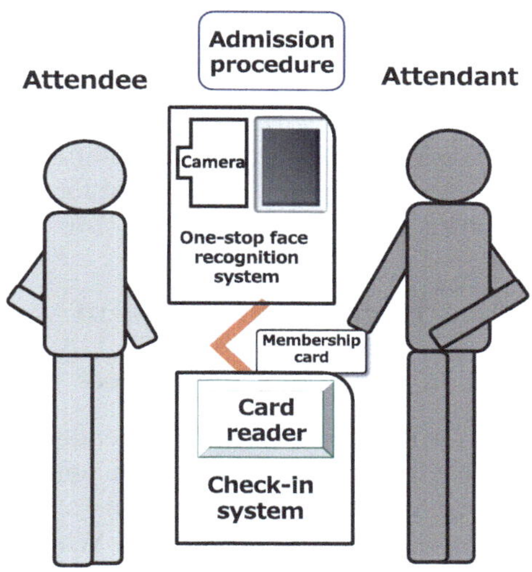

Fig. 2. One-stop face-recognition system

3.2 Problems with One-Stop System

The average time for identity verification from check-in to entry admission was 7 seconds per person, and the average accuracy of face recognition was 90%. It is necessary to improve face-recognition accuracy by securing clear facial photos because face recognition fails when unclear facial photos are obtained, i.e., when event attendees have their eyes closed, not looking directly forward, or have their faces covered with hair or items such as masks and mufflers. Examples in which recognition was not successful are shown in Fig. 3. The recognition failed because the individuals had their eyes closed (Fig. 3 left), were not looking directly forward (Fig. 3 middle), or had their faces covered with hair (Fig. 3 right).

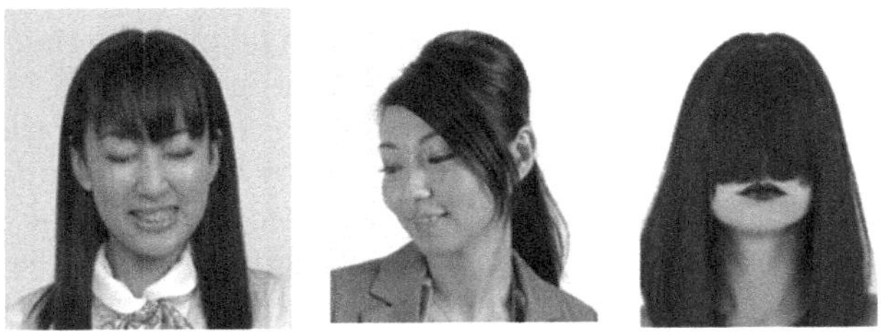

Fig. 3. Examples of recognition failure

When face recognition fails, venue attendants have to verify attendees carefully by direct visual inspection. This increases the mental and physical burden on attendants, which makes attendees have an unreliable impression of the system. When face-recognition accuracy is 90%, two attendees are successively verified without face recognition failure with a probability of 81%. This means that 19% of attendees may experience face-recognition failure or observe it in front of them. Improving face-recognition accuracy is critical for decreasing attendants' stress and attendees' waiting time.

4 Continuous-Face-Recognition System

4.1 Managing Check-in Behavior of Attendees

We propose an identity-verification system for attendees of large-scale events using continuous face recognition improved by managing check-in behavior of the attendees. The proposed system enables attendees to check in themselves (check-in doers are not attendants, but attendees). While the previous system is equipped with a card reader, the proposed system verifies attendees with a QR code reader set up at the same position for recognizing faces of attendees standing still in front of a venue attendant, as shown in Fig. 4.

Managing facial directions and eye contact are two major issues regarding facial recognition. The proposed system addresses these issues with the following methods:

(1) Managing facial direction

The proposed system secures facial photos of attendees directly facing a camera by leading them to scan their QR codes just before executing face recognition. We found most people spontaneously look at the code-reader, i.e., turn

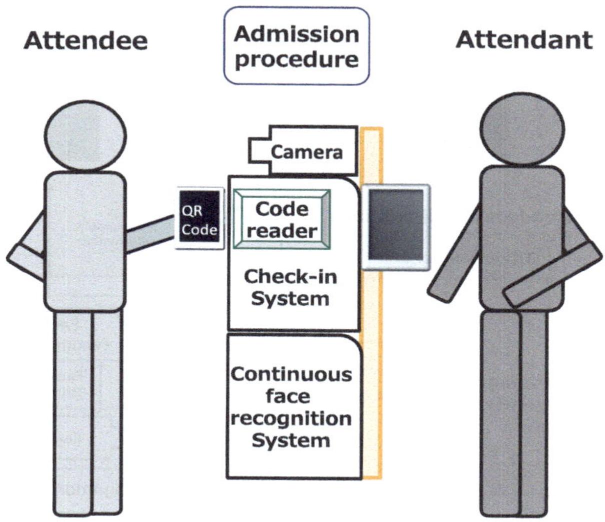

Fig. 4. Continuous face-recognition system

their faces to the reader during check-in. The face-recognition camera of the proposed system is placed at the same position as the code-reader, as shown in Fig. 4, which makes it possible to take an attendee's photo when directly facing the camera when the photo is taken just after check-in.

(2) Managing eye contact

The proposed system uses a continuous-face-recognition system for accepting two photos of attendees successively taken with a single camera after an interval of 0.5 s to obtain facial photos with their eyes open. Few people spontaneously keep their eyes closed longer than 0.5 s because human blink duration is on average between 0.1 and 0.4 s [13]. Few people spontaneously blink twice in 0.5 s because human blink rate is between 7 and 17 per minute [14]. It is possible to manage eye contact of attendees when we take the first photo at the same time of them scanning a QR code and then take the second photo after an interval of 0.5 s with a single camera.

4.2 Configuration of Proposed System

Figure 5 shows a configuration of the proposed system including event-attendee control platform and continuous-face-recognition system.

The configuration is almost the same as that of the previous system [3] except for a check-in doer and a QR code reader. While check-in doers of the previous system are attendants, those of the proposed system are attendees. While the previous system uses a card reader to scan membership cards, the proposed system uses a QR-code reader to scan tickets with QR codes as shown in Fig. 6.

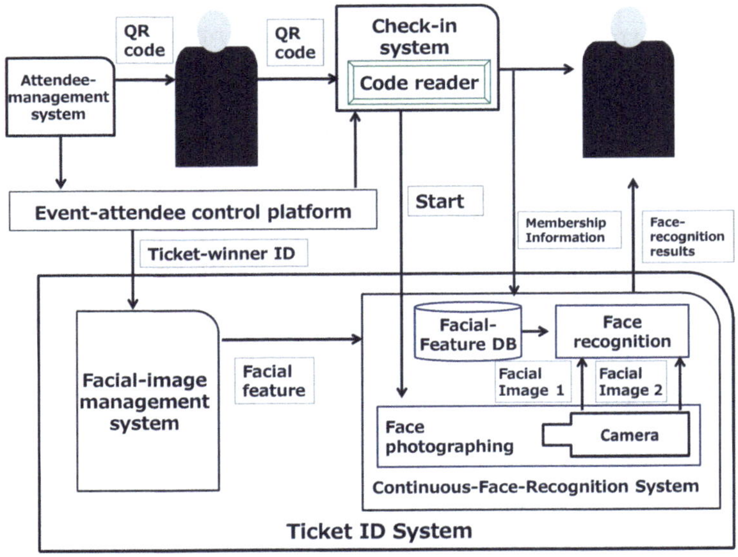

Fig. 5. Configuration of proposed system

Fig. 6. Example of ticket with QR code

When attendees check in at a location that has the proposed system installed, they scan their QR-coded tickets. The attendee-management system provides the attendees the tickets in advance of the event day. Attendees can obtain tickets with QR codes with their smartphones. The ticket has the concert name, date and time, venue, QR code containing attendee's member-ship information, his/her name, seat number, registered photos, and so on.

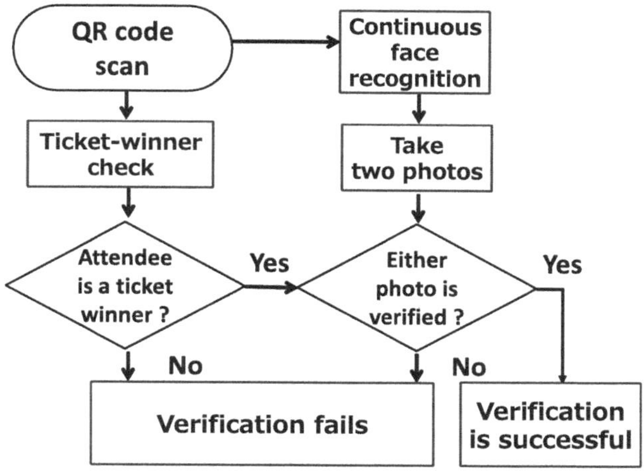

Fig. 7. Flowchart of verification

4.3 Identity-Verification Procedure

An attendee's identity is verified with the procedure shown in Fig. 7. When attendees scan their QR codes, a check-in system performs ticket-winner check as well as showing the attendants the member information of the attendees, which is retrieved from the ticket-winner database with search keys of membership numbers obtained through a QR code reader. Scanning a QR code automatically activates continuous face recognition by taking two photos of the attendee after an interval of 0.5 s. When either photo is verified with the registered photo of the attendee, face recognition is successful. When attendees are ticket winners and face recognition is successful, the verification is successful. Otherwise, verification fails.

4.4 Operational Steps

The proposed system has the following operational steps from ticket application be-fore the event day to admission on the event day:

Step 1: Ticket application is the same as that of the one-stop face-recognition system described in Section 2.1.
Step 2: Event organizers notify ticket winners, i.e., successful applicants that have been selected. They can obtain attendee's tickets including their QR codes and registered facial photos.
Step 3: At the check-in site on the event day, attendees scan their QR codes with the code reader according to attendant's instruction.
Step 4: Attendants can confirm that attendees are successful applicants who applied for the tickets.

Step 5: If identity is verified, the attendee is admitted entry. Otherwise, identity is verified by an attendant with direct visual inspection of the facial photo on the ticket.

5 Demonstration of Proposed System

5.1 Results

Six sets of the proposed system were used for a popular concert on November 17, 2018 in Tottori prefecture, Japan. Face recognition was carried out for 1,547 attendees. Face-recognition accuracy was 93%, and identity-verification time was 2.7 s on average in cases in which face recognition was not successful. No cases of attendees impersonating others were reported for the concert. The false reject rate (FRR) was 7% and the false accept rate (FAR) was 0%. There were two reasons for recognition failure: The first was that faces of incorrect attendees were detected when photos contained other attendees behind the correct attendee. The second was that the attendees had their faces covered with hair or items such as facemasks and mufflers. Failure was not observed due to the fact that attendees had their eyes closed or were not directly facing a camera.

5.2 Discussion

Table 1 compares the results of our previous and proposed systems. The average identity-verification time was 4.3 s shorter than that of the previous system because of changing check-in doers and improving face-recognition accuracy. The proposed system does not require handing over a membership card between attendants and attendees. The face-recognition accuracy of the proposed system was 3% higher than that of the previous system. This resulted in decreasing the number of direct visual inspections, which increases identity-verification time. There was a problem with face detection in that faces of incorrect attendees were detected when photos contained other people behind an attendee. This can be solved by choosing the face with the largest face area among all the detected faces. Face recognition failed when attendees had their faces covered with hair or items. There were no cases in which attendees had their eyes closed or were not directly facing a camera in the two successive photos, i.e., in the first photo and second photo taken after 0.5 s. Managing facial direction and eye contact of attendees worked as expected. It is difficult to solve the problem of when the faces of attendees are covered with hair or items because there would be no differences between the first and second photos on the attendee's covered faces. The attendee's cooperation is necessary for solving this problem.

Surveys were conducted for attendees who were admitted with the previous or proposed system. They were asked to respond to two questions about its convenience in verification. The 241 survey respondents are broken down by age and gender in Fig. 8 on the previous system.

The 109 survey respondents are broken down by age and gender in Fig. 9 on the proposed system.

Table 1. Results of our previous and proposed systems

<table>
<tr><th></th><th>Previous system</th><th>Proposed system</th></tr>
<tr><td>Identity- verification time</td><td>7 s</td><td>2.7 s</td></tr>
<tr><td>Face-recognition accuracy</td><td>90%</td><td>93%</td></tr>
<tr><td>Check-in doer</td><td>Attendant</td><td>Attendee</td></tr>
<tr><td rowspan="4">Reasons for recognition failure</td><td>Attendees had their eyes closed</td><td></td></tr>
<tr><td>Attendees were not directly facing camera</td><td></td></tr>
<tr><td></td><td>Incorrect attendee's faces were detected</td></tr>
<tr><td colspan="2">Attendees had their faces covered with their hair or items</td></tr>
</table>

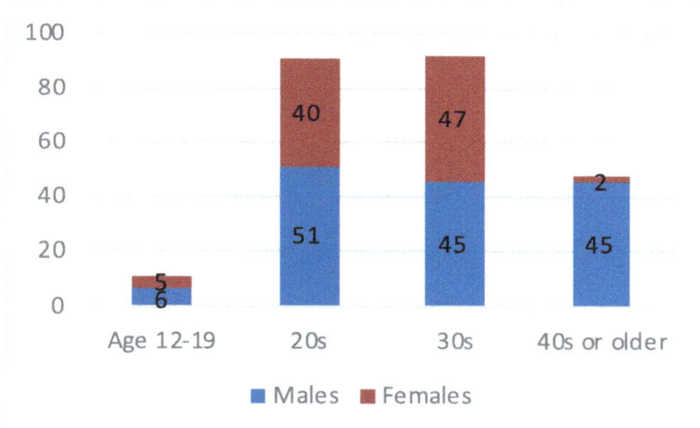

Fig. 8. Survey respondents by age and gender on the previous system

The survey results are shown as percentages in Tables 2 and 3. Regarding the question, "Is the system more convenient than having the attendant verify ID cards and the like visually?" (Table 2), 83% of the respondents of the previous system said either "Definitely yes" or "Yes, I think so." 87.1% of the respondents of the proposed system said so. The proposed system obtained higher ratio of positive responses than the previous one.

Regarding the question, "Why do you think the system is more convenient?" (Table 3), the offered responses included, "It streamlines the admission procedure", "It makes showing an ID card unnecessary", and "It shortens the waiting time." The first and the last responses are increased on the proposed system.

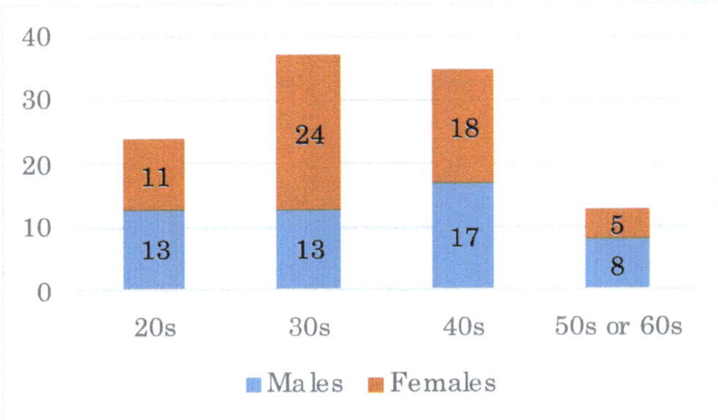

Fig. 9. Survey respondents by age and gender on the proposed system

Table 2. Is the system more convenient than having the attendant verify ID cards and the like visually?

Responses	Previous system	Proposed system
Definitely yes	36.9	48.6
Yes, I think so	46.1	38.5
No, I don't think so	11.6	11.9
Definitely no	5.4	0.9

Table 3. Why do you think the system is more convenient?

Responses	Previous system	Proposed system
It streamlines the admission procedure	63.0	74.0
It makes showing an ID card unnecessary	42.0	42.0
It shortens the waiting time	38.0	41.0

6 Future Issues

Faces of incorrect attendees were detected when they stood behind a correct attendee. This can be solved by choosing the face with the largest face area among all the detected faces. If this improvement does now work, we are preparing a partitioning screen to be placed behind the correct attendee to prevent incorrect attendees from being photographed. The largest obstacle remaining to improving face-recognition accuracy is that of covered faces. This problem could be solved with attendee's cooperation. We have been developing an identity-verification system using face recognition from selfies taken by attendees with their smartphone cameras [15]. Self-photographing is regarded as helpful for securing clear facial photos because attendees can control intrinsic parameters such as their expressions, facial hair, and facial directions. We are planning to use of this system with the proposed system for solving the problem of covered faces. The proposed system has been widely reported in the mass media. The system is highly regarded from reviews on the Internet [16]. It was used to carry out face recognition for more than 100,000 attendees in 2018. Though no cases of attendees impersonating others were reported for any of these events, i.e., the FAR was 0%, the FAR should be more carefully examined from the view-point of preventing impersonation. It is necessary to evaluate the robustness against impersonation with pseudo attack tests. These tests should include disguise and lookalike tests. A disguise test makes people's facial appearances as similar to each other as possible by using facial paraphernalia such as facial hair, glasses, and makeup. A lookalike test is conducted for those, such as twins or similar looking siblings, with similar facial features. A disguise test will reveal considerable disguise methods and help in creating operational manuals for venue attendants to detect these methods. A lookalike test will disclose the technical limitations of current face-recognition methods and help in establishing next-generation technology.

7 Conclusion

We proposed an identity-verification system for attendees of large-scale events using continuous face recognition improved by managing check-in behavior of the attendees. The proposed system could secure facial photos of attendees directly facing a camera by leading them to scan their QR codes on a QR-code reader placed close to the camera just before executing face recognition. The system took two photos of attendees with this one single camera after an interval of 0.5 s to obtain facial photos with their eyes open. The system achieved 93% face-recognition accuracy with an average identity-verification time of 2.7 seconds per person when it was used for verifying 1,547 attendees at a concert of a popular music singer. The system made it possible to complete identity verification with higher accuracy with shorter average time than the previous system. We plan to improve our system to further streamline the verification procedure.

Acknowledgments. Thanks are expressed to all the personnel related to our systems, especially to TAIPIRS Inc. for their operation and Mr. Shinji Nakamura, Executive Vice President of NEC Solution Innovators, Ltd. for his encouragement and support.

References

1. Chapple, J.: Ticket resale? NO, says Japanese music business, 23 August 2016. https://www.iq-mag.net/2016/08/ticket-resale-no-says-japanese-live-business-resaleno/#.W_01Jk8Un3g
2. JE fandom: Johnny's Tracks Illegally Sold Tickets for Arashi's Japonism Tour, 25 October 2015. available from https://jnewseng.wordpress.com/2015/10/25/johnnys-tracks-illegally-sold-tickets-for-arashis-japonism-tour/

3. Okumura, A., Hoshino, T., Handa, S., Nishiyama, Y., Tabuchi, M.: Identity verification of ticket holders at large-scale events using face recognition. J. Inf. Process. **25**, 448–458 (2017)
4. Schilling, M., Komon, M.: The Encyclopedia of Japanese Pop Culture, pp. 135–138. Weatherhill, New York (1997)
5. Takeda, I., Miyoshi, S., Keene, D.: Chushingura: The Treasury of Loyal Retainers, p. 170 (1971)
6. Wikipedia, the free encyclopedia: Tōyama Kagemoto. https://en.wikipedia.org/wiki/T%C5%8Dyama_Kagemoto
7. Imaoka, H., Mizoguchi, M., Hara, M.: Biometrics technology to preserve safety and security. Inf. Process. **51**(12), 1547–1554 (2010). (in Japanese)
8. Soto, M.: Using biometric authentication technology in Japanese financial institutions. Inf. Process. **47**(6), 577–582 (2006). (in Japanese)
9. Sakamoto, S.: Present status and prospects of biometric products and solutions. NEC Technical Report, vol. 5, no. 3, October 2010. http://www.nec.com/en/global/techrep/journal/g10/n03/pdf/100303.pdf
10. Zhao, W., Chellappa, R., Rosenfeld, A., Phillips, P.J.: Face recognition: a literature survey. ACM Comput. Surv. **35**, 399–458 (2003)
11. Jafri, R., Arabnia, H.R.: A survey of face recognition techniques. J. Inf. Process. Syst. **5**(2), 41–67 (2009)
12. NEC: Face Recognition. https://www.nec.com/en/global/solutions/safety/face_recognition/index.html
13. Bentivoglio, A.R., Bressman, S.B., Cassetta, E., Carretta, D., Tonali, P., Albanese, A.: Analysis of blink rate patterns in normal subjects. Mov. Disord. **12**(6), 1028–1034 (1997)
14. Nosch, D.S., Pult, H., Albon, J., Purslow, C., Murphy, P.J.: Relationship between corneal sensation, blinking, and tear film quality. Optom. Vis. Sci. **93**(5), 471–481 (2016)
15. Okumura, A., Hoshino, T., Handa, S., Yamada, E., Tabuchi, M.: Identity verification for attendees of large-scale events using face recognition of selfies taken with smartphone cameras. J. Inf. Process. **26**, 779–788 (2018)
16. Hachima: Concerts of Hikaru Utada were successfully operated, November 2018. http://blog.esuteru.com/archives/9218587.html. (in Japanese)

Author Index

Y. Ohsawa et al. (Eds.): JSAI 2019, AISC 1128, pp. 305–306, 2020.
https://doi.org/10.1007/978-3-030-39878-1

The manufacturer's authorised representative in the EU is Springer Nature Customer Service Centre GmbH, Europaplatz 3, 69115 Heidelberg, Germany. If you have any concerns regarding our products, please contact ProductSafety@springernature.com

Printed and bound by CPI Group (UK) Ltd, Croydon, CR0 4YY
25/09/2026
02222891-0008